Pictorial Laboratory Guide for Anatomy and Physiology

Pictorial Laboratory Guide for Anatomy and Physiology

Dennis Strete, Ph.D.

McLennan Community College

Consultant, Department of Pathology and Laboratory Medicine

Waco Veterans Affairs Medical Center

An imprint of Addison Wesley Longman, Inc.

Menlo Park, California • Reading, Massachusetts • New York • Harlow, England
Don Mills, Ontario • Sydney • Mexico City • Madrid • Amsterdam

Acquisitions Editor: Bonnie Roesch
Developmental Editor: Cyndy Taylor
Project Coordination: Electronic Publishing Services Inc.
Cover Designer: Yvo Riezebos
Art Studio: Electronic Publishing Services Inc.
Full-Service Production Manager: Valerie L. Zaborski
Manufacturing Manager: Helene G. Landers
Electronic Page Makeup: Electronic Publishing Services Inc.
Printer and Binder: Courier/Stoughton, Inc.
Cover Printer: Courier/Stoughton, Inc.

Library of Congress Cataloging-in-Publication Data
Strete, Dennis.
Pictorial laboratory guide for anatomy and physiology / Dennis Strete.
p. cm.
Includes index.
ISBN 0-673-99225-X
1. Anatomy--Atlases. 2. Physiology--Atlases. I. Title.
QM25.S77 1996 96-45953
611' .0022'2--dc20 CIP

ISBN 0-673-99225-X
12345678910–CRS–99989796

DEDICATED TO THE STUDENTS
WHO WILL CONTINUE TO SEEK KNOWLEDGE
BEYOND THE CONFINES OF THIS BOOK

CONTENTS

PREFACE		ix
CHAPTER 1	◆ Body Organization, Anatomical Position, and Body Planes	1
CHAPTER 2	◆ The Cell	12
CHAPTER 3	◆ Mitosis, Meiosis, and Chromosomes	17
CHAPTER 4	◆ Epithelium	21
CHAPTER 5	◆ Connective Tissue	26
CHAPTER 6	◆ Specialized Connective Tissue: Cartilage	31
CHAPTER 7	◆ Specialized Connective Tissue: Bone and the Skeleton System	34
CHAPTER 8	◆ Articulation	65
CHAPTER 9	◆ Specialized Connective Tissue: Blood and Bone Marrow	78
CHAPTER 10	◆ Muscle	82
CHAPTER 11	◆ Nervous Tissue and the Central Nervous System	94
CHAPTER 12	◆ Integumentary System	107
CHAPTER 13	◆ Cardiovascular System	112
CHAPTER 14	◆ Lymphatic System	120
CHAPTER 15	◆ Digestive System	126
CHAPTER 16	◆ Respiratory System	138
CHAPTER 17	◆ Urinary System	144
CHAPTER 18	◆ Endocrine System	150
CHAPTER 19	◆ Male Reproductive System	155
CHAPTER 20	◆ Female Reproductive System	162
CHAPTER 21	◆ Fetal Development	167
CHAPTER 22	◆ Organs of Special Sense	173
CHAPTER 23	◆ Cat Dissection	182
CHAPTER 24	◆ Fetal Pig Dissection	191
CREDITS		193
INDEX		195

PREFACE

This anatomy and physiology atlas is the result of my years of teaching experience in the aforestated disciplines. I have tried to supplement the micrographs and photographs with line drawings wherever possible. This, I feel, helps the students to relate the organs of a given system.

This atlas is specifically designed for use by either graduate or undergraduate students enrolled in allied health, biology, medical, dental, veterinary medicine, anatomy, or anatomy and physiology programs.

Among the outstanding features of this atlas are the following:

- Introduction to the cell structure using light and electron micrographs
- Body positioning using a live model
- Extensive coverage of the skeletal system and bone articulation
- All organ systems covered in detail
- Body systems supported by well-labeled histological slides and their magnification
- A cat dissection chapter that includes coverage of the important musculature, blood circulation, glands, and developmental process
- Section on fetal pig dissection
- Carefully written, brief captions pointing to key structures included with all figures

I wish to express my gratitude and appreciation to Bonnie Roesch and Cyndy Taylor, to Charles Hickman, who was instrumental in establishing contact between myself and the publishers, and to Kathy Johnson, who typed the manuscript.

My sincere thanks and appreciation go to Dr. Kelly Sexton for his input on the skeleton system, and to Dr. Charles Conley, pathologist, and Dr. Keith Young of Waco Veterans Affair Medical Center for their encouragement and suggestions. In addition, the following reviewers lent their expertise to the development of this atlas: Clifford Barnes, University of Colorado; Gary Iwamoto, University of Illinois; John Moore, Parkland College; and Bruce Sundred, Harrisburg Area Community College.

Finally, I would like to thank Benjamin Cummings for their financial support and publication of this atlas.

I hope students will find this atlas beneficial to their course of study. If you have suggestions or questions, please feel free to contact me.

Dennis Strete, Ph.D.

Body Organization, Anatomical Position, and Body Planes

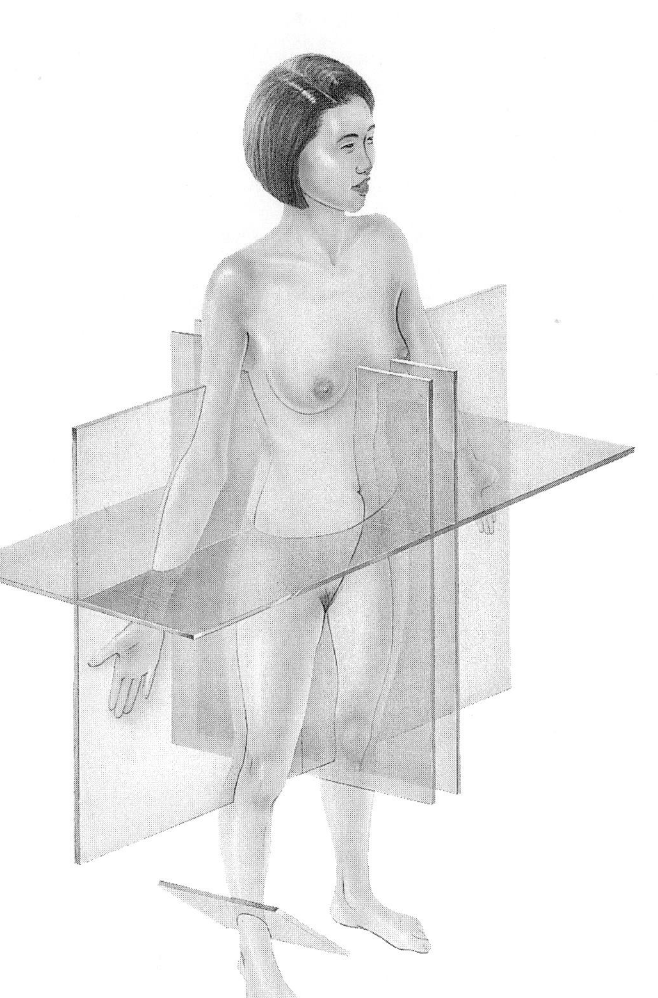

The human body consists of the head, trunk (neck, chest, abdomen, and pelvis), and limbs (arms and legs). The lower part of the trunk and pelvis is also called the perineum. The central axis of the body includes the head and a long vertebral column that consists of seven cervical, twelve thoracic, five lumbar, five sacral (fused), and five fused coccyx vertebrae. The sacrum is fused with two large pelvic bones.

The upper limbs of the body consist of the shoulders, arms, forearms, and hands. The lower limbs include the thighs, legs, and feet.

The **anatomical position** of the body is described by assuming that the individual is standing upright, eyes and head facing forward, and arms relaxed and straight, parallel to the body with palms of the hands facing toward the front. Once the anatomical position has been established, the body structures are described relative to this standard position, irrespective of the body being in an upright position, lying in bed, or lying on a table in the laboratory.

The body planes are imaginary sections passing through the body. The **median sagittal plane** passes vertically and divides the body into two identical halves. The **coronal** or **frontal plane** is described by imaginary sections progressing from the anterior to the posterior part of the body. Coronal and sagittal sections are widely used in studying the anatomy of the head and brain. The **transverse plane** is a section passing at right angles to the sagittal and coronal planes.

Other anatomical terms commonly used in describing relative positions are anterior and posterior, which mean front and back of the body, respectively; cephalic or cranial means toward the head, whereas caudal means toward the tail. Proximal and distal mean nearer or further away, respectively. Superficial means close to the skin surface; deep means below the skin surface. Superior means above and inferior means below. The anterior or ventral surface of the arm is referred to as palm or palmar. The posterior of the hand is called the dorsal surface or dorsum. The undersurface of the foot is called the plantar or sole of the foot.

Anatomical Regions

FIGURE 1.1
Facial, anterior view.

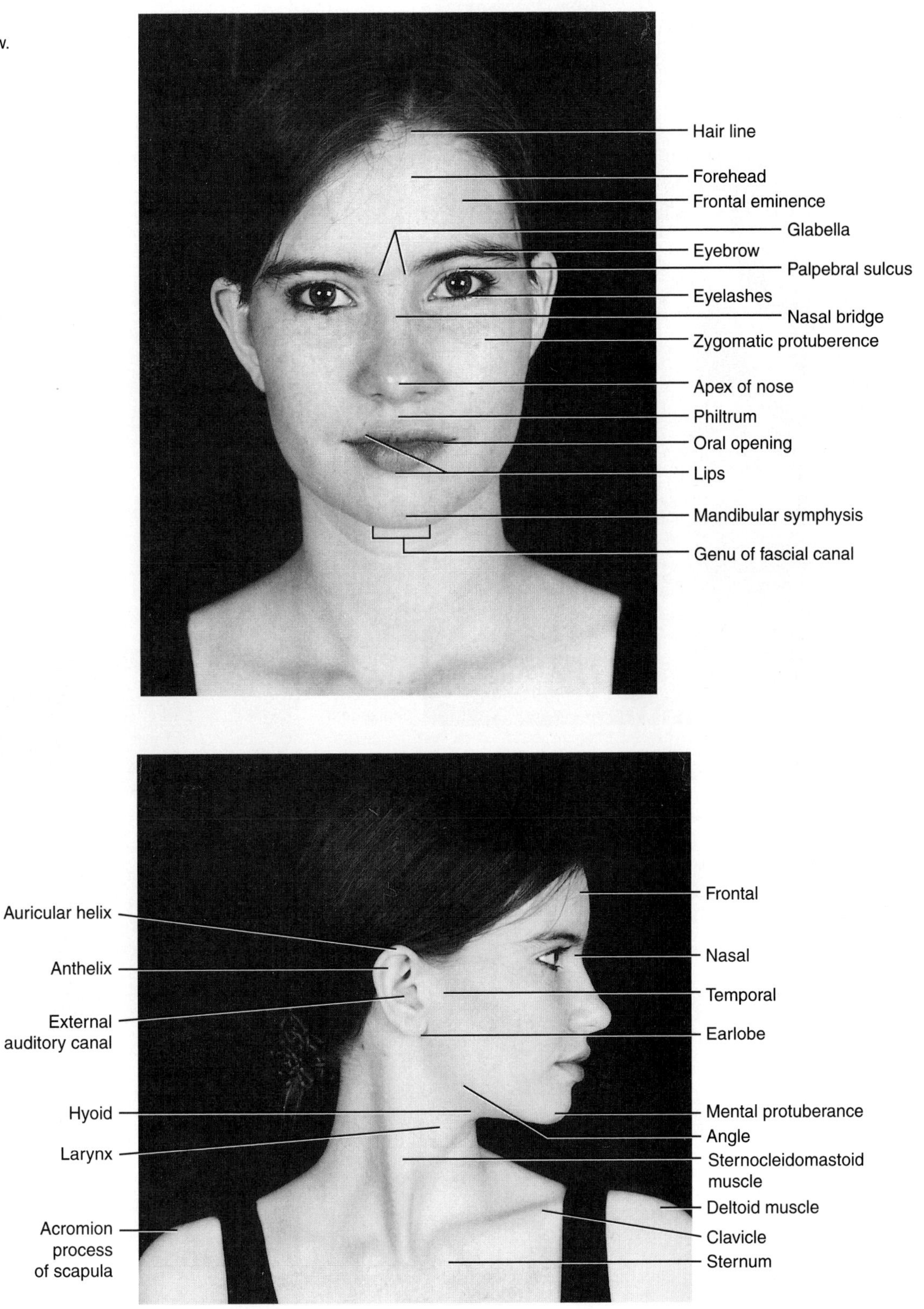

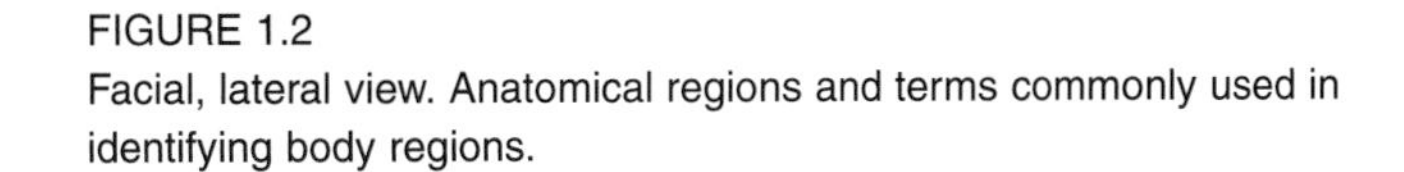

FIGURE 1.2
Facial, lateral view. Anatomical regions and terms commonly used in identifying body regions.

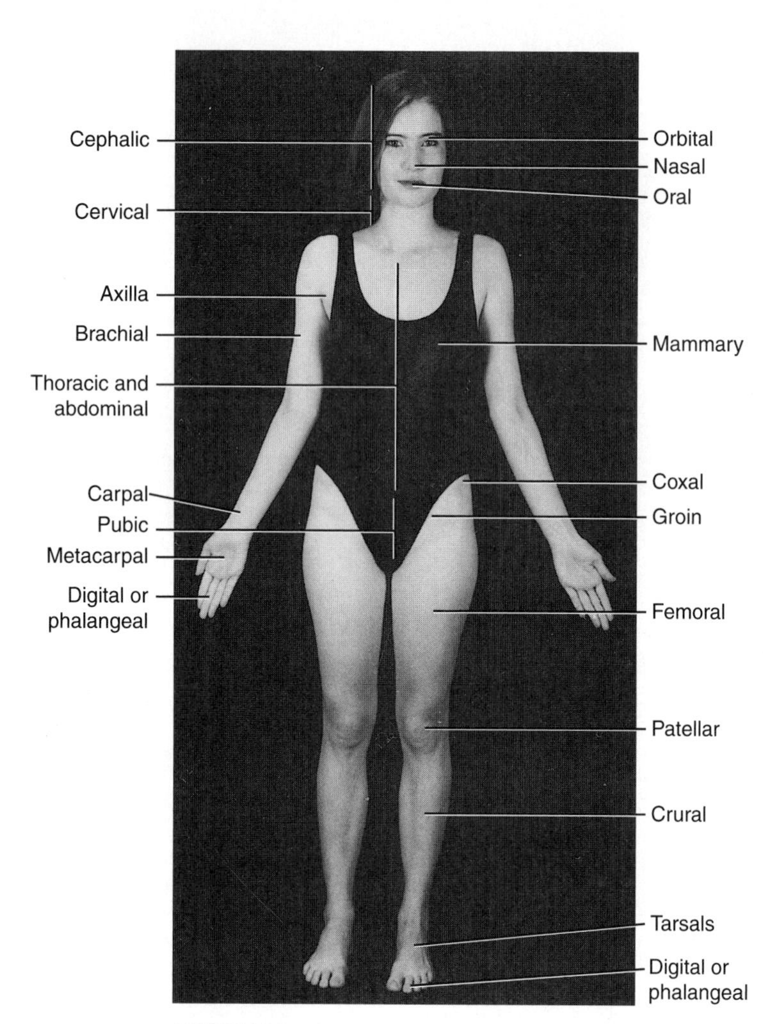

FIGURE 1.3
Anterior view.

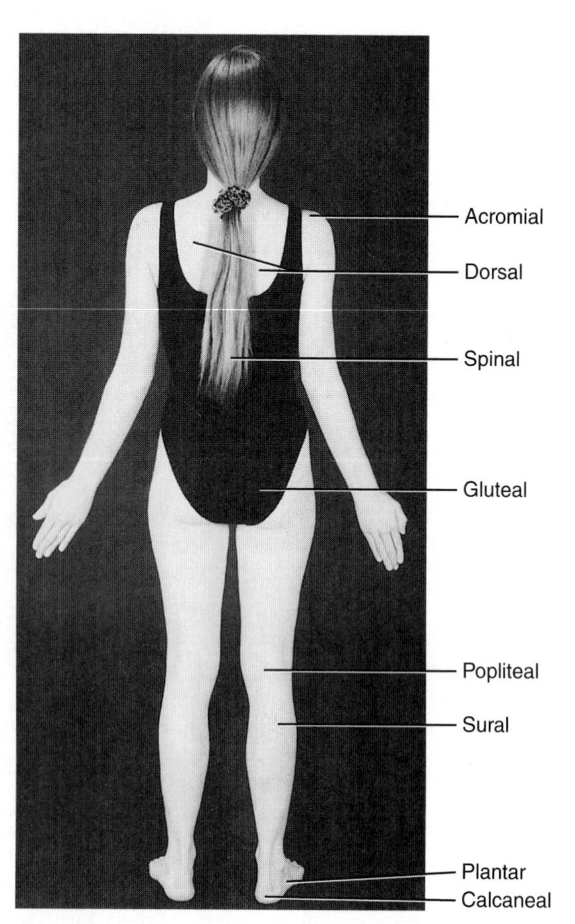

FIGURE 1.4
Posterior view.

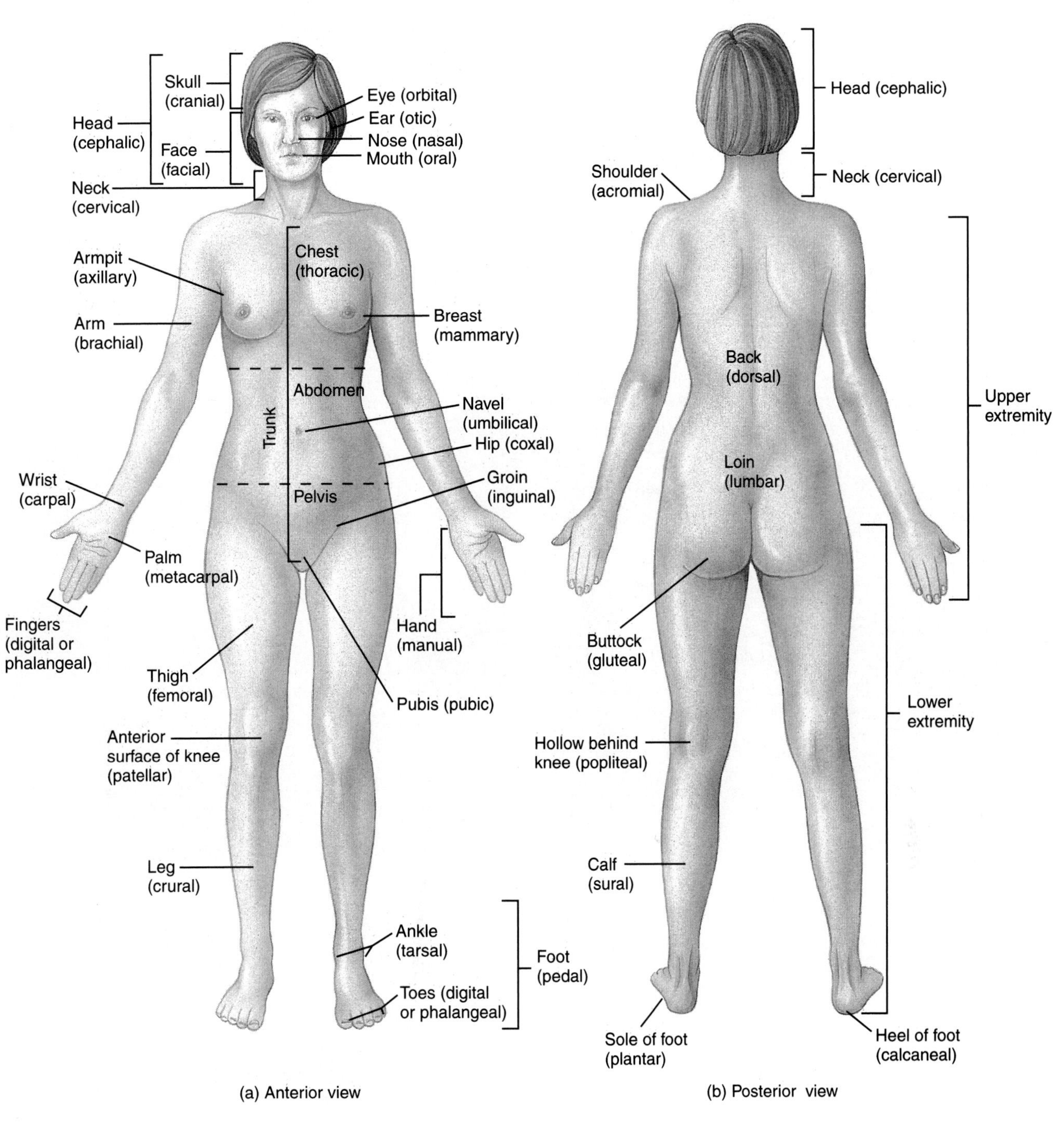

(a) Anterior view

(b) Posterior view

FIGURE 1.5
Diagrammatic presentation of anatomical position and body regions.

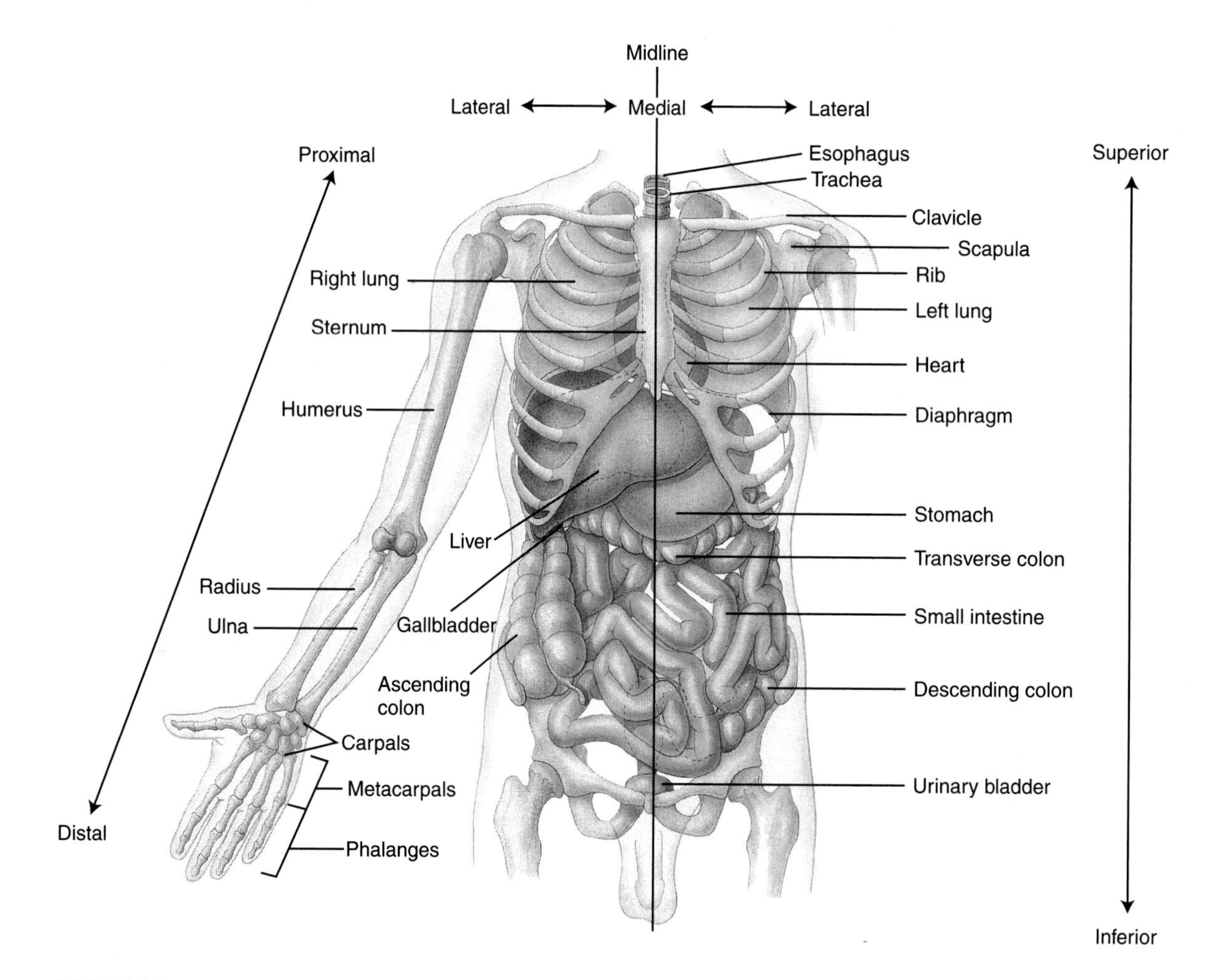

FIGURE 1.6
Diagrammatic presentation of anatomical structures, their location, and understanding of directional terms.

FIGURE 1.7
Diagrammatic presentation of the nine regions and the anatomical structures associated with the regions.

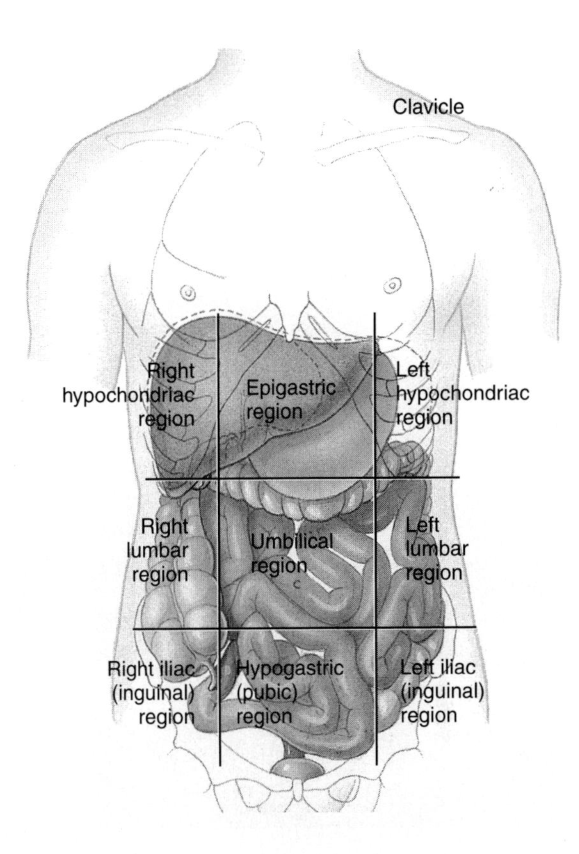

Location of abdominopelvic regions, anterior view

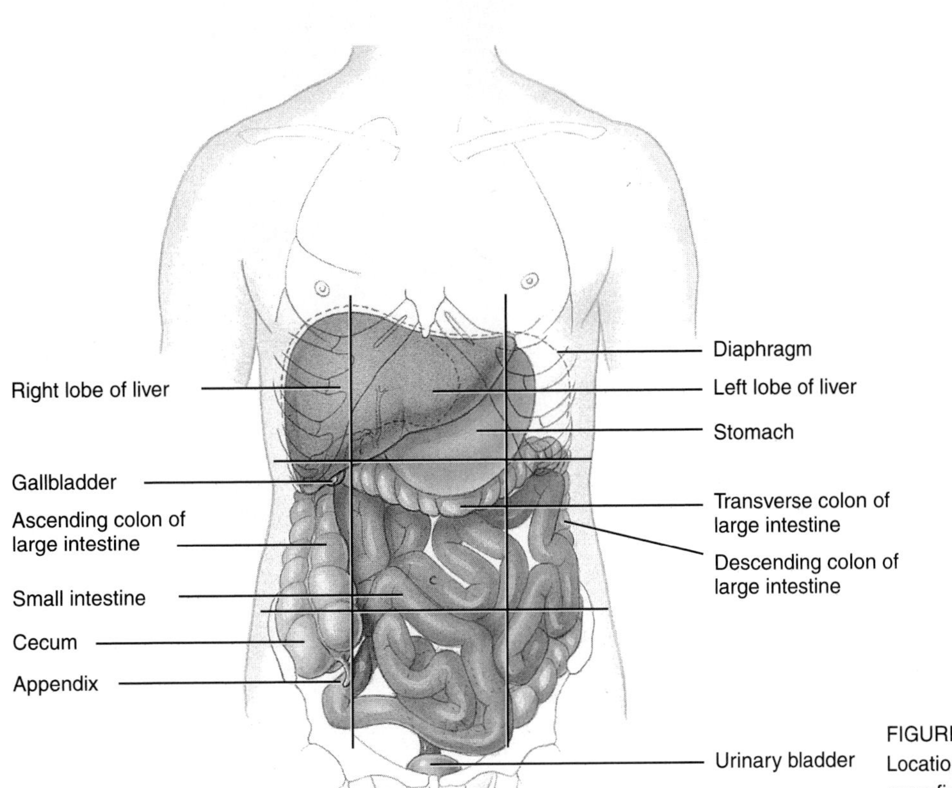

FIGURE 1.8
Location of abdominopelvic regions, superficial view.

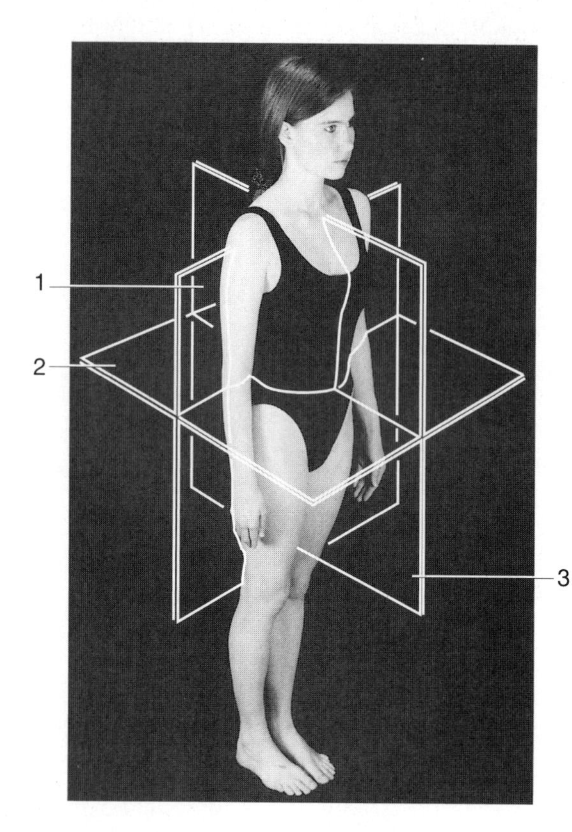

FIGURE 1.9
Body planes superimposed on the model.

1. Frontal or coronal plane.
2. Transverse or cross-sectional plane.
3. Midsagittal plane.

Body organization based on some of the subcutaneous muscles and supporting structures.

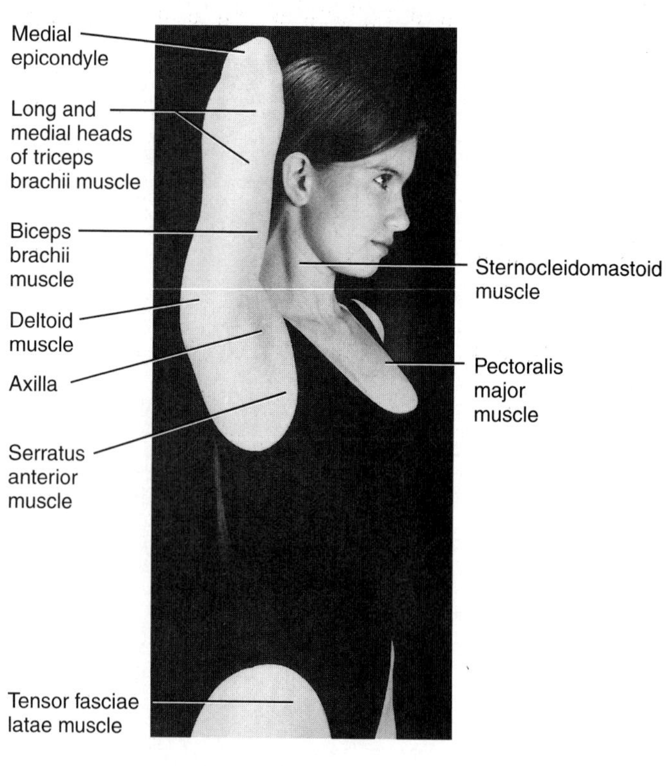

FIGURE 1.10
Anterolateral view.

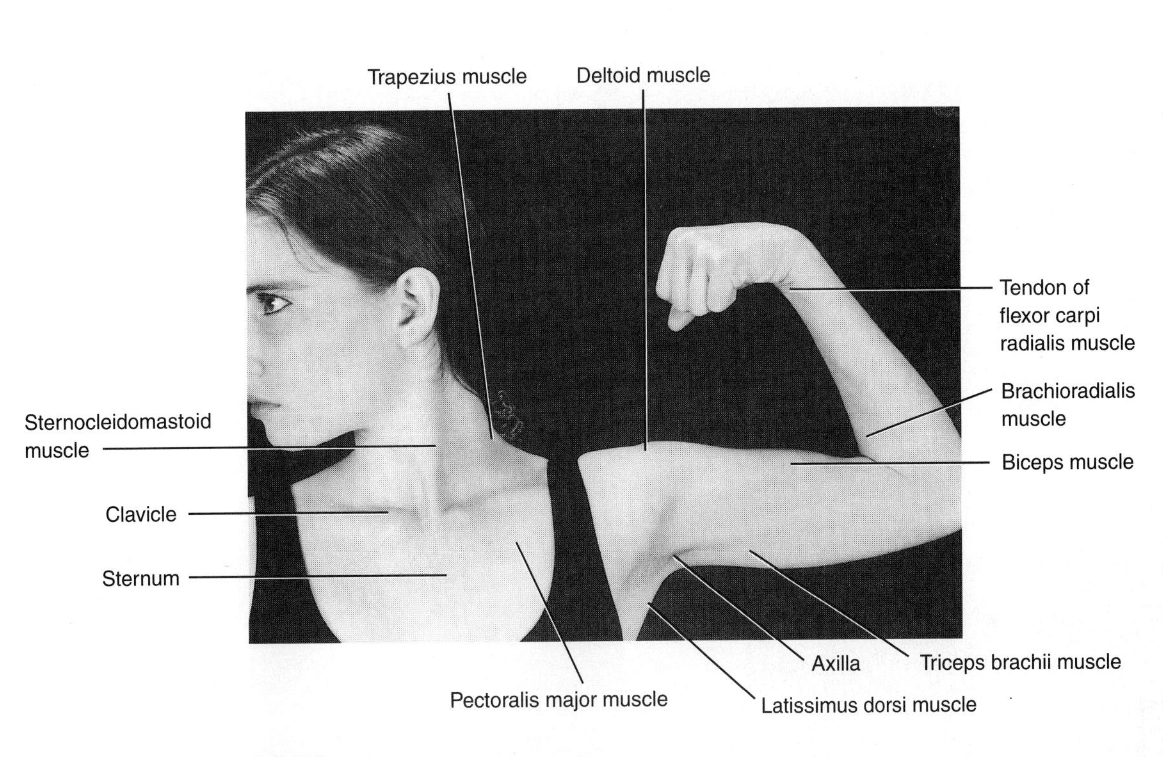

FIGURE 1.11
Left shoulder, axilla, and upper appendage.

FIGURE 1.12
Anteroposterior view.

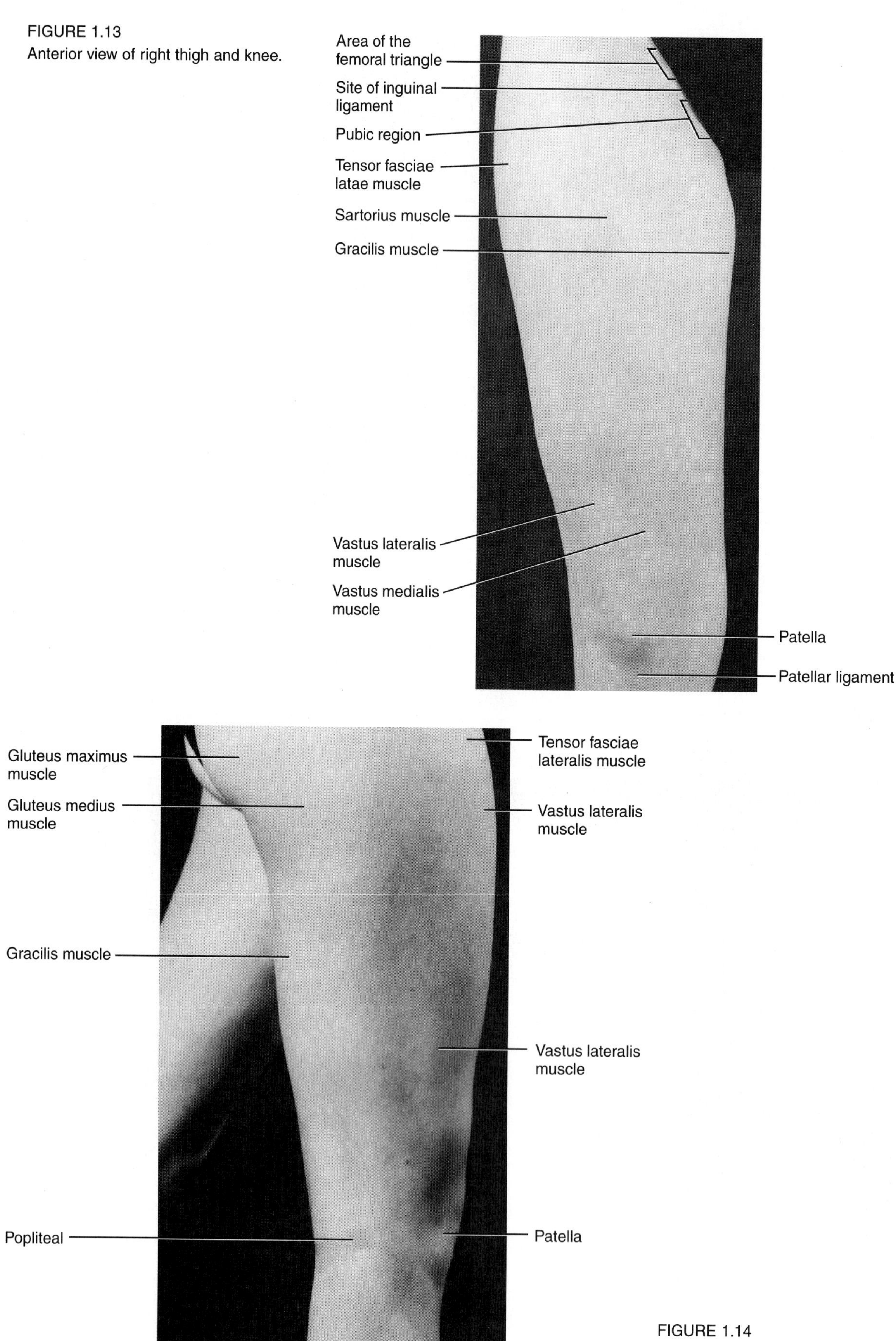

FIGURE 1.13
Anterior view of right thigh and knee.

FIGURE 1.14
Lateral view of right thigh and knee.

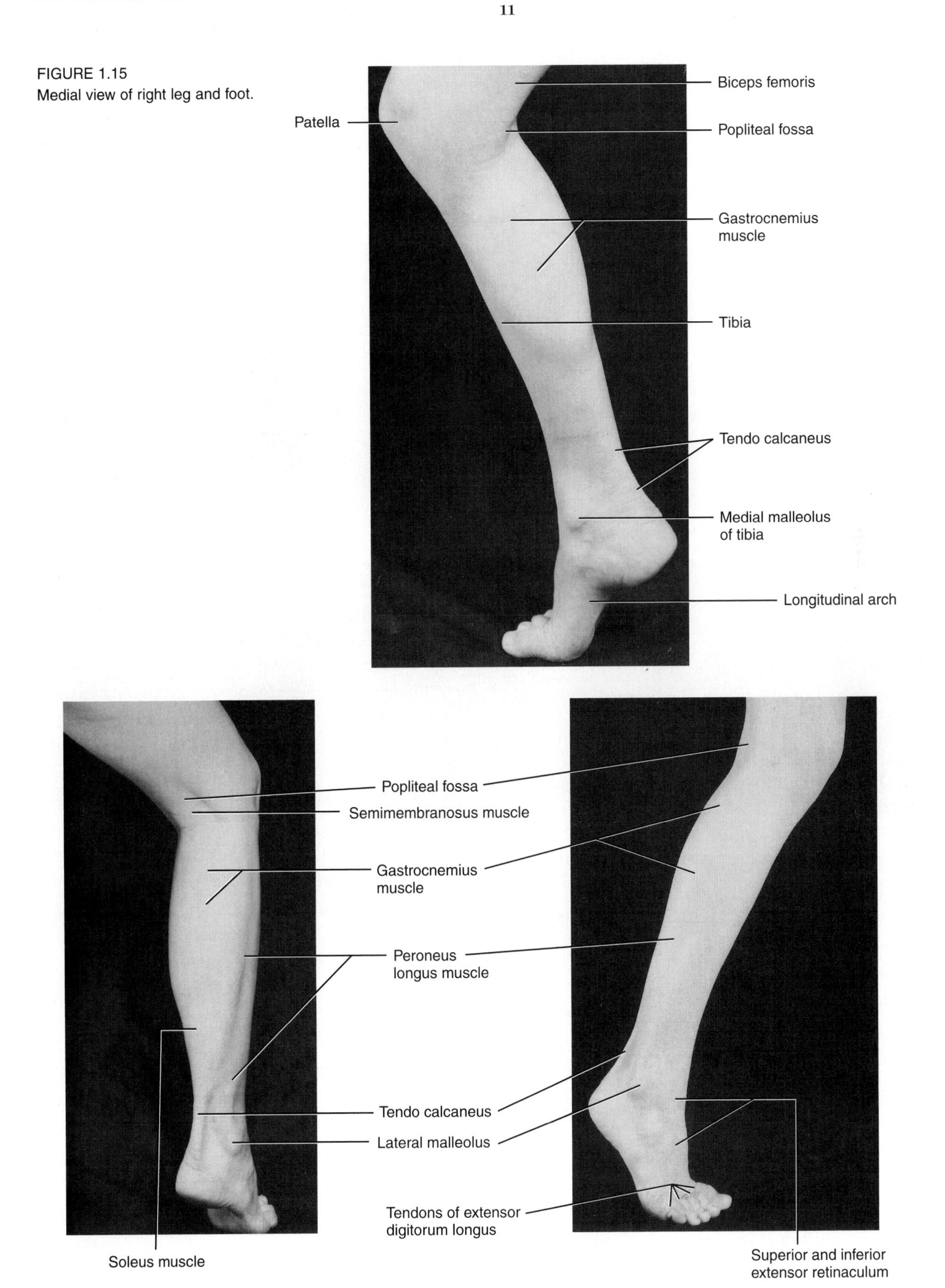

FIGURE 1.15
Medial view of right leg and foot.

FIGURE 1.16
Posterior view of right leg and foot.

FIGURE 1.17
Lateral view of right leg and foot.

The Cell

The **cell** forms the basic structural and functional unit of a tissue. The **tissues,** in turn, form four morphological and functional divisions of the body: **epithelial tissue, connective tissue, nervous tissue,** and **muscular tissue.** Every mature, living cell contains a **nucleus, cytoplasm, organelles, inclusions,** and a highly specialized limiting membrane called the **plasma membrane** or **plasmalemma.** The nucleus stores the genetic information and coordinates all organelle and cellular metabolic activities.

Cell Membrane (Plasmalemma) The plasmalemma is a trilaminar membrane structure (a lipid layer between two protein layers) that measures approximately 8 to 10 nm in thickness. The external protein layer of the membrane may be lined by polysaccharide units that form the outer, uneven glycocalyx coating. The thickness of the glycocalyx coating may vary, depending on the type and function of the cell.

Protoplasm The protoplasm displays physiological characteristics such as conductivity, contractility, irritability, absorption, secretion, excretion, and metabolism. These characteristics of the protoplasm are important if the cell is to survive, divide, and engender cell-to-cell communication.

Cell Organelles Cytoplasmic organelles are specific functional units of the cell. Included in the organelles are such diverse structures as agranular and granular endoplasmic reticula, plasmalemmas, Golgi complexes, ribosomes, cilia, lysosomes, peroxisomes, mitochondria, centrioles, filaments, and microtubules. Most of the organelles are surrounded by a limiting membrane that is morphologically similar to the plasma membrane. Intermediate filaments in the cell matrix can be of many types (nonfibrillary, desmoid, keratin, acidic, and protein) depending on the function of the cell. Other cytoplasmic components of the cells may include pigments, glycogen and lipid molecules, vacuoles, and undifferentiated granules.

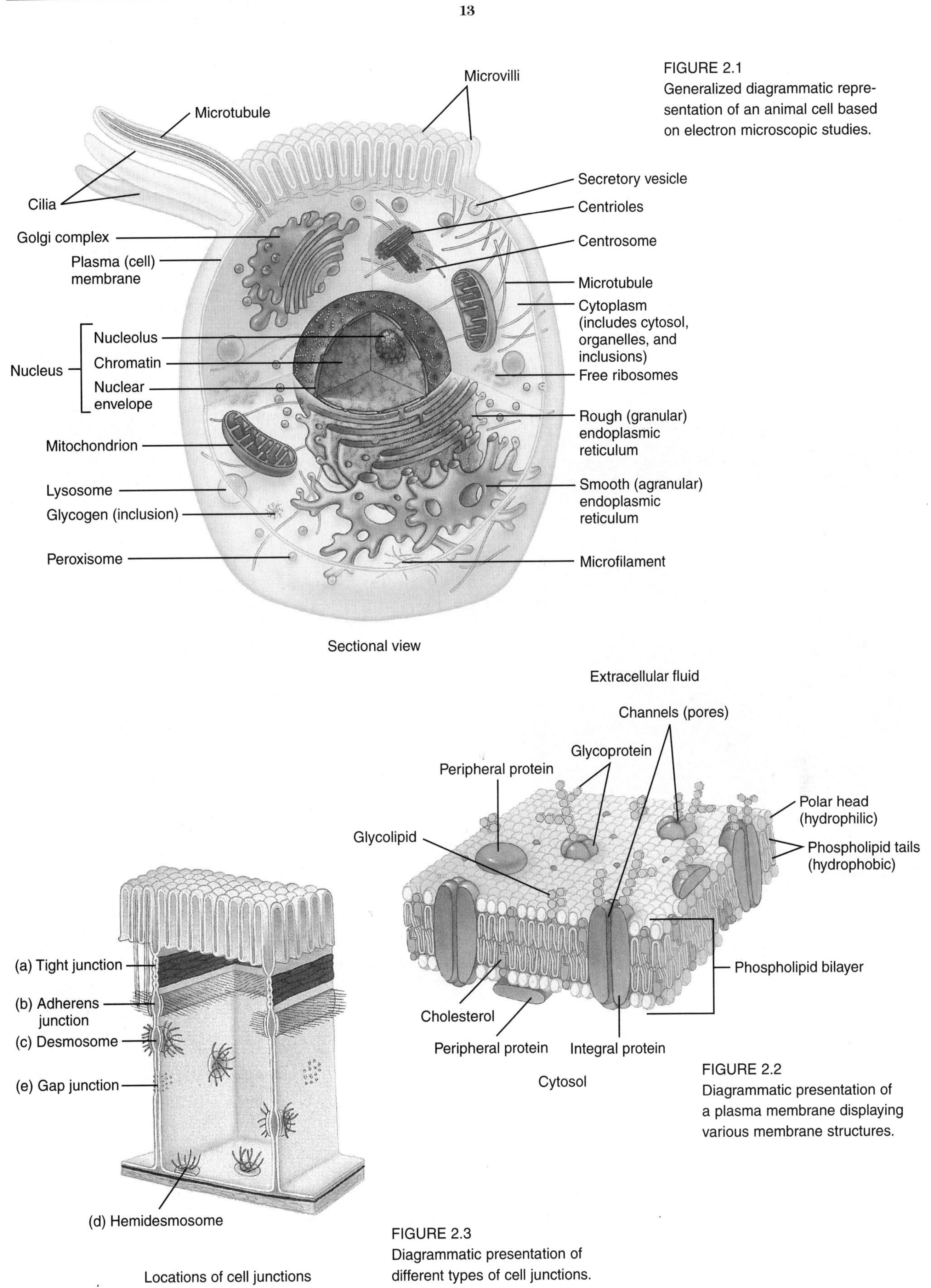

FIGURE 2.1
Generalized diagrammatic representation of an animal cell based on electron microscopic studies.

FIGURE 2.2
Diagrammatic presentation of a plasma membrane displaying various membrane structures.

FIGURE 2.3
Diagrammatic presentation of different types of cell junctions.

FIGURE 2.4
TEM of a lymphocyte exhibiting a large nucleus (which is characteristic of lymphocytes), mitochondria, vacuoles, and a pair of centrioles

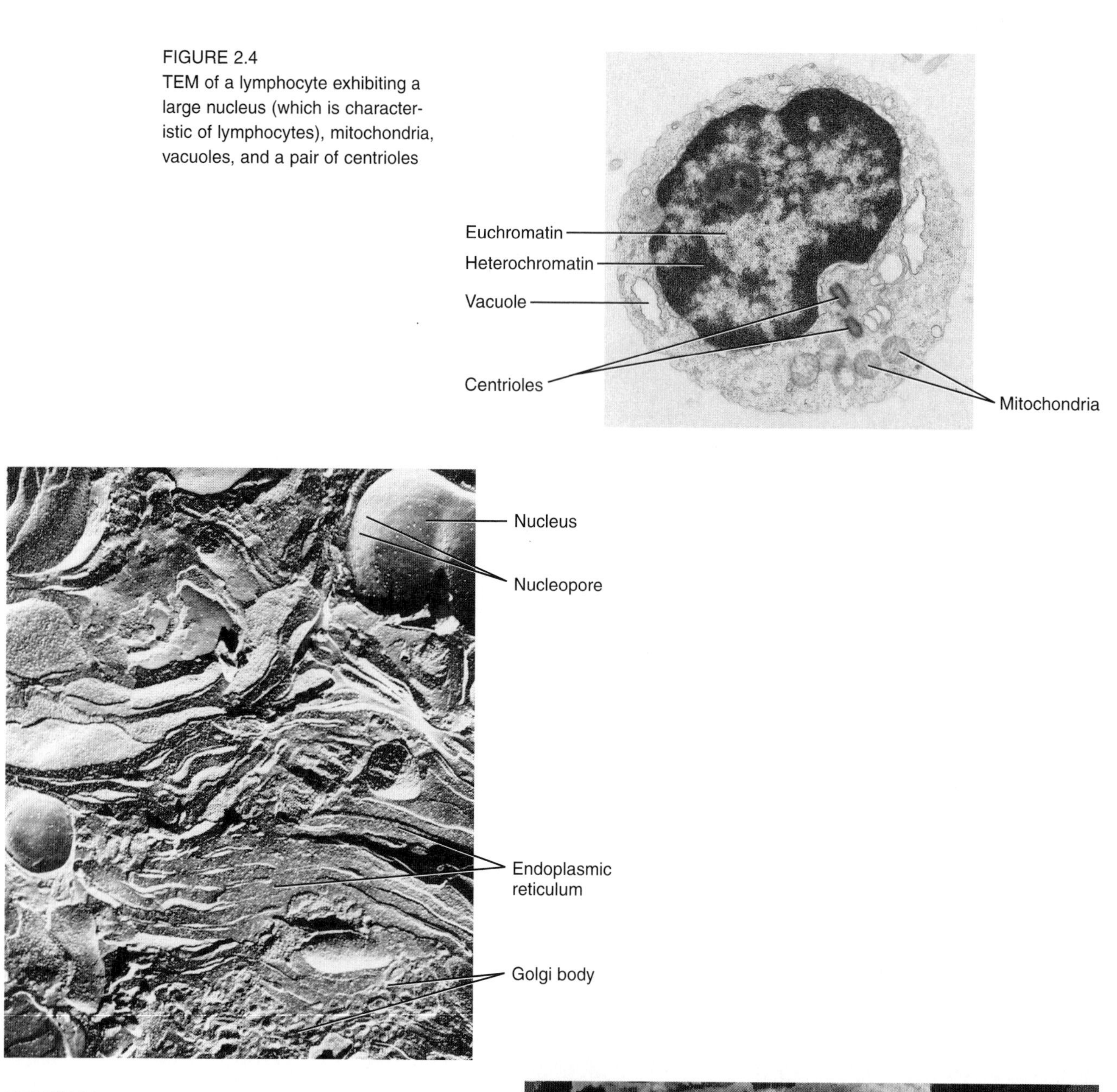

FIGURE 2.5
TEM of a freeze-fractured cell. The nucleus reveals pores in the membrane. Rough endoplasmic reticulum and the Golgi complex can also be resolved in the micrograph. (30,000×)

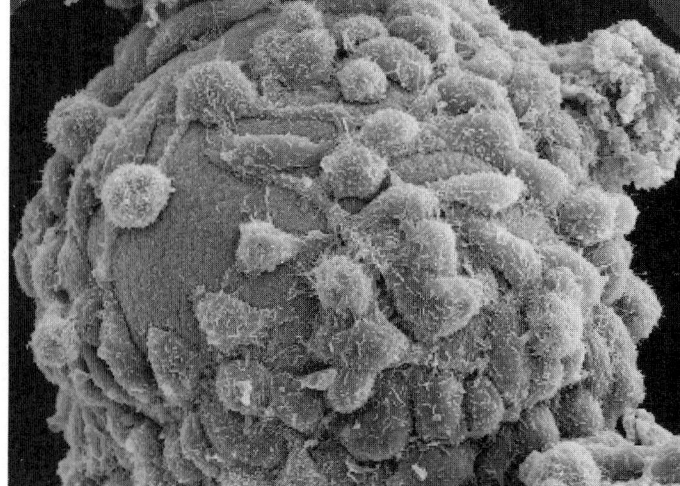

FIGURE 2.6
Scanning electron micrograph (SEM) of transformed rat granulosa cells (ovarian follicular cells) that have been grown on collagen-coated cytodex micro-carrier beads. (1000×)

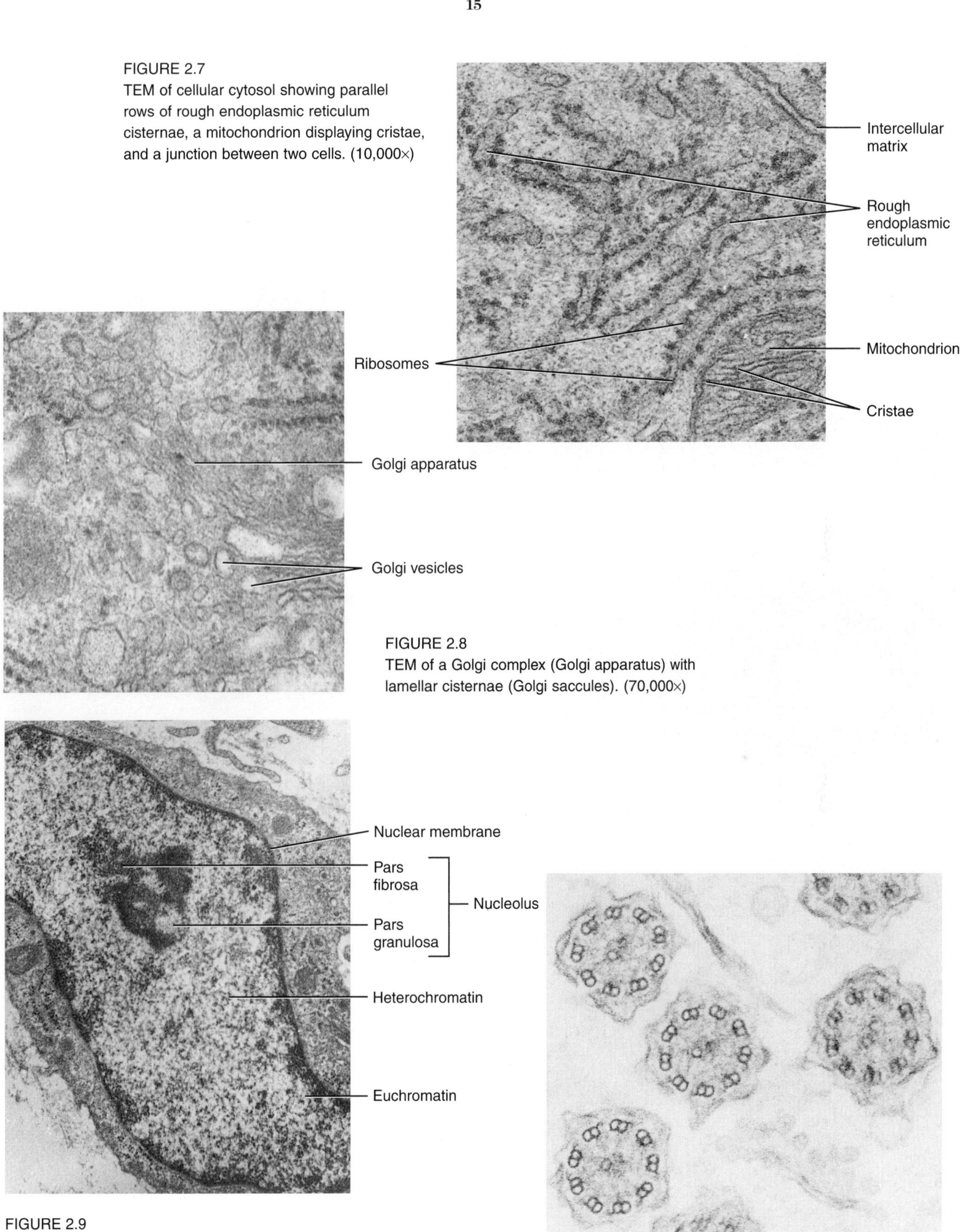

FIGURE 2.7
TEM of cellular cytosol showing parallel rows of rough endoplasmic reticulum cisternae, a mitochondrion displaying cristae, and a junction between two cells. (10,000×)

FIGURE 2.8
TEM of a Golgi complex (Golgi apparatus) with lamellar cisternae (Golgi saccules). (70,000×)

FIGURE 2.9
TEM of a cell nucleus exhibiting the characteristic appearance of a nuclear membrane, granular heterochromatin, dispersed euchromatin, and a nucleolus. (6000×)

FIGURE 2.10
TEM of cilia, which are motile appendages of specialized cells. (50,000×)

FIGURE 2.11
TEM of collagen fibers displaying unit fibrils. The fibrils exhibit repeating bands of fibrils spaced at 640 nm/interval. (36,000×)

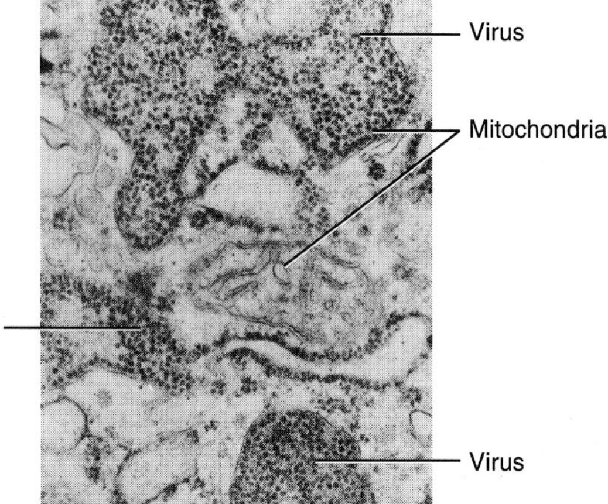

FIGURE 2.12
TEM of mitochondria and endoplasmic reticulum infiltrated with a virus. The mitochondria change shape and disintegrate as the virus multiplies within the cell. (150,000×)

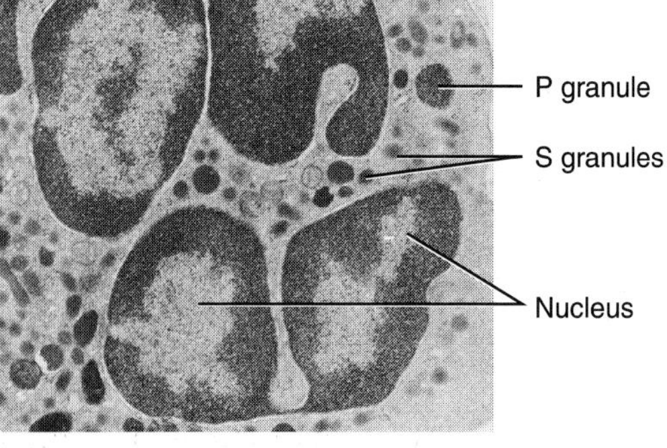

FIGURE 2.13
TEM of a neutrophil displaying the characteristic multilobed nucleus. (2000×)

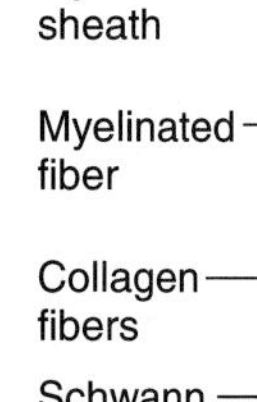
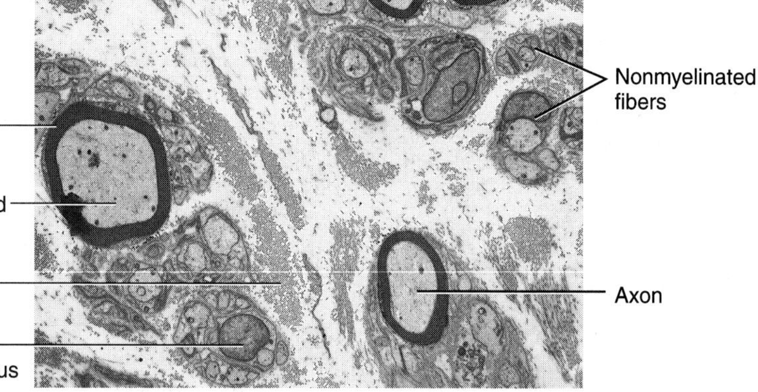

FIGURE 2.14
TEM of a cross section through a small area of a nerve showing myelinated and non-myelinated nerve fibers (axons) of various sizes embedded in Schwann cells. (6000×)

FIGURE 2.15
TEM of skeletal muscle demonstrating the myofibrils with alternating actin and myosin myofilaments. (10,500×)

Mitosis, Meiosis, and Chromosomes

The Cell Cycle

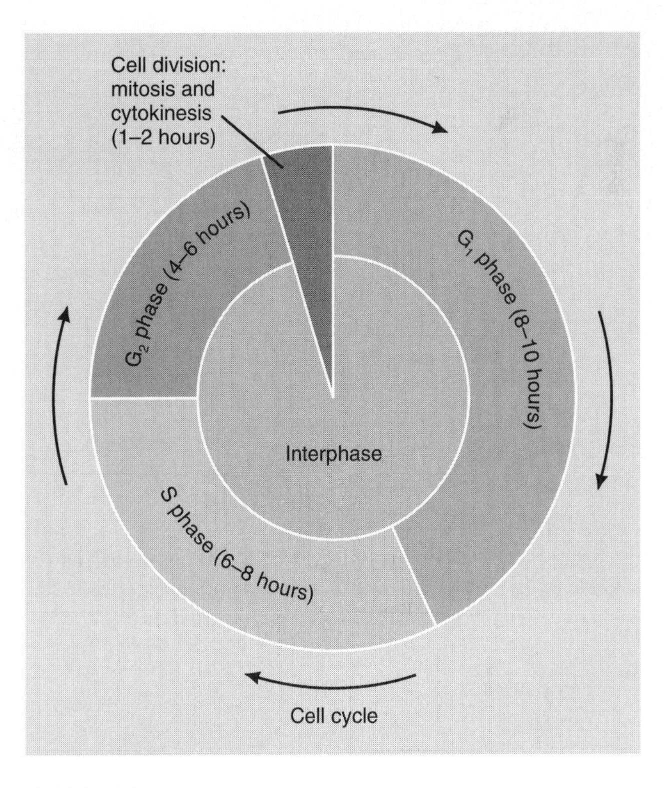

FIGURE 3.1
Diagrammatic presentation of the cell cycle displaying length of relative time it takes for each phase of the cell cycle to be completed.

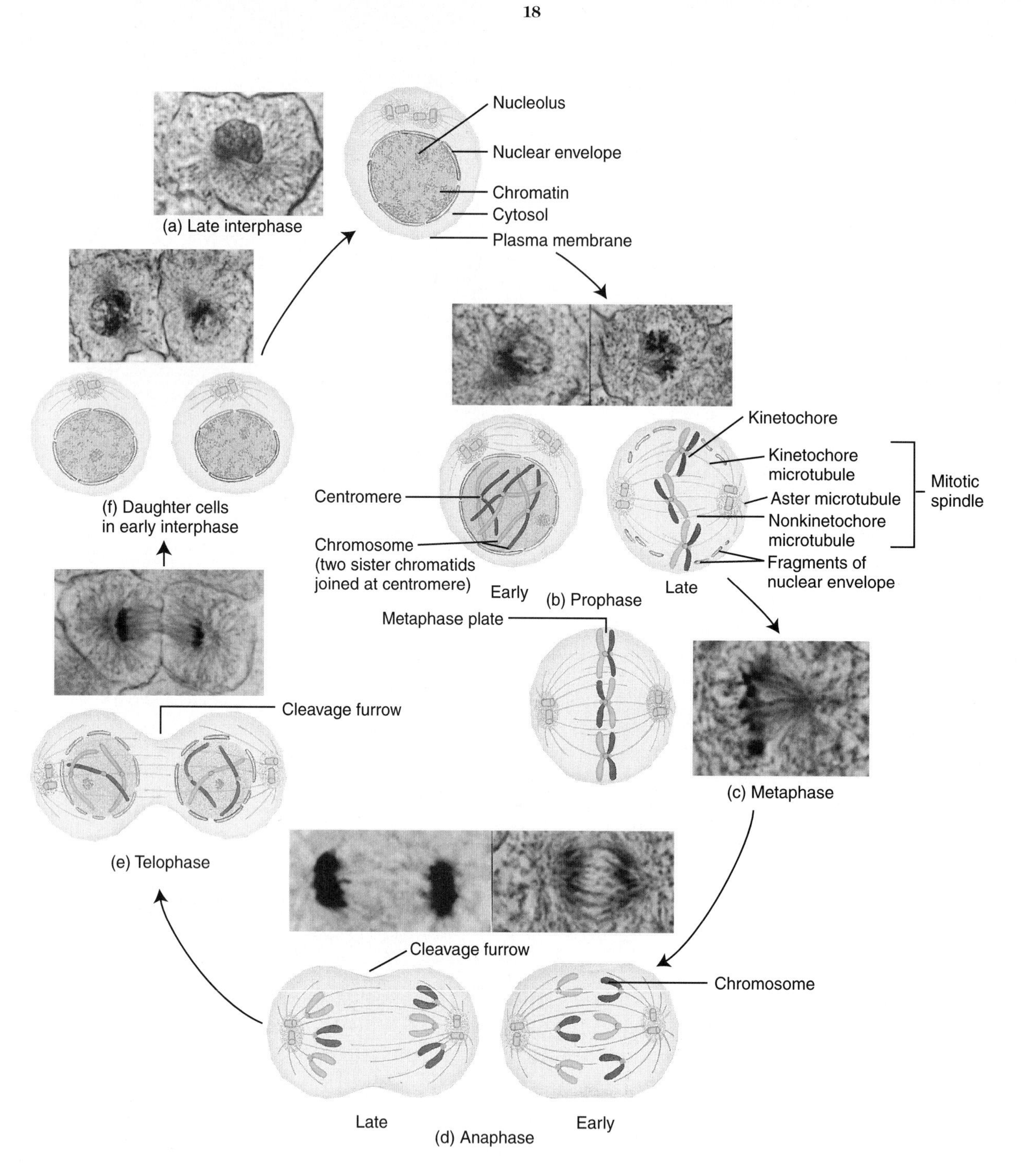

FIGURE 3.2
Diagrammatic representation of cell division—mitosis and cytokinesis—as seen in animal cells: (a) cell in the late interphase stage of the cell cycle; (b) cell in the early and late prophase stages; (c) cell in the metaphase stage; (d) cell in the anaphase stage; (e) cell cytoplasm going through cytokinesis (cytoplasmic division); and (f) formation of new daughter cells.

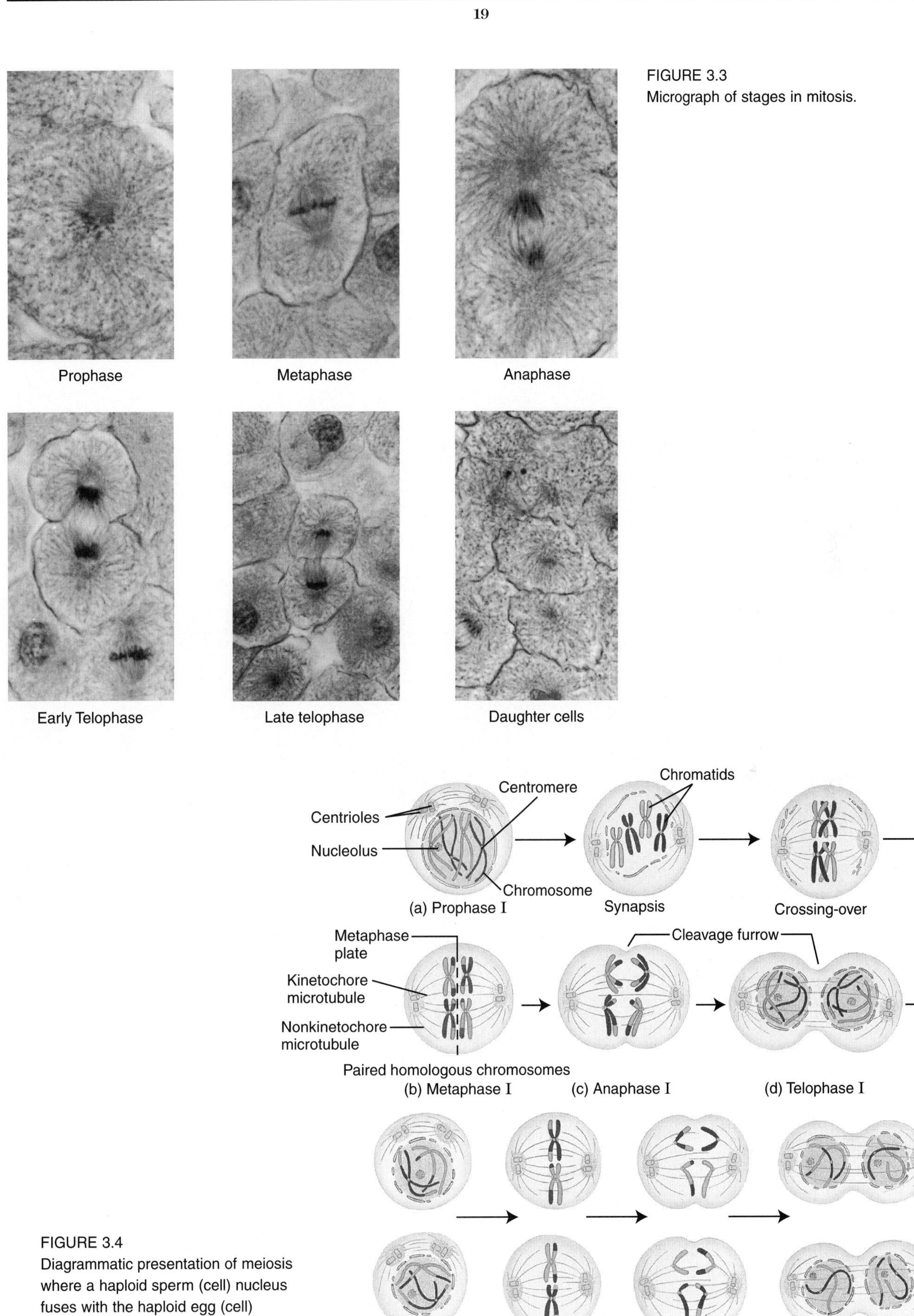

FIGURE 3.3
Micrograph of stages in mitosis.

FIGURE 3.4
Diagrammatic presentation of meiosis where a haploid sperm (cell) nucleus fuses with the haploid egg (cell) nucleus, thus initiating the reduction division (meiosis I).

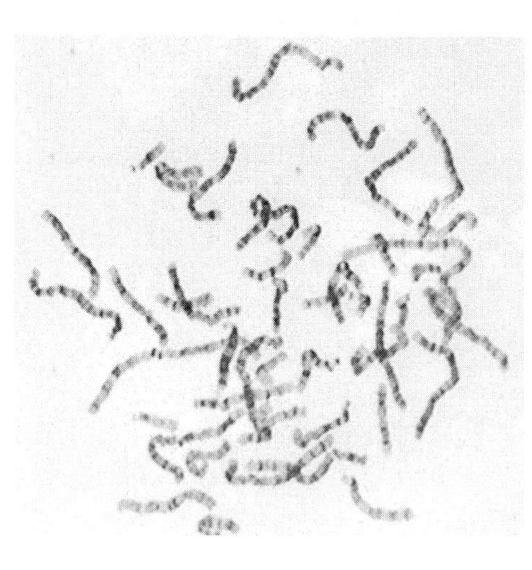

FIGURE 3.5
Light micrograph (LM) of human chromosomes isolated from *in vitro* cell culture at the onset of mitosis.

FIGURE 3.6
LM illustrating the karyotype of human female chromosomes arrested in the metaphase stage of mitosis, *in vitro,* by colchicine treatment.

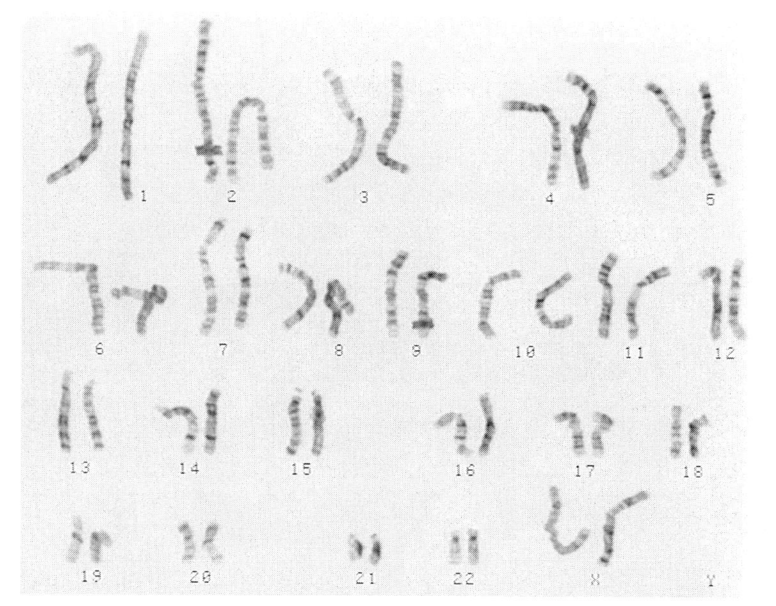

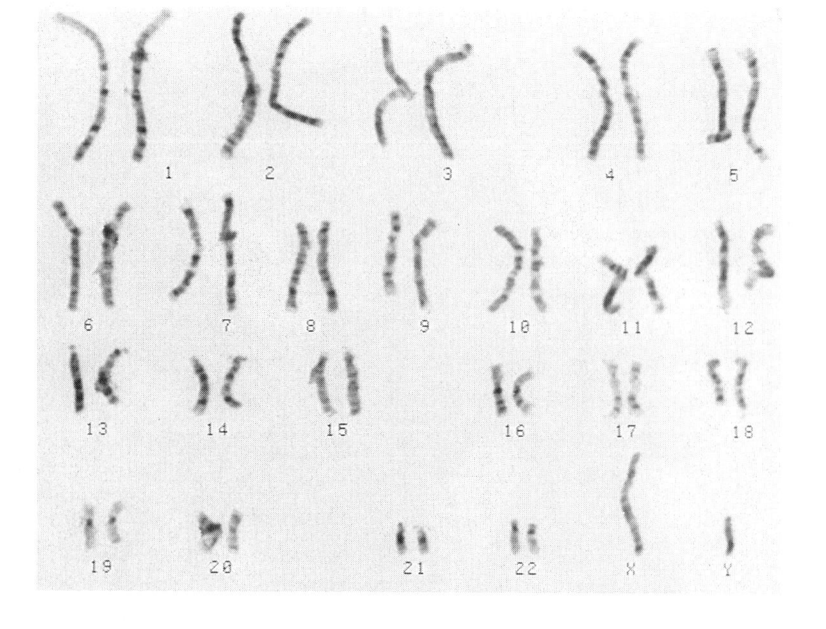

FIGURE 3.7
LM showing the karyotype of human male metaphase chromosomes, prepared from white blood cells cultured *in vitro.*

Epithelium

There are four basic types of body tissues: **epithelial, connective, muscular,** and **nervous.** These tissues are found in all organs of the body and function in close association with each other. Epithelial tissue covers free surfaces, lines body cavities and tubular structures, assumes many secretory and lubrication functions, and protects organs from invasion by microorganisms.

The classification of epithelial tissue is based on the shape and arrangement of cells that overlie a **basement lamina** or membrane. The lamina also separates the epithelial tissue from the underlying connective tissue. A group of epithelial cells that forms a single layer of flat, irregular cells on the lamina is known as **simple squamous epithelium.** In the lymphatic and cardiovascular systems, this layer is the endothelium. In the lining of the peritoneal, pleural, and pericardial cavities, the epithelium is called mesothelium. Owing to the thin, flat shape of simple squamous cells, passive transport of fluid metabolites, nutrients, and gases can readily take place between the vascular system and the epithelial cells.

Simple columnar epithelial cells have greater height than width. They are often found in the gastrointestinal tract, where they function as secretory and absorptive cells. They are also found in the bronchi and as secretory cells in the oviduct and uterus.

The **simple cuboidal cells** are cubic in shape. They function as secretory or absorptive cells (kidney). They also line the ducts of glandular tissue.

The **stratified squamous epithelium** forms several cell layers in the skin, oral cavity, pharynx, esophagus, anal canal, and vagina. The epithelium forms a protective covering with the capacity to replace dead cells periodically. The stratified squamous epithelium may be keratinized or nonkeratinized. Nonkeratinized epithelium lines most cavities (mouth, pharynx, vagina, and anal canal).

Keratinized epithelium forms the upper layer of the skin and the gingival lining of the teeth. **Stratified cuboidal epithelium** and **stratified columnar epithelium** are not common. They are generally found in the ducts of the pancreas, sweat glands, and salivary glands. The stratification of cells does not exceed more than two or three layers.

Transitional epithelium changes shape when stretched. The surface cells take on a dome-shaped appearance when relaxed and a squamous shape when stretched. Transitional epithelium lines the urinary bladder and urinary tract.

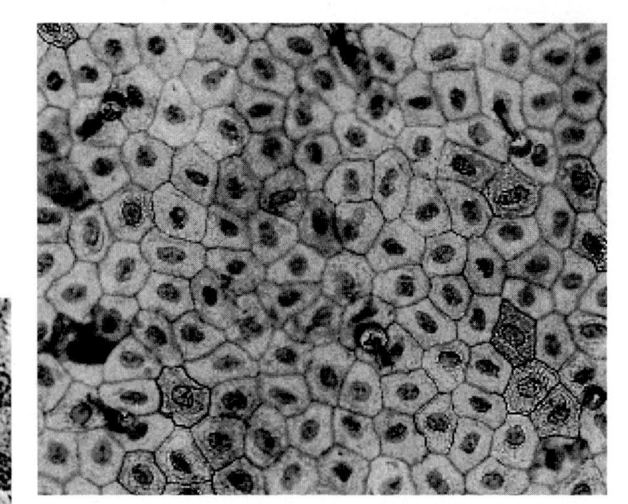

FIGURE 4.1
Light micrograph (LM) of simple squamous epithelial cells in a surface view of mesothelium. (400×)

Parietal cells
Visceral cells
Glomerulus
Capsular space

FIGURE 4.2
LM of a Bowman's capsule located at the vascular pole of the renal corpuscle. (400×)

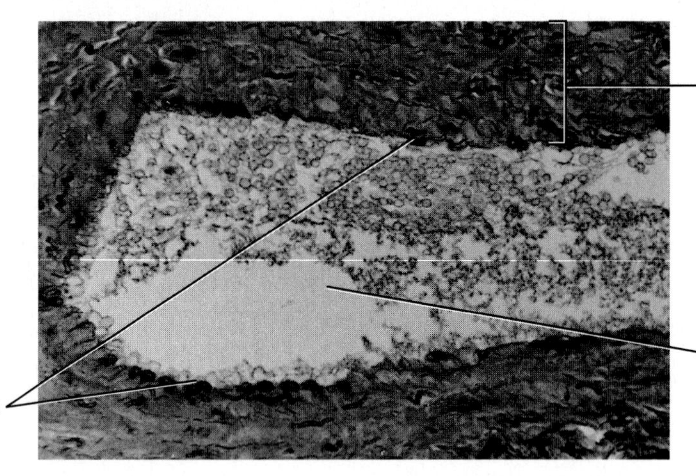

FIGURE 4.3
LM of a blood vessel in cross section. The lumen of the blood vessel is lined with simple squamous epithelium. (400×)

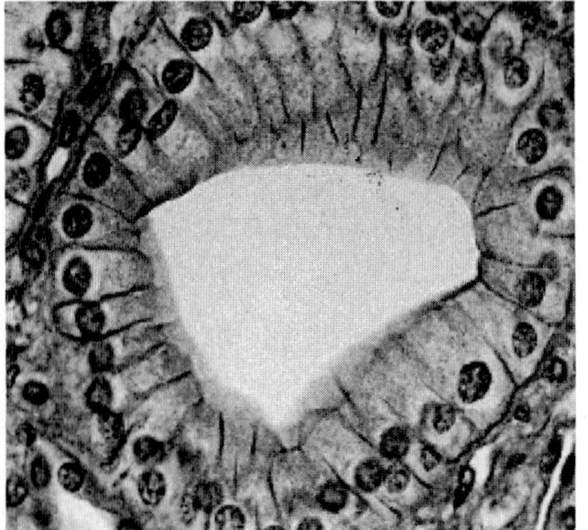

FIGURE 4.4
LM of simple columnar cells lining the collecting ducts in the medulla of the kidney. (400×)

Microvilli
Goblet cell
Columnar cells
Lamina propria

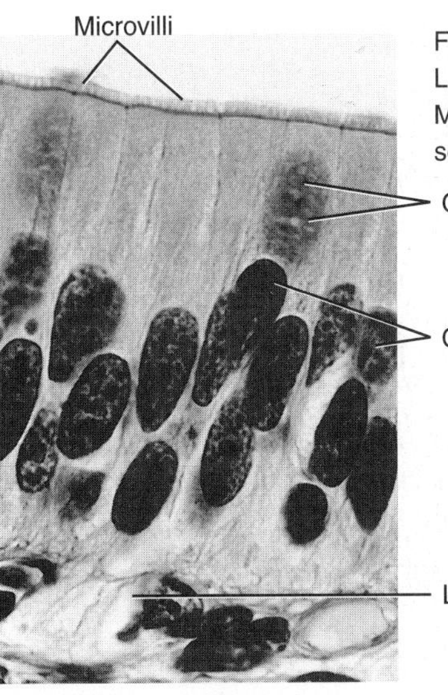

FIGURE 4.5
LM of simple columnar epithelium at a higher magnification. Microvilli border the columnar cells. Goblet cells can also be seen in the micrograph. (400×)

Cuboidal cells
Lumen

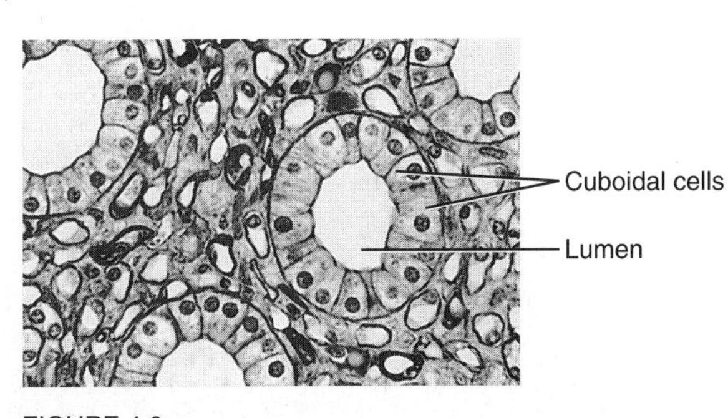

FIGURE 4.6
LM of simple cuboidal epithelium in a collecting tubule of the kidney medulla. (400×)

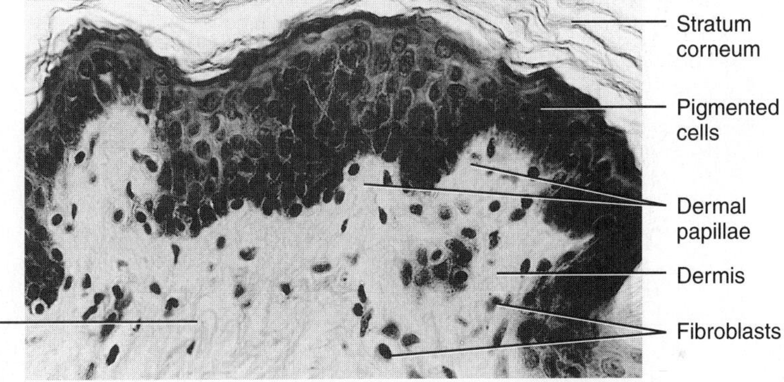

FIGURE 4.7
LM of a highly pigmented stratified squamous epithelium lining the skin. (200×)

Stratum corneum
Pigmented cells
Dermal papillae
Dermis
Fibroblasts
Connective tissue

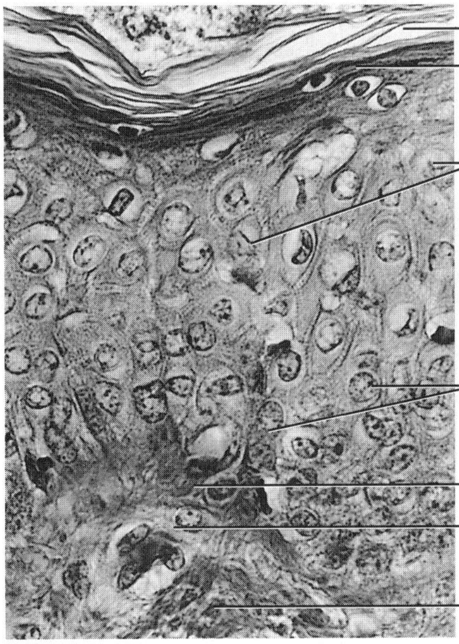

Stratum corneum
Stratum lucidum
Stratum granulosum
Stratum spinosum
Stratum germinativum
Dermal papilla
Connective tissue of dermis

FIGURE 4.8
LM of keratinized, stratified squamous epithelium as seen in a thick skin. (200×)

FIGURE 4.9
LM of upper region of esophagus, showing an internal lining of nonkeratinized, stratified squamous epithelium. (200×)

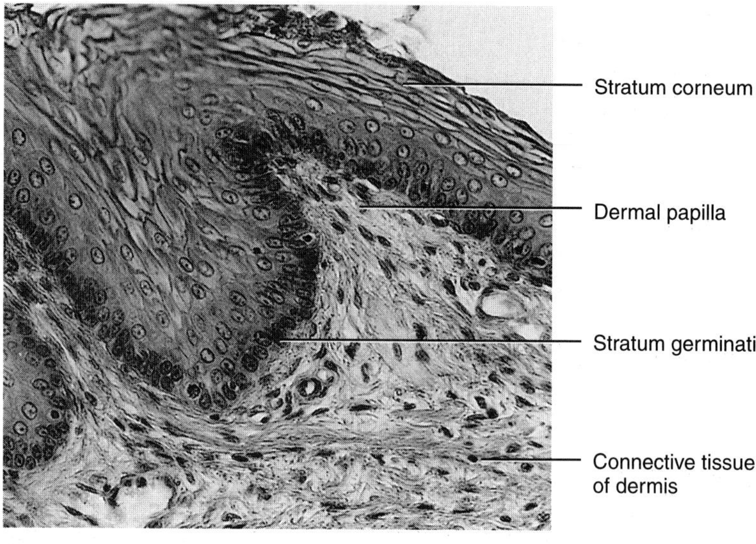

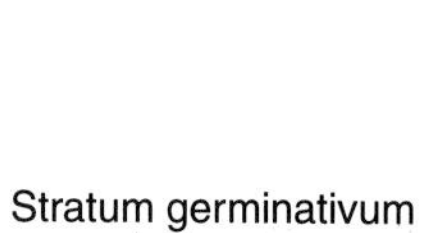

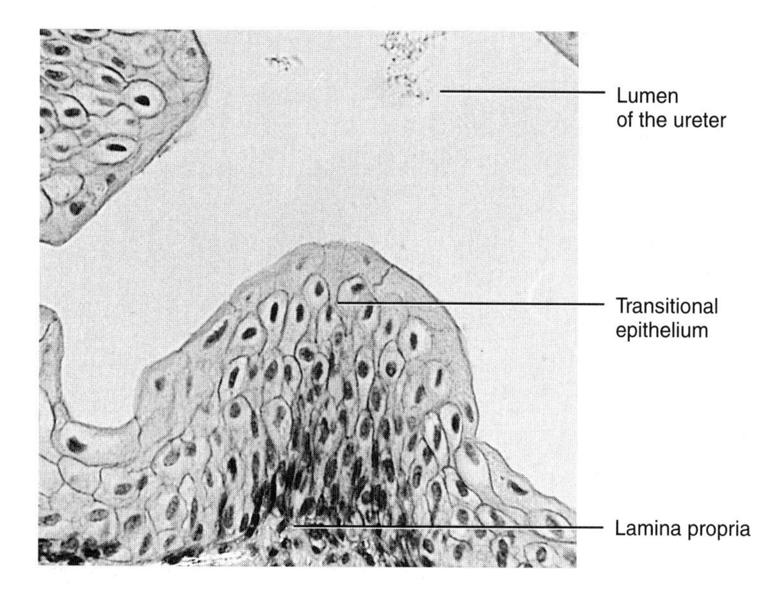

FIGURE 4.10
LM of transitional epithelium lining the ureters. (200×)

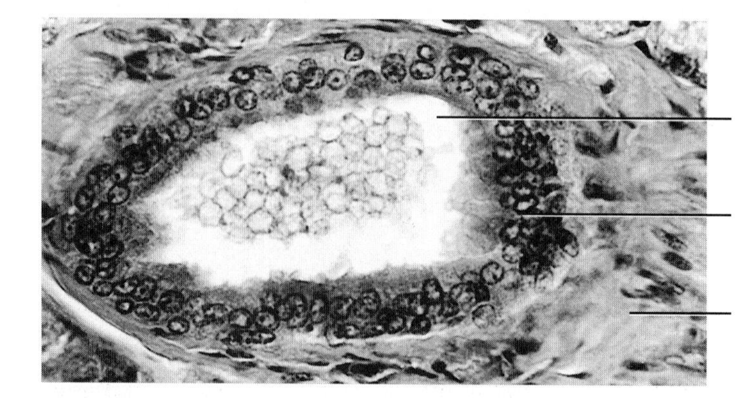

FIGURE 4.11
LM of stratified cuboidal epithelium in the ducts of a submaxillary gland. (400×)

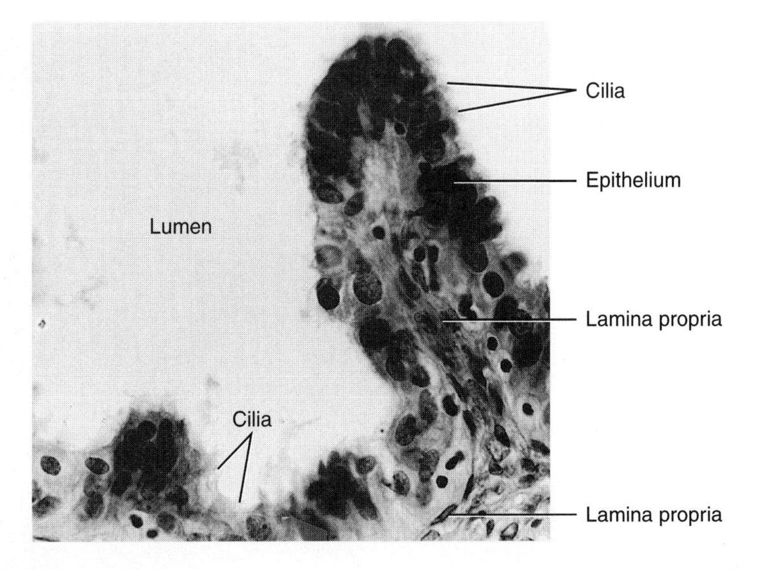

FIGURE 4.12
LM of simple ciliated columnar epithelium of the oviduct magnified to delineate tufts of cilia bordering the columnar cells. (1000×)

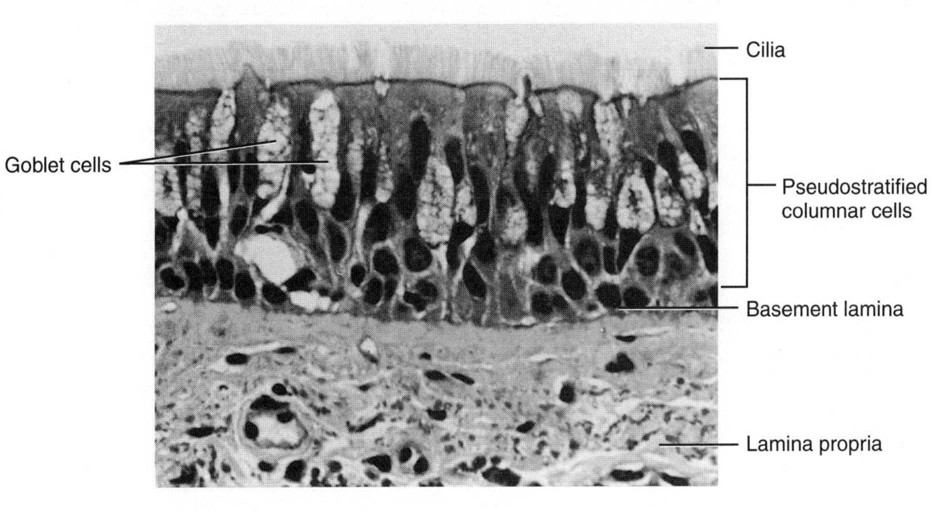

FIGURE 4.13
LM of pseudostratified ciliated columnar epithelium with interspersed goblet cells in the human trachea. (1000×)

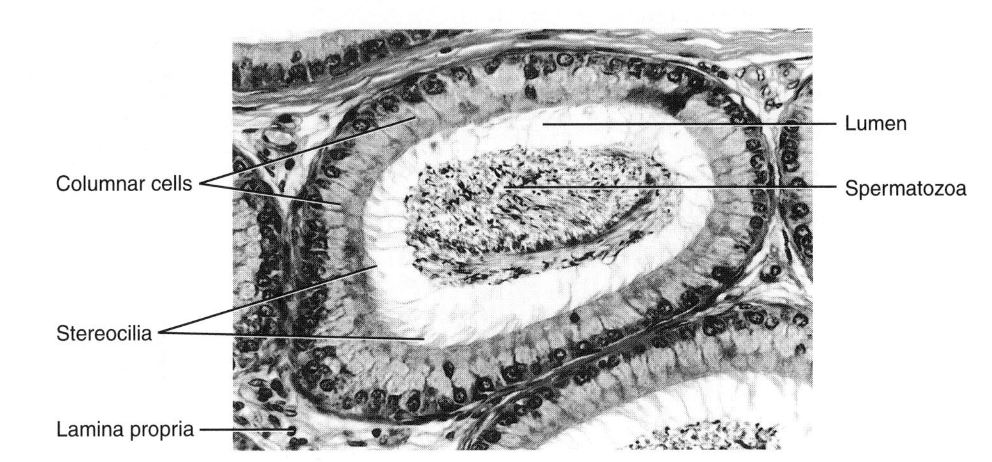

FIGURE 4.14
LM of tall pseudostratified columnar epithelial cells in the lining of the epididymis. (400×)

Connective Tissue

In a developing embryo, the majority of the connective tissue is derived from mesenchyme, a derivative of the mesodermal germinal layer. The connective tissue is a combination of a diverse group of tissues that are responsible for performing several functions. Even though these tissues are structurally and functionally diverse, they share common qualities and for this reason are considered collectively. Functionally, connective tissues provide support, transport, defense, repair, storage, packing material, and insulation. Some of the functions may be general, whereas other functions may be delegated to specialized connective tissues such as blood, cartilage, and bone.

I. **Fetal Connective Tissues**
   A. **Mucous** (e.g., umbilical cord)
   B. **Mesenchymal** (e.g., developing embryo and fetus)

II. **Adult Connective Tissues**
   A. **Loose or areolar:** packing tissue associated with most organs
   B. **Reticular:** tissue found in bone and lymph nodes
   C. **Adipose:** storage tissue in subcutaneous and omentum areas
   D. **Dense irregular:** found in dermis, periosteum, perichondrium, and capsules of organs
   E. **Dense irregular:**
      1. **Collagenous** (e.g., tendons, aponeurosis, cornea, and ligaments)
      2. **Elastic** (e.g., ligamentum nuchae, ligamentum flavum, and suspensory ligament of penis)

III. **Specialized Connective Tissues**
   A. **Supporting tissue**—cartilage:
      1. **Hyaline** (e.g., in trachea and costal cartilage)
      2. **Elastic** (e.g., in epiglottis and external ear)
      3. **Fibrous** (e.g., symphysis pubis, intervertebral disk)
   B. **Supporting tissue—bone:**
      1. **Cancellous,** or spongy bone
      2. **Compact,** or dense bone

IV. **Transport Tissues** (e.g., blood, cardiovascular system and hematopoietic tissue)

Connective Tissue Cells

A diverse group of cells is found in the connective tissue. Cells of the loose or areolar tissue are organized into six basic types: **fibroblast, macrophage, plasma, mast cells, adipose cells,** and **leukocytes** under certain pathological conditions.

Connective Tissue Fibers

Based on their morphological differences, connective tissue fibers can be divided into three groups: collagen, elastic, and reticular fibers.

Collagen fibers are the most common fibers of the connective tissues and are found in almost all types of connective tissue. Collagen fibers give great tensile strength to organs and parts of the body where skeletal movement occurs.

Elastic fibers are less common than collagen fibers. They are thin, branched, and relatively short, and they have lower tensile strength. Elastic fibers exhibit elasticity. These fibers can be seen in the lungs, large blood vessels, urinary bladder, skin, and elastic cartilage.

Reticular fibers are the least common of all the connective tissue fibers. They are found in the lymph nodes, spleen, liver, and hematopoietic tissue.

Ground Substance of Connective Tissue

The ground substance forms a permeable amorphous matrix in which fibers and cells are embedded. Chemically, the matrix is a mixture of proteoglycans, which create a gel-like consistency, and glycosaminoglycans (carbohydrate polymers). It also contains hyaluronic acid, chondroitin 6-sulfate, chondroitin 4-sulfate, dermatan sulfate, and heparin sulfate. Glycoprotein and fibronectin may also be present in the connective tissue proper.

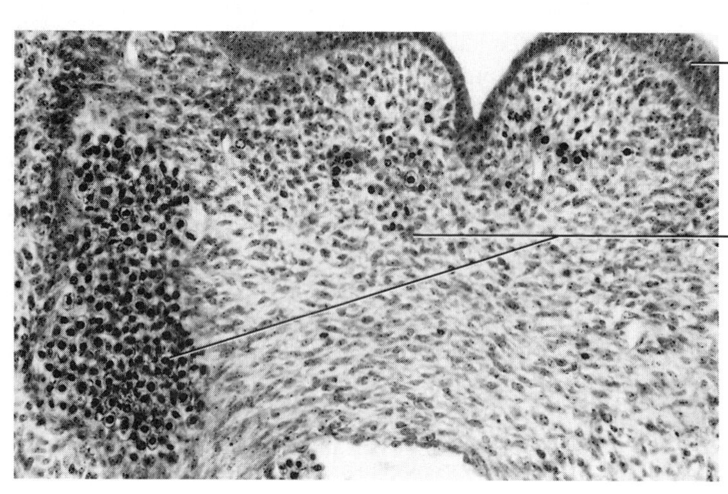

FIGURE 5.1
Light micrograph (LM) of mesenchymal connective tissue in a human fetus. The tissue is still at an immature state and is partially cellular. (100×)

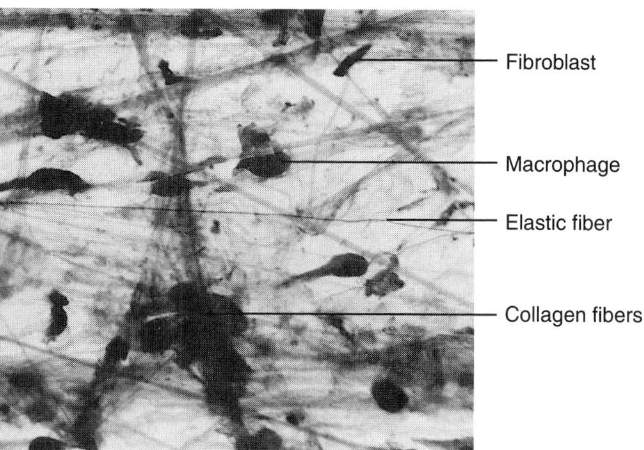

FIGURE 5.2
LM of areolar connective tissue in a spread of omentum. The collagen fibers are thick and run an undulating course in the field. (400×)

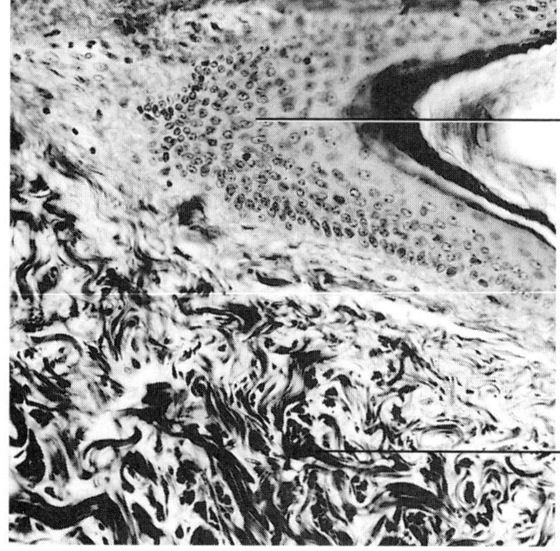

FIGURE 5.3
LM of dense irregular connective tissue as seen in the dermis of the axilla. (200×)

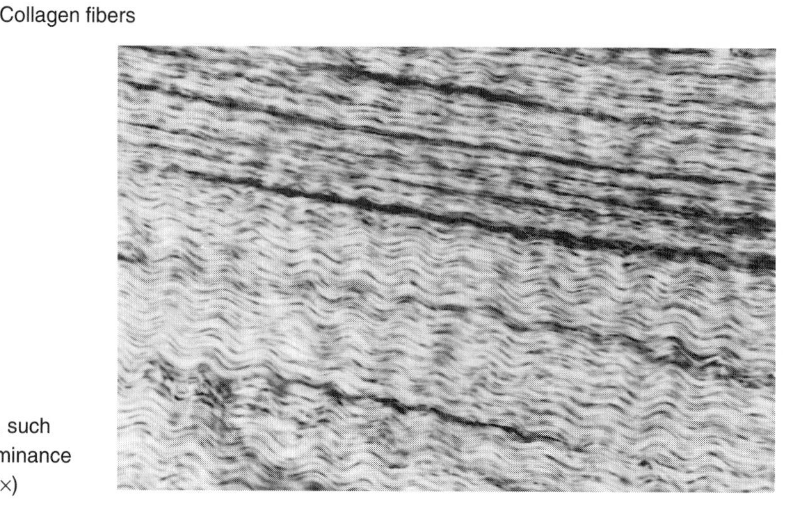

FIGURE 5.4
LM of dense regular connective tissue, such as in tendons, which contains a predominance of collagen fibers in parallel rows. (200×)

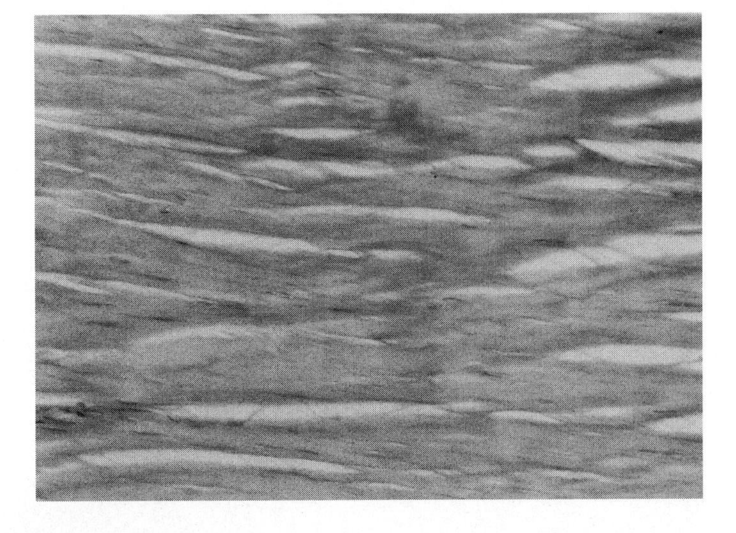

FIGURE 5.5
LM of dense regular connective tissue of ligamentum nuchae, as seen at a higher magnification. (200×)

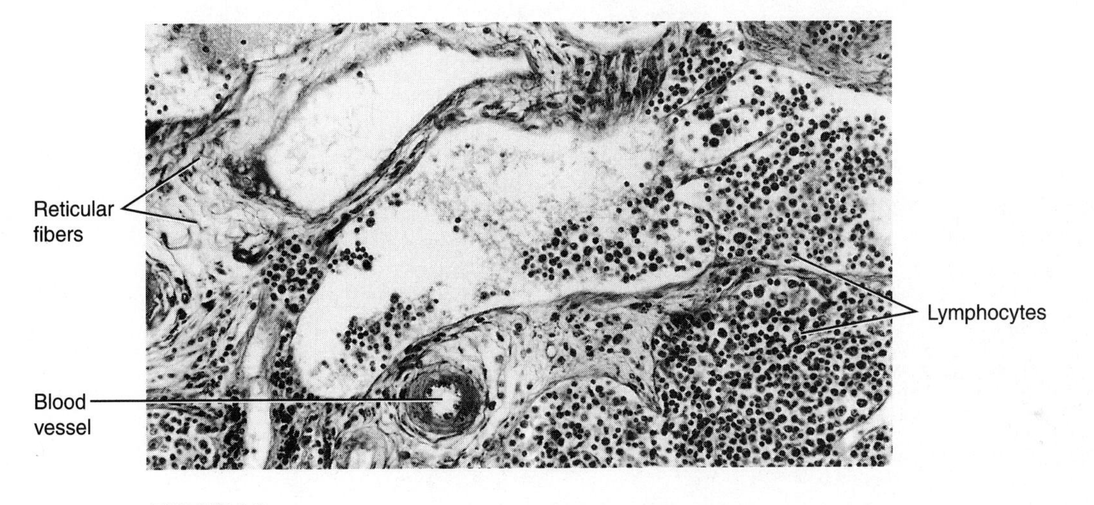

FIGURE 5.6
LM of reticular connective tissue in the medulla of a lymph node. (200×)

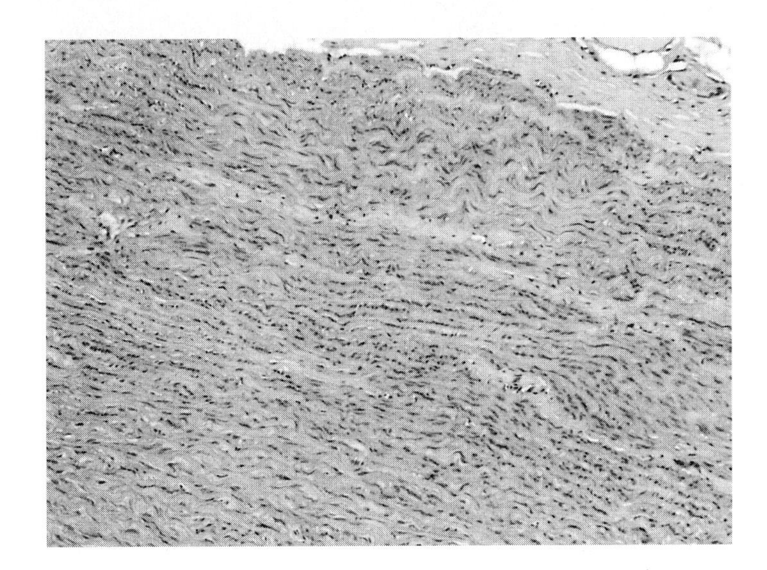

FIGURE 5.7
LM of elastic connective tissue as seen in the body wall of the aorta. (200×)

FIGURE 5.8
LM of adipose connective tissue showing adipocytes, or fat cells, without any cytoplasmic details. (400×)

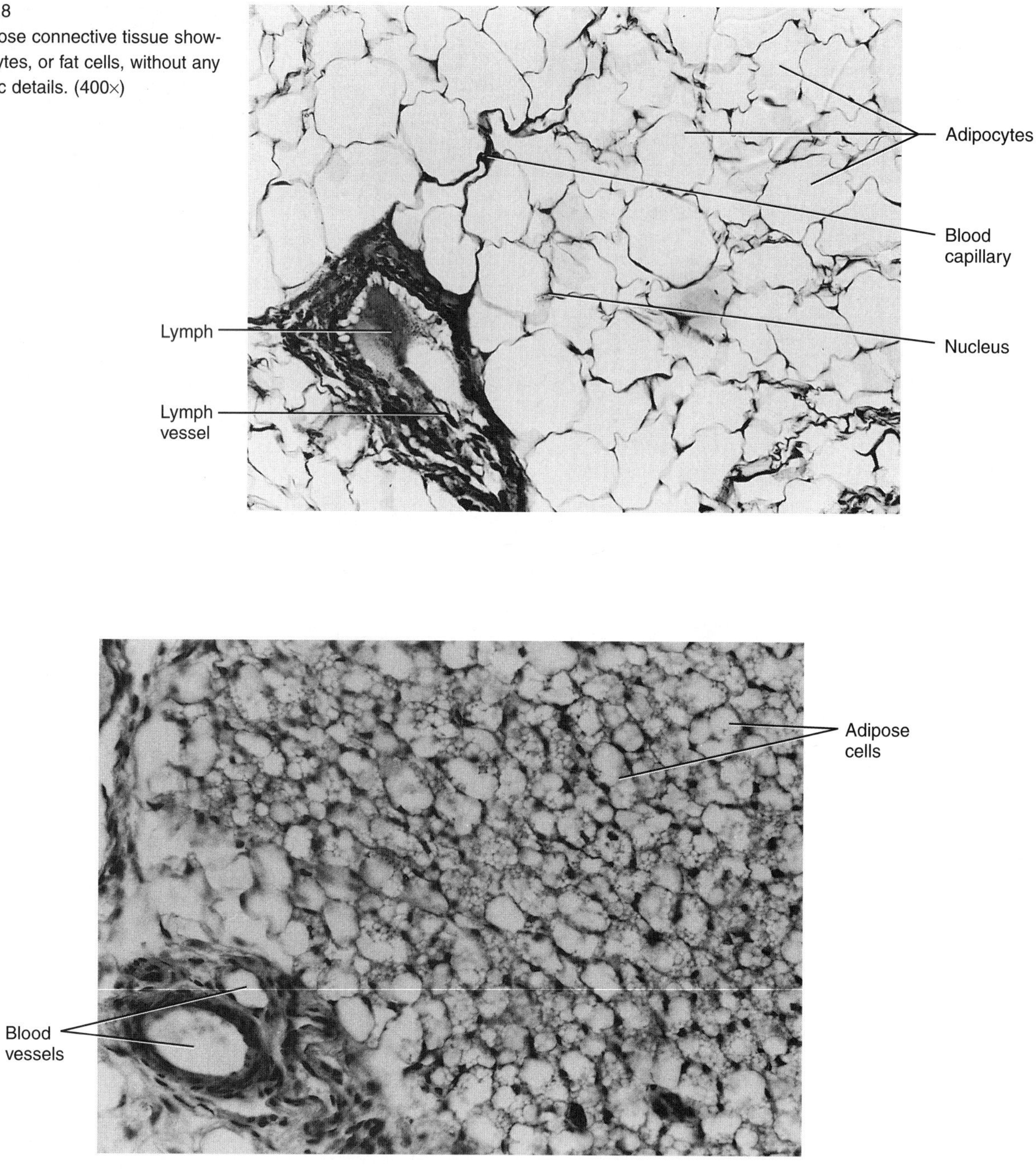

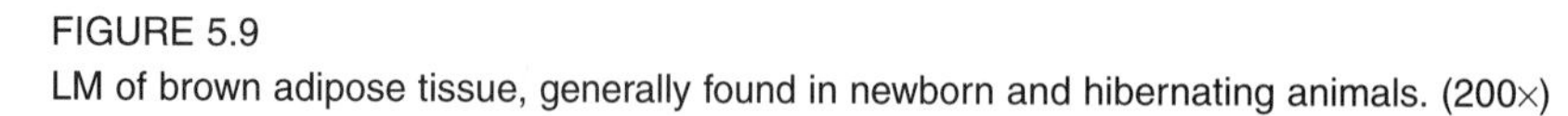

FIGURE 5.9
LM of brown adipose tissue, generally found in newborn and hibernating animals. (200×)

Specialized Connective Tissue: Cartilage

Skeletal tissue, which includes **cartilage** and **bone,** is specialized connective tissue composed of fibers, cells, and ground substance, the latter two of which form the intercellular matrix. Chemically, the ground substance in cartilage is composed of chondromucoids that have a high concentration of sulfated proteoglycans and hyaluronic acid. The combination of the two forms chondronectin and aggregates of proteoglycans. In comparison, collagenous fibers in bones are impregnated with an amorphous ground substance composed predominantly of calcium hydroxide crystals.

The **cartilage matrix** is composed of elastic and collagen fibers that increase the flexibility and strength of the tissue so that it can adapt to stress and the mechanical requirements of different regions of the body.

Cartilage, unlike other forms of connective tissue, has secretory cells called **chondrocytes,** which are embedded in the intercellular matrix. Surrounding the cells are empty spaces called **lacunae.** However, the space around the cells in reality is a fixation artifact caused by the shrinking of the matrix at the time of tissue processing. The cells found in the peripheral region of the cartilage, below the perichondrium, are called **chondroblasts.** The chondroblasts/chondrocytes create an elaborate extracellular matrix. The cartilage can be classified into three types: **hyaline cartilage, elastic cartilage,** and **fibrous cartilage.** The three types of cartilage are distributed in different parts of the body according to function.

Hyaline cartilage is one of the most common types of cartilage and is found as costal cartilage between the ribs and the sternum, articular cartilage between bone joints, and tracheal rings of the trachea and bronchi. It is also the cartilage of the nose and larynx and the cartilage that proliferates in elongation of areas of the long bones. Surrounding the hyaline cartilage is a dense mass of connective tissue that forms the covering or the perichondrium.

Elastic cartilage, which is a type of modified hyaline cartilage, has a predominance of branched elastic fibers that almost obscure the matrix. Just below the perichondrium, the elastic fibers are loosely arranged and form a network of fibers that blends with the overlying perichondrium. Owing to the nature of elastic fibers, elastic cartilage is more flexible than hyaline and fibrous cartilage. The elastic cartilage can be seen in such places as the auditory tube, external ear, epiglottis, and the smaller cartilages associated with the larynx.

Unlike hyaline and elastic cartilage, **fibrous cartilage,** or **fibrocartilage,** is not a derivative of mesenchymal tissue. It is a product of dense connective tissue that has differentiated into cartilage as a result of stress and the weight-bearing demands of the body. Fibrous cartilage is found in the intervertebral disks, symphysis pubis of the pelvic girdle, menisci, and ligaments associated with joints and junctional areas between bones, tendons, and ligaments.

Hyaline Cartilage

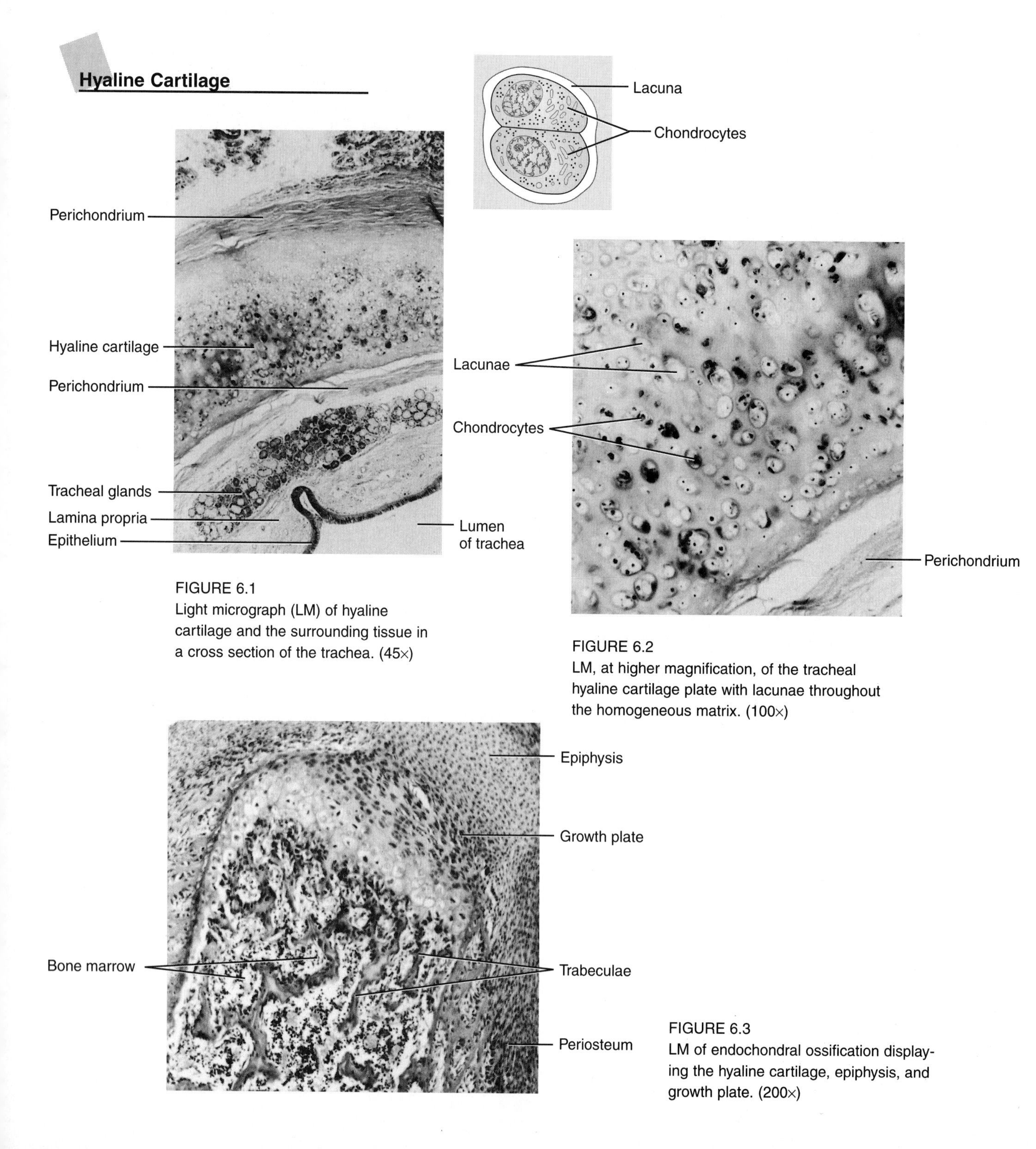

FIGURE 6.1
Light micrograph (LM) of hyaline cartilage and the surrounding tissue in a cross section of the trachea. (45×)

FIGURE 6.2
LM, at higher magnification, of the tracheal hyaline cartilage plate with lacunae throughout the homogeneous matrix. (100×)

FIGURE 6.3
LM of endochondral ossification displaying the hyaline cartilage, epiphysis, and growth plate. (200×)

Elastic Cartilage

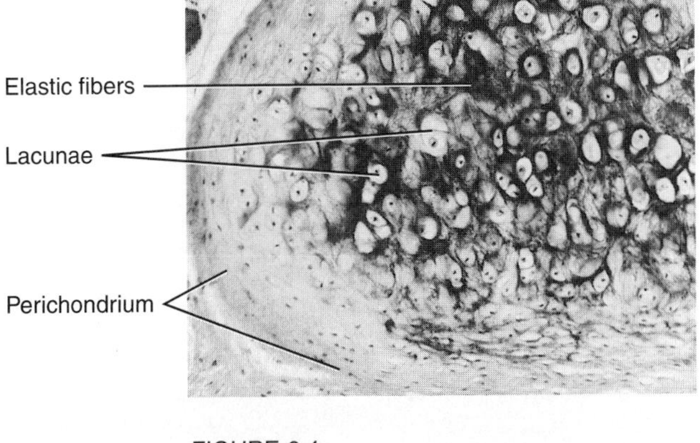

FIGURE 6.4
LM of elastic cartilage taken from the epiglottis. (100×)

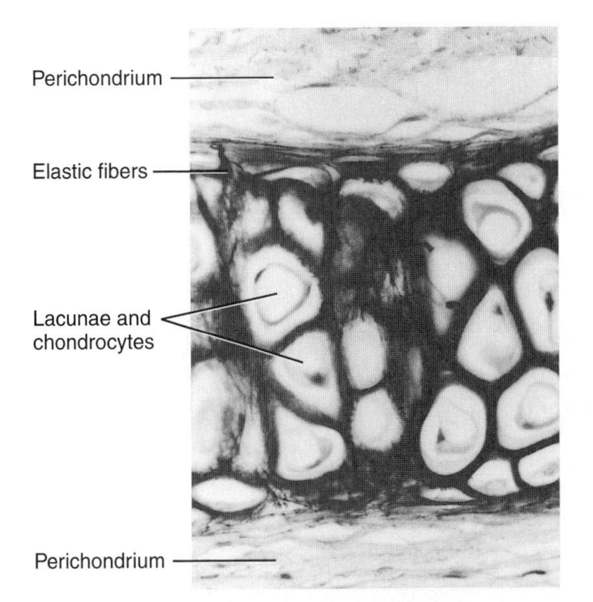

FIGURE 6.5
LM of a cross section of elastic cartilage displaying a dense network of elastic fibers surrounding the lacunae. Perichondrium can also be seen in the micrograph. (200×)

Fibrous Cartilage

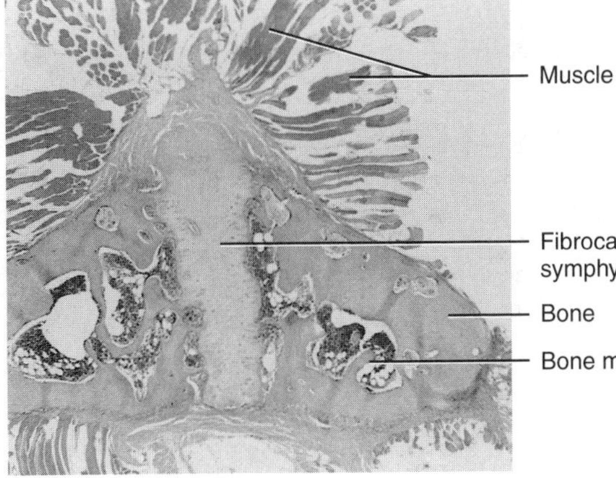

FIGURE 6.6
LM of fibrocartilage as seen in the symphysis pubis of the pelvis. (40×)

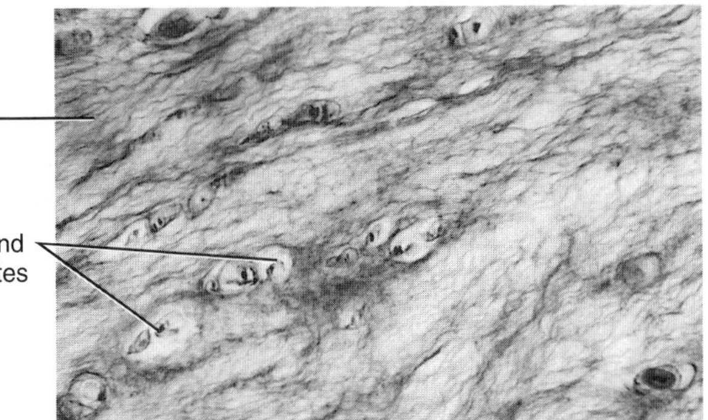

FIGURE 6.7
LM of fibrocartilage in the intervertebral disk. (100×)

Specialized Connective Tissue: Bone and the Skeleton System

CHAPTER 7

Bone, as in the case of **cartilage,** is a modified connective tissue. However, the ground substance secreted by the bone cells is mineralized, resulting in a dense, hard, nonpliable, weight-bearing, and high-compression-strength substance. Bone is a complex living tissue that is highly vascularized and is constantly being deposited and reabsorbed throughout life. **Hormones** play an important role in the process of bone deposition and bone reabsorption.

Bone matrix possesses a predominance of collagen fibers that are infiltrated with a heavily mineralized amorphous substance. Embedded in the bony matrix are bone cells or **osteocytes.** The osteocytes in bone, like the **chondrocytes** in cartilage, are surrounded by spaces or **lacunae.** Bone tissue also possesses a covering of dense connective tissue called **periosteum.** Bone cells, **osteoblasts,** and osteocytes are direct derivatives of **mesenchyme,** as is cartilage.

Another unique feature of bone is that its cells are nourished by blood capillaries that lie approximately 0.2 nm or less from the bone cells. The bone matrix is traversed by fine channels, known as canaliculi, which are filled with nutritious fluid from the blood capillaries. The canaliculi connect the osteocytes, which are in the lacunae, to the source of nutrients, the blood capillaries, thus providing a constant supply of oxygen and nutrition to all the osteocytes in the bone tissue.

Cells associated with the bone remodeling of calcified bones are called **osteoclasts.** These cells are multinucleated and are formed by the fusion of numerous **osteoprogenitor** cells. The osteoclasts are generally present in bone depressions called **Howship's lacunae.** The periosteum that covers the external surface of the bone is heavily anchored to the bone by bundles of dense collagenous connective tissue fibers called **Sharpey's fibers.** The connective tissue is infiltrated by fibroblasts and osteoprogenitor cells.

Morphologically, bone can be classified into two types: **compact** or **dense** bone, and **cancellous** or **spongy** bone. In compact bone, the bone may be deposited either in layers parallel to one another or in a concentric manner around a blood vessel, thus presenting a lamellar arrangement of bone. This concentric lamellar deposition of bone around a blood vessel forms a microscopic unit comprised of a blood vessel, a lymph vessel, and possibly a nerve. This unit of bone is known as a Haversian system.

Dense bone contains longitudinally oriented channels; these channels are remnants of blood vessels and are called **Haversian canals.**

Cancellous (or spongy) bone is in many ways similar to dense (or compact) bone; however, the ossified **trabeculae** and **spicules** are thin and are surrounded by bone marrow and blood vessels. Osteons are mostly absent in spongy bone. The spaces between the spicules and trabeculae are filled with **hematopoietic** tissue.

In a typical long bone, the shaft, or **diaphysis,** is comprised predominantly of compact bone that surrounds the medullary or bone marrow cavity. The terminal ends (**epiphysis**) of the long bones are essentially cancellous bone covered with a thin shell of compact bone. The marrow cavity of the long bone (diaphysis) is continuous with the marrow cavity of the cancellous bone.

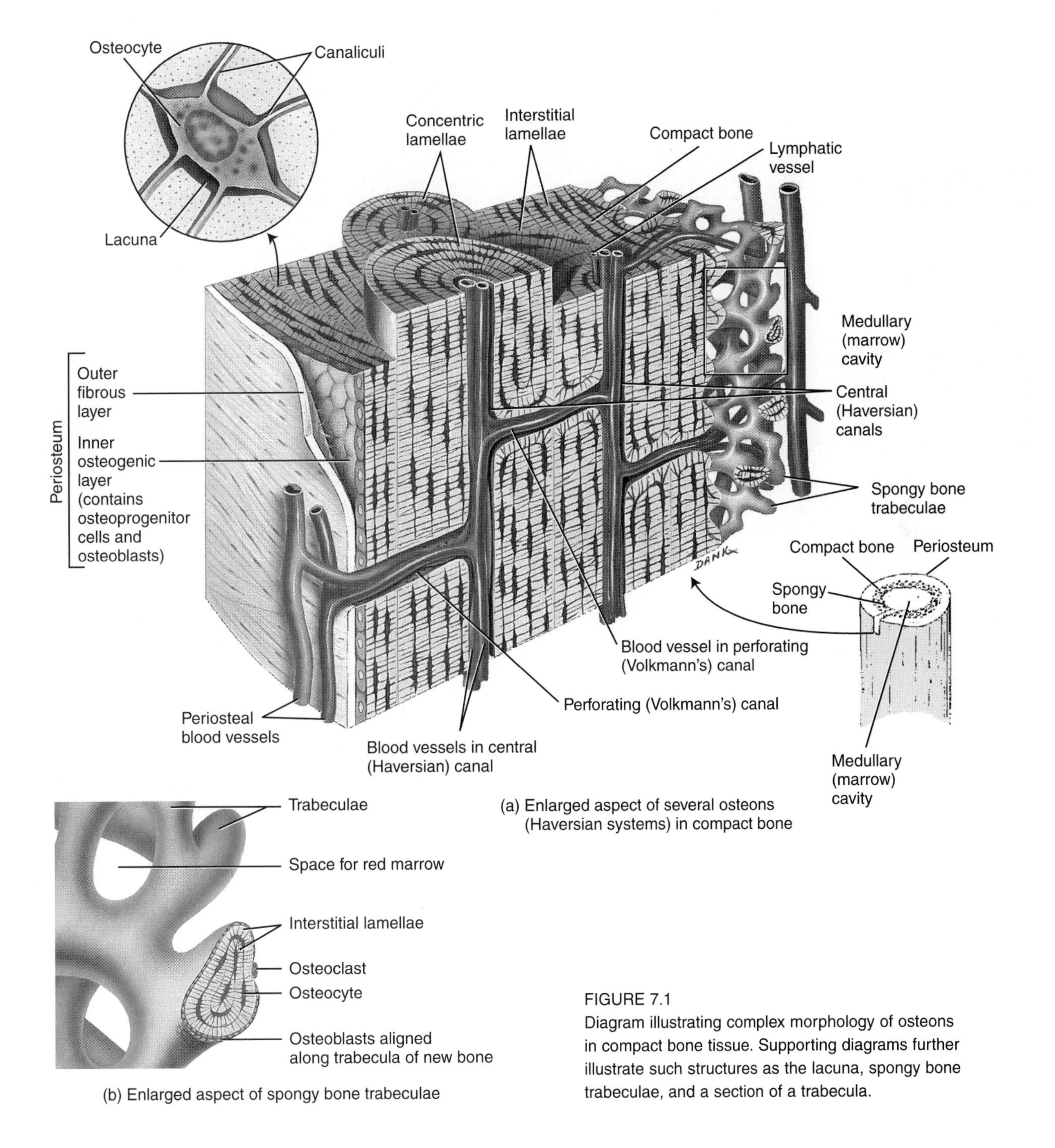

(a) Enlarged aspect of several osteons (Haversian systems) in compact bone

(b) Enlarged aspect of spongy bone trabeculae

FIGURE 7.1
Diagram illustrating complex morphology of osteons in compact bone tissue. Supporting diagrams further illustrate such structures as the lacuna, spongy bone trabeculae, and a section of a trabecula.

FIGURE 7.2
Light micrograph (LM) of ground bone that has been made paper thin by grinding with abrasives. The canaliculi and almond-shaped lacunae with osteocytes are clearly demonstrated. (200×)

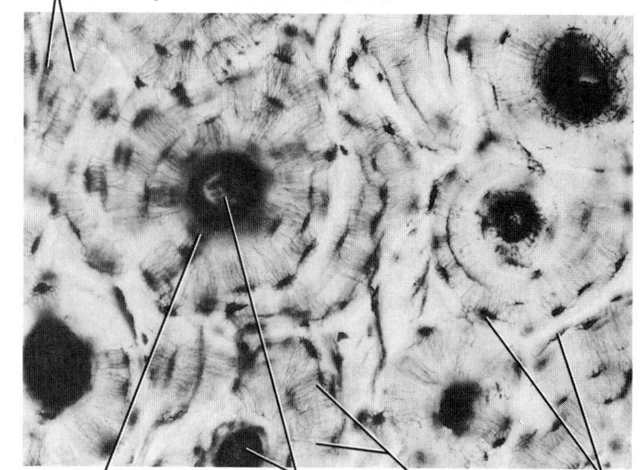

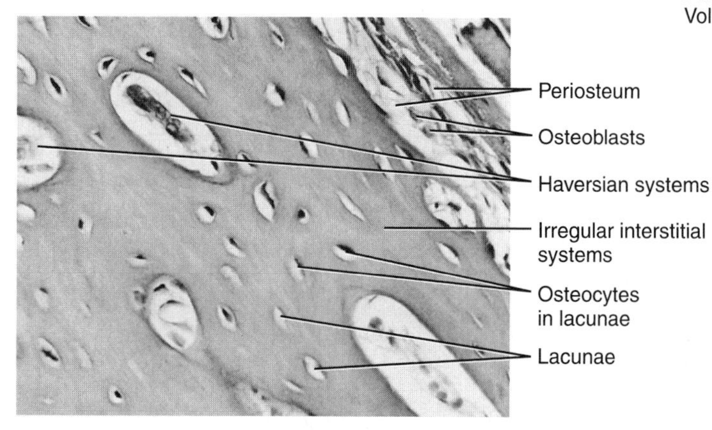

FIGURE 7.3
LM of dense bone at a higher magnification. (400×)

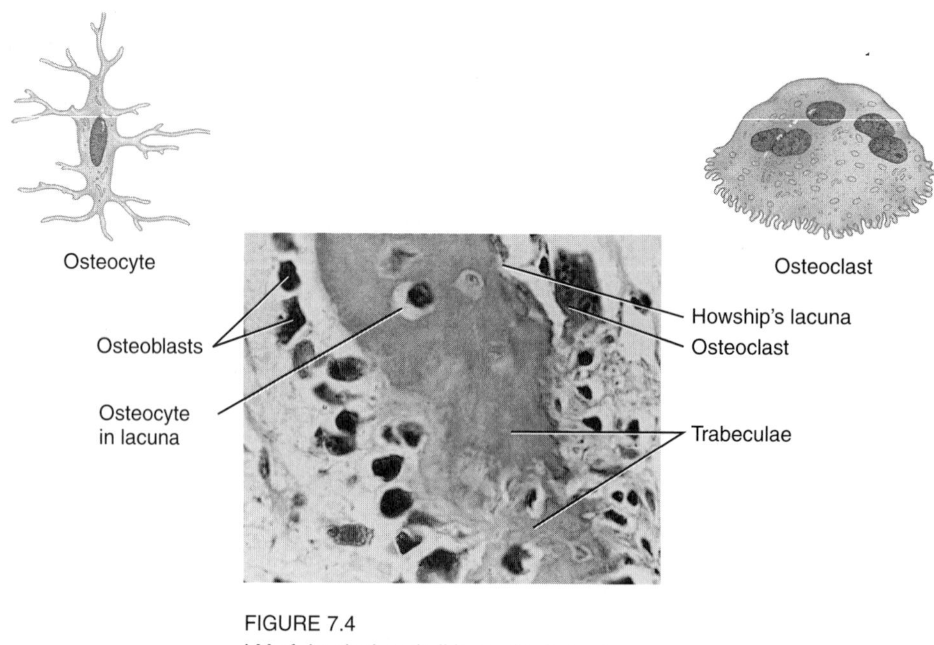

FIGURE 7.4
LM of developing skull bone displaying bone deposition by osteoblasts and bone reabsorption by osteoclast cells. (400×)

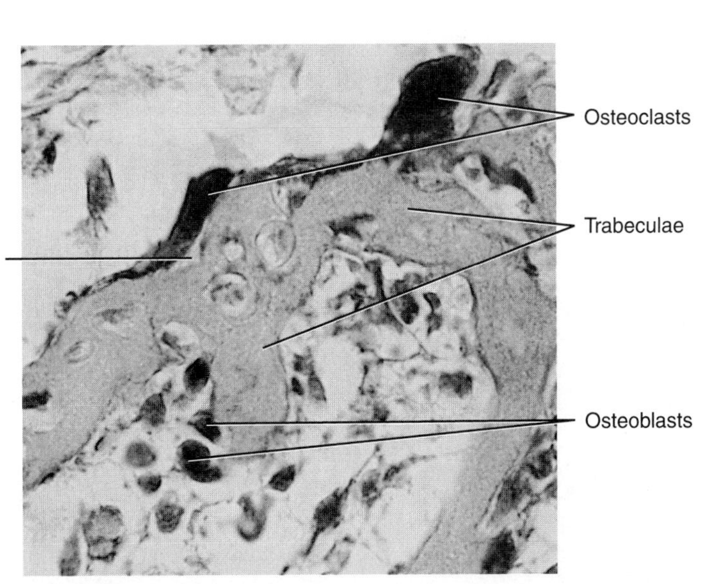

FIGURE 7.5
LM of osteoclast cells adhering to the bony trabeculae. Osteoclasts are actively involved in bone reabsorption. (200×)

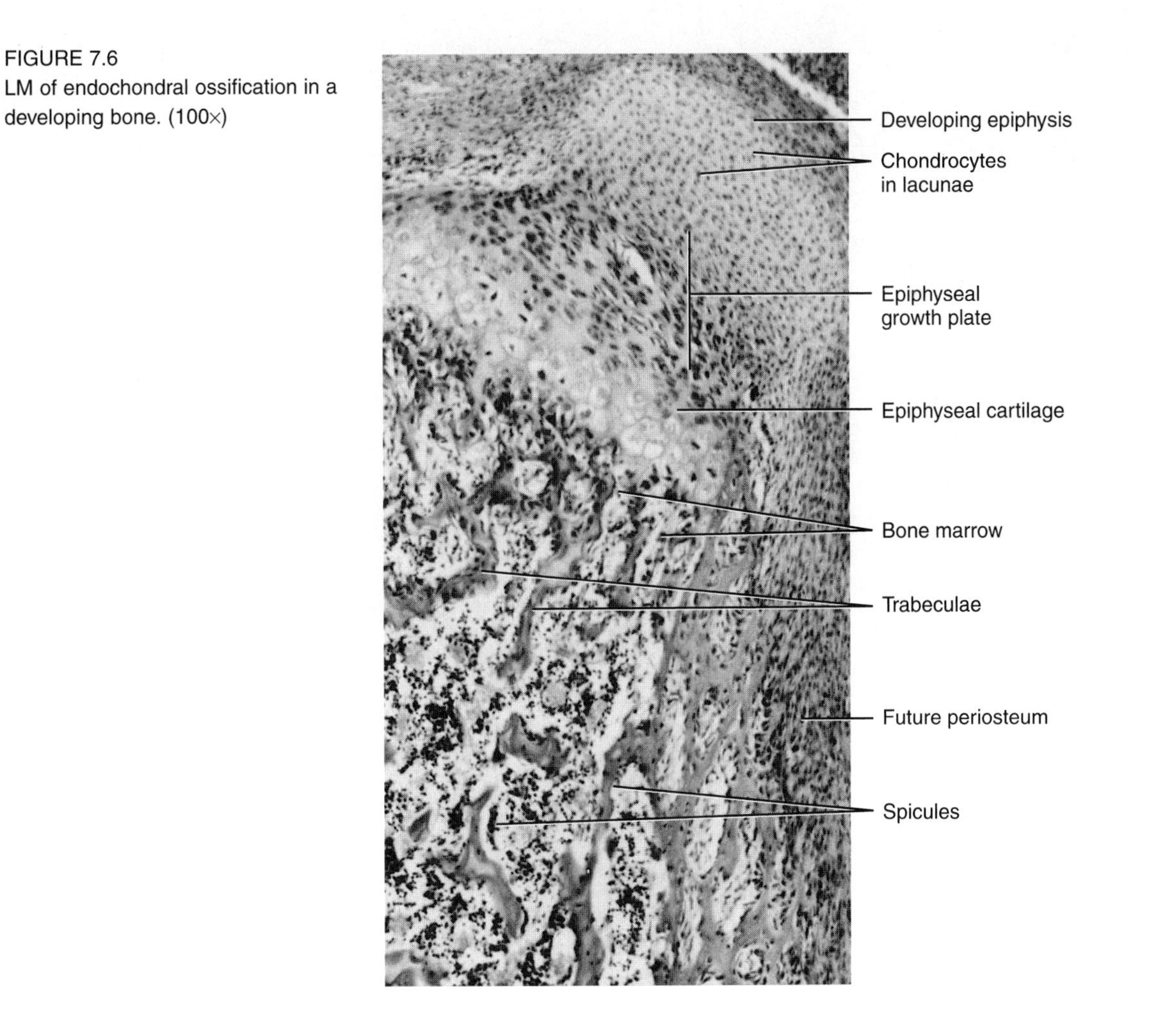

FIGURE 7.6
LM of endochondral ossification in a developing bone. (100×)

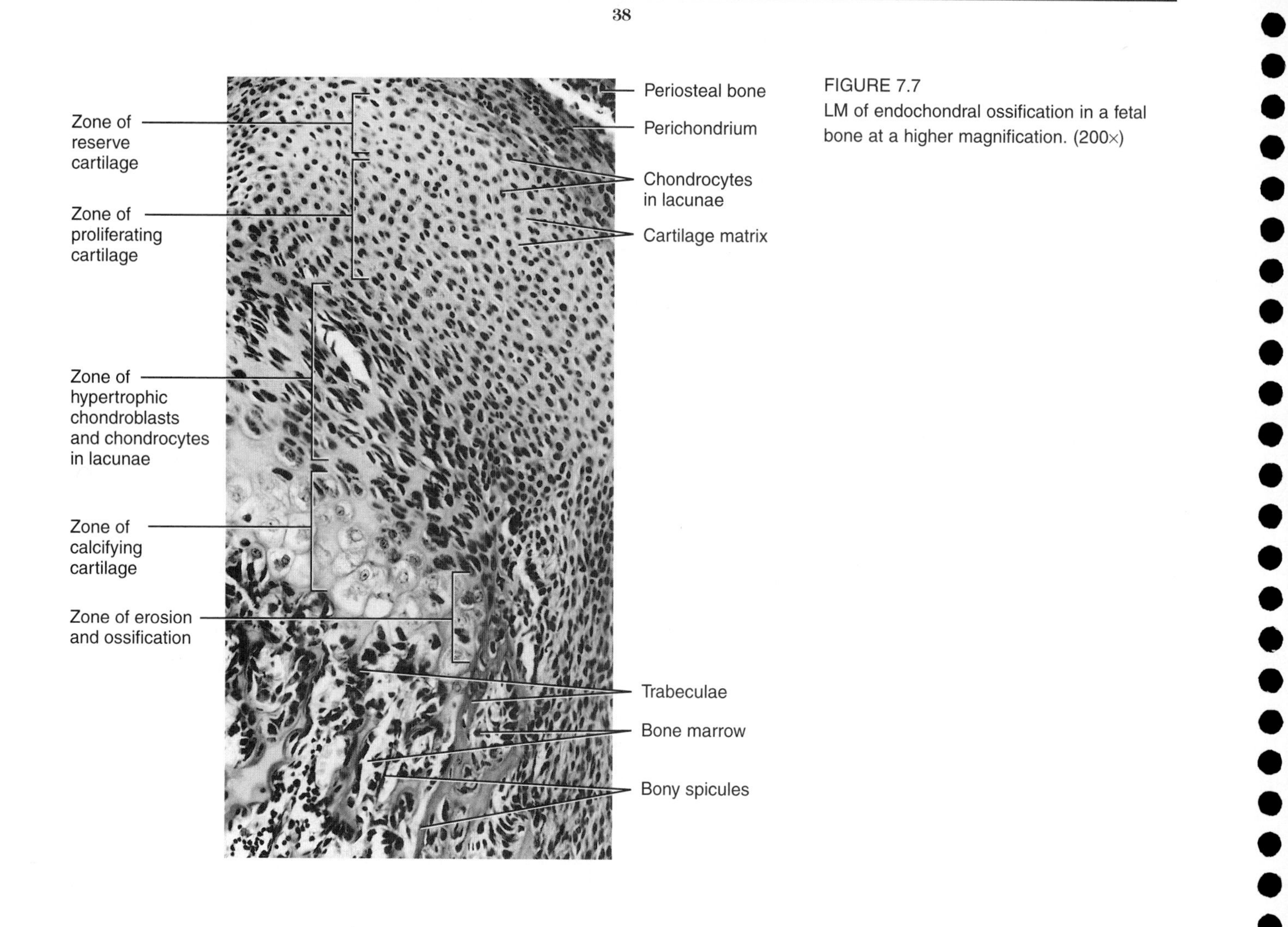

FIGURE 7.7
LM of endochondral ossification in a fetal bone at a higher magnification. (200×)

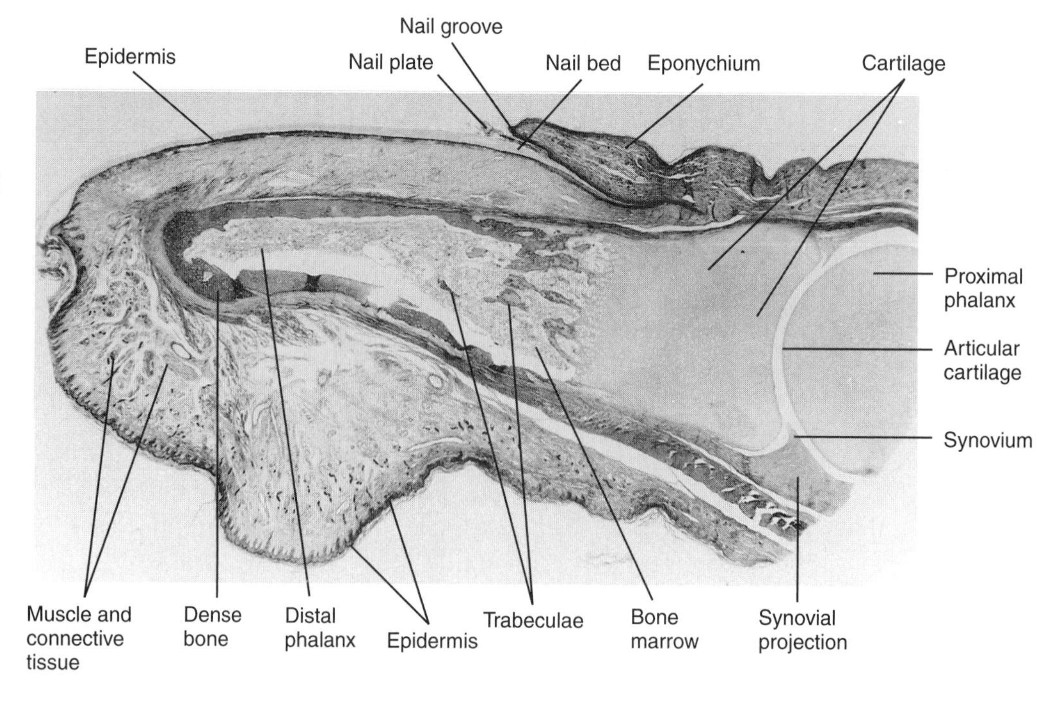

FIGURE 7.8
LM of fetal finger showing the fingernail, the synovial joint between the distal and medial phalanges, muscle, connective tissue, and the epithelium. (20×)

The Skeletal System

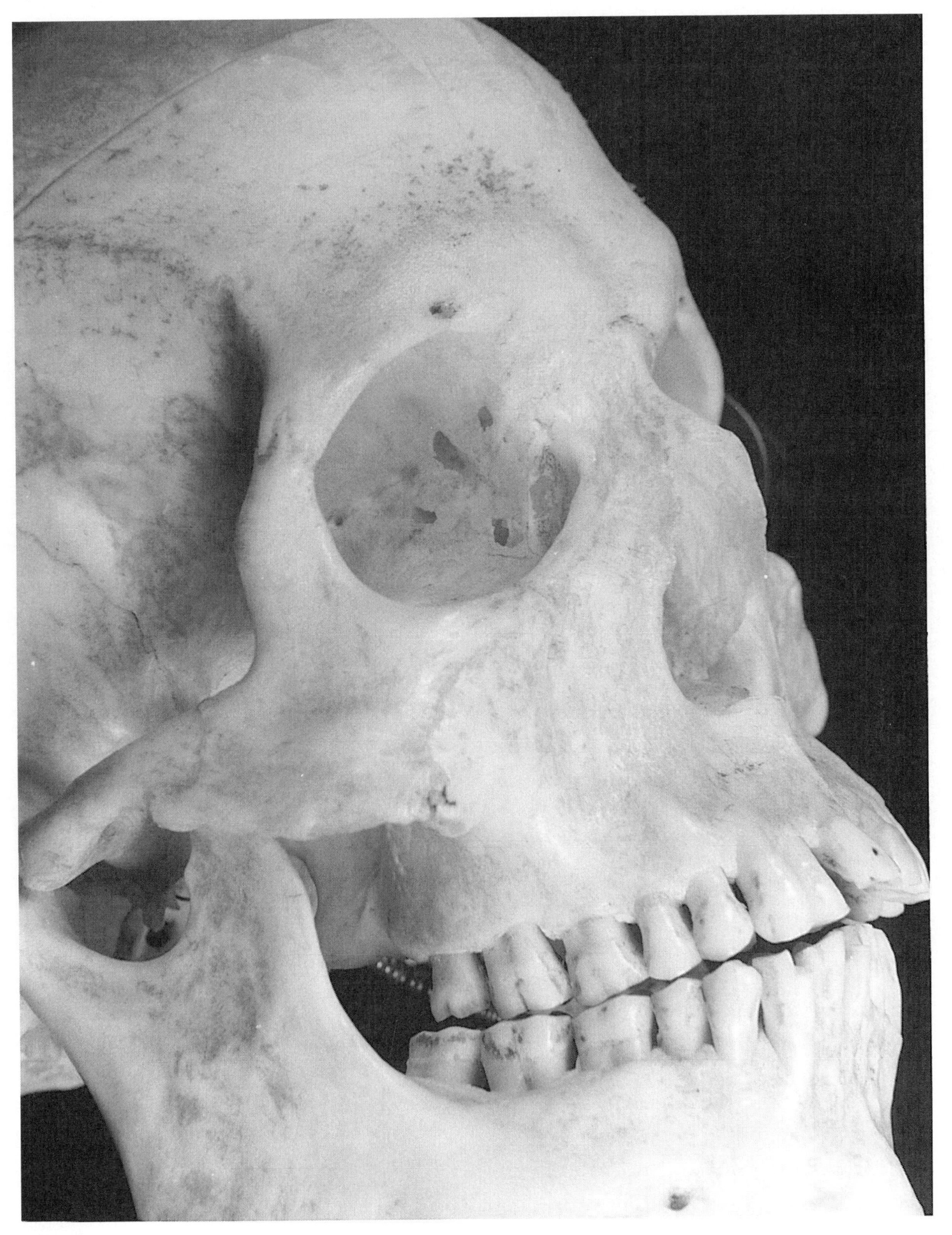

FIGURE 7.9
The skull.

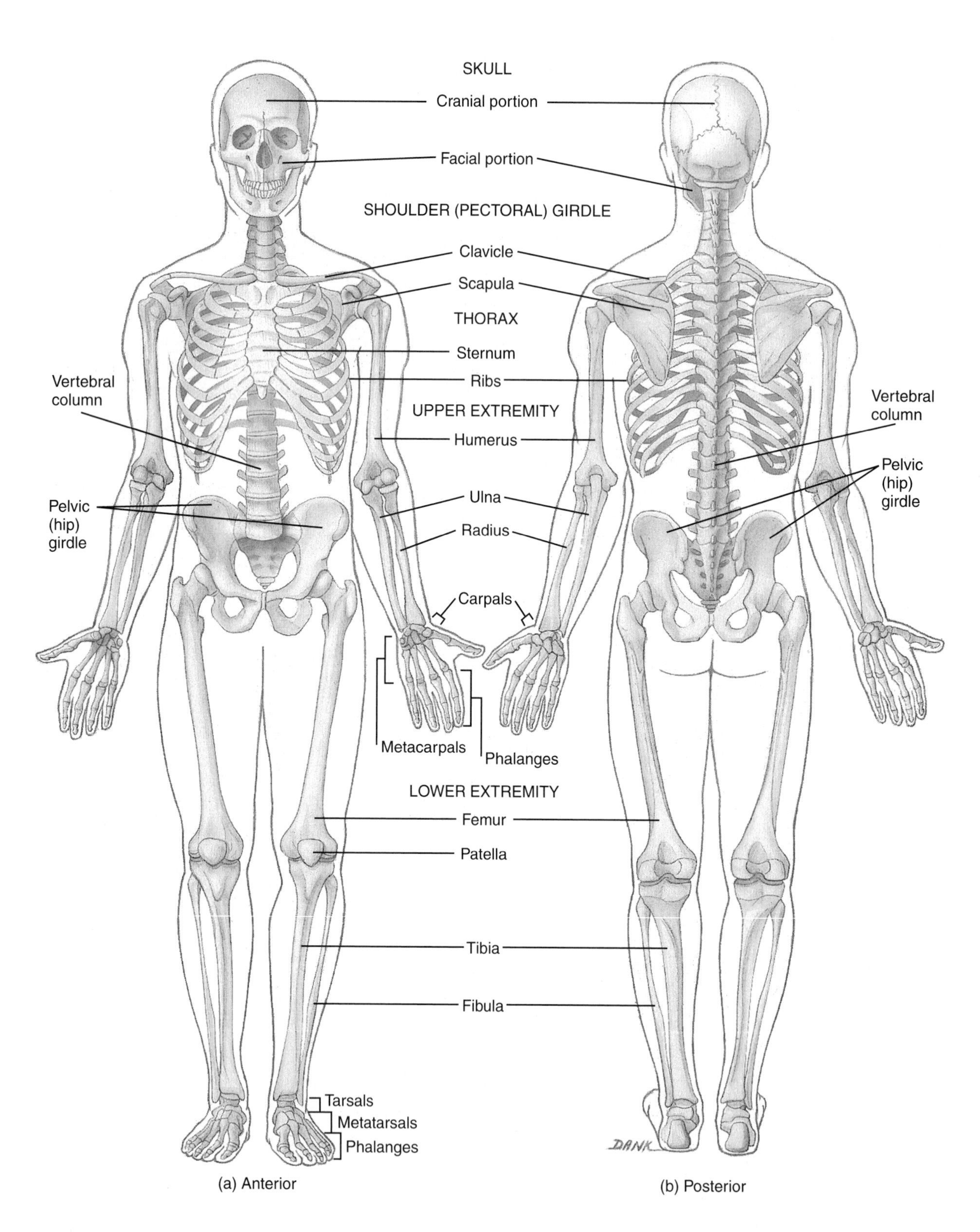

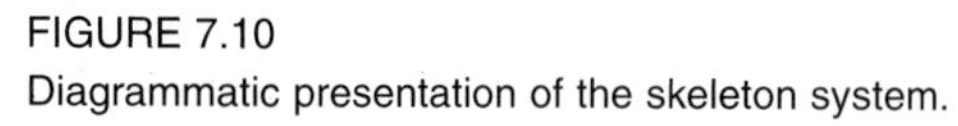
FIGURE 7.10
Diagrammatic presentation of the skeleton system.

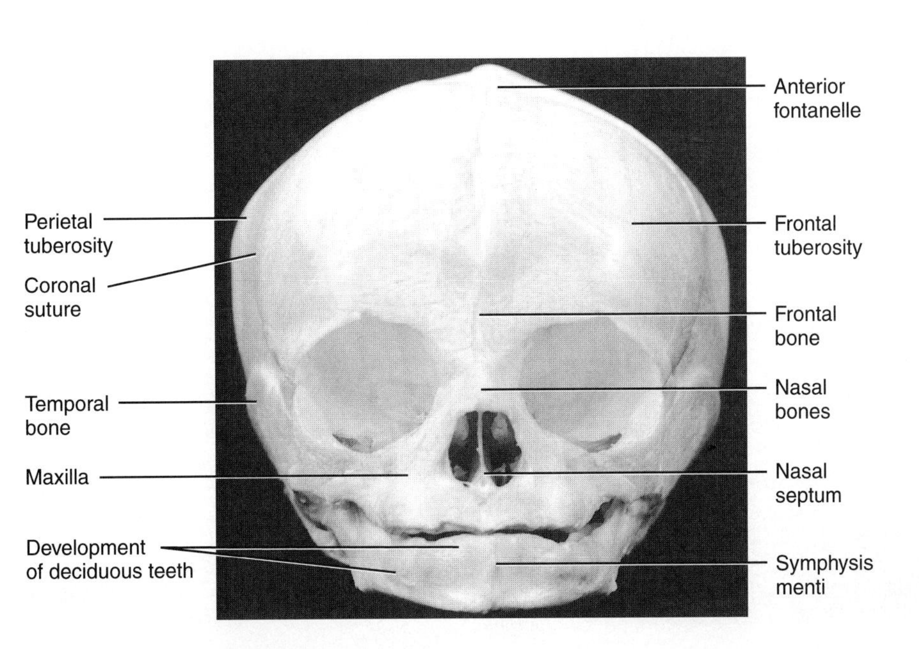

FIGURE 7.11
Anterior view of fetal skull.

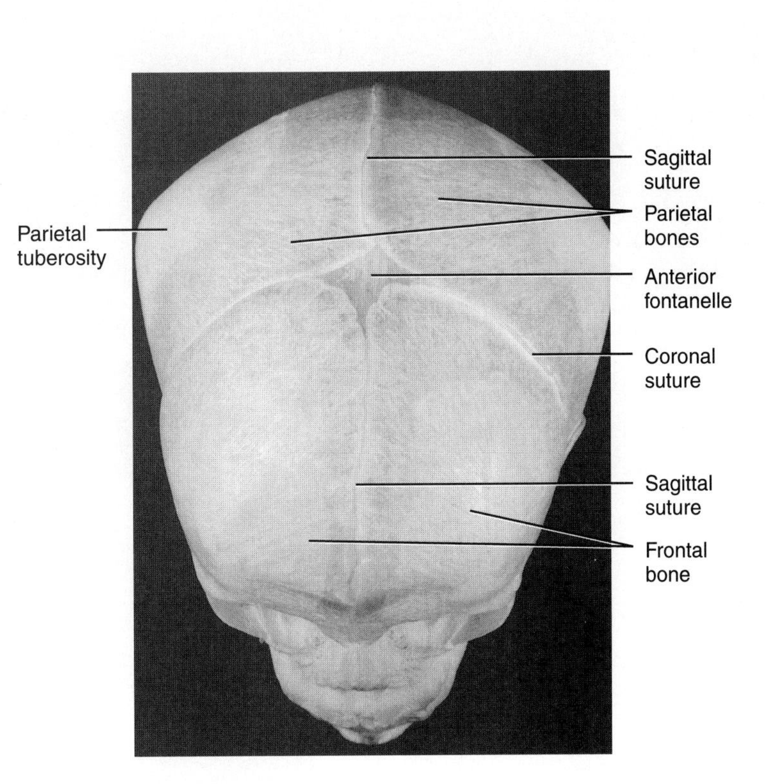

FIGURE 7.12
Superior view of fetal skull.

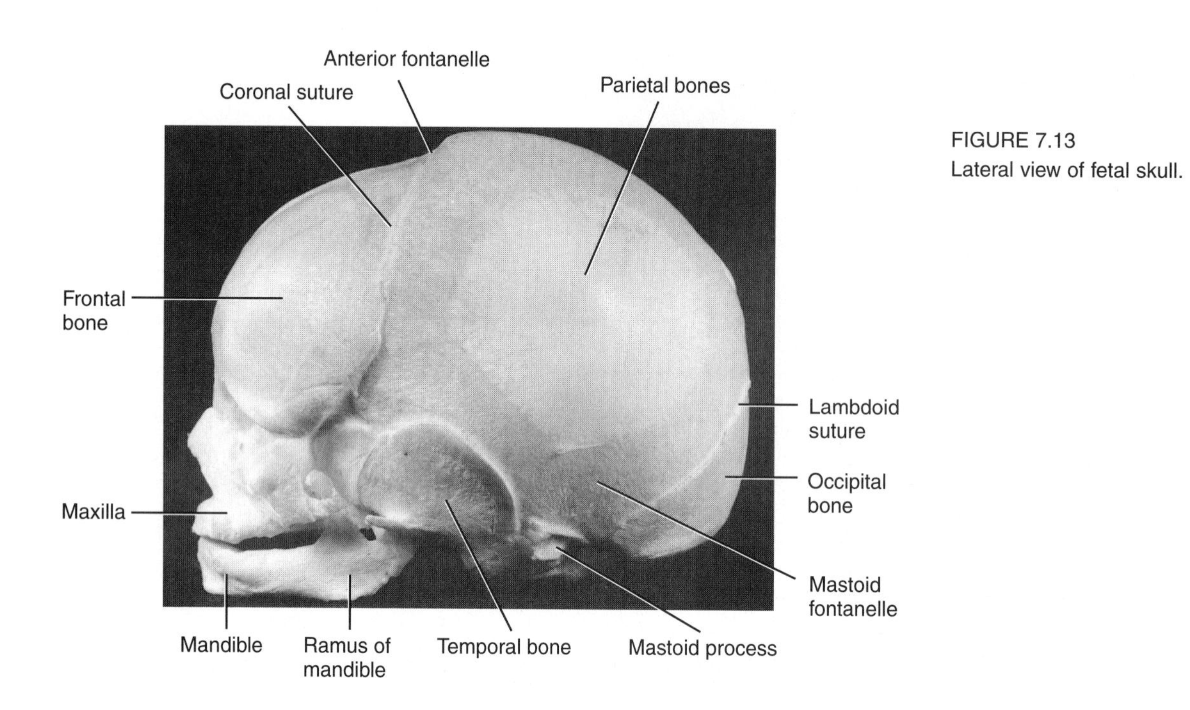

FIGURE 7.13
Lateral view of fetal skull.

FIGURE 7.14
Anterior view, human skull.

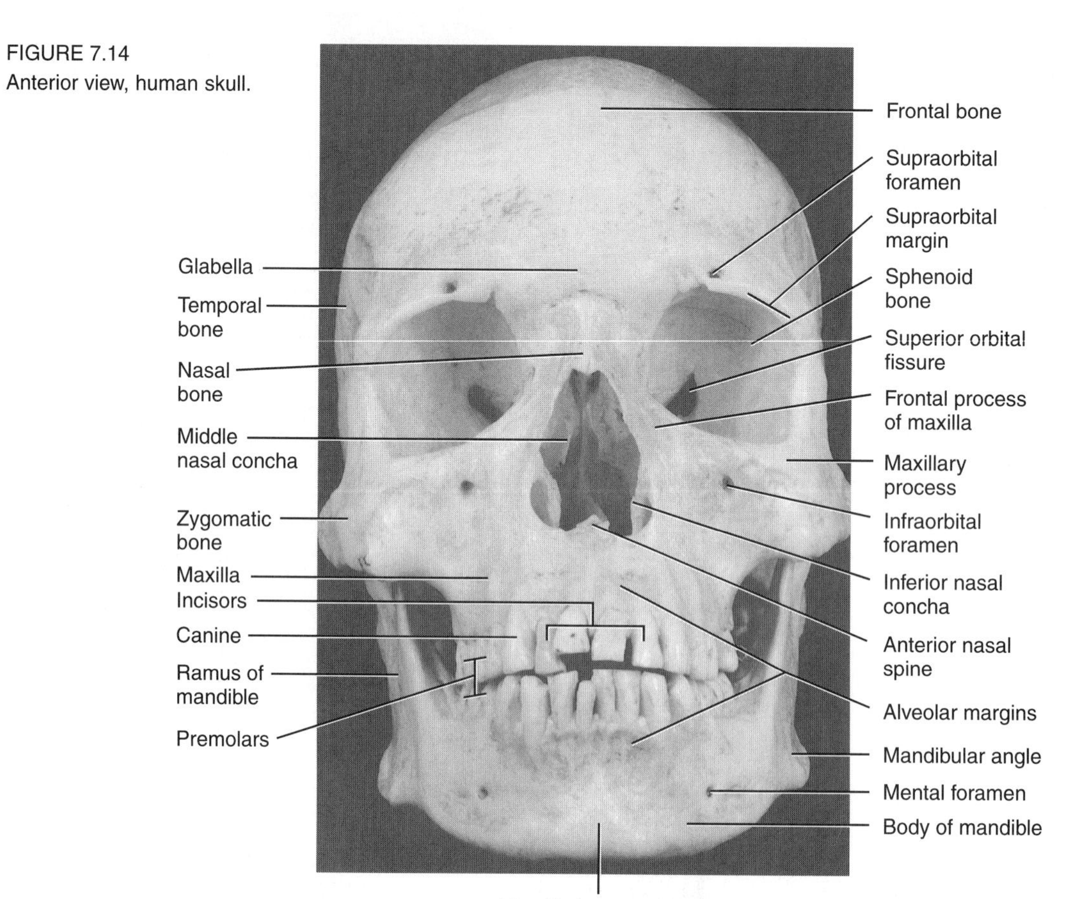

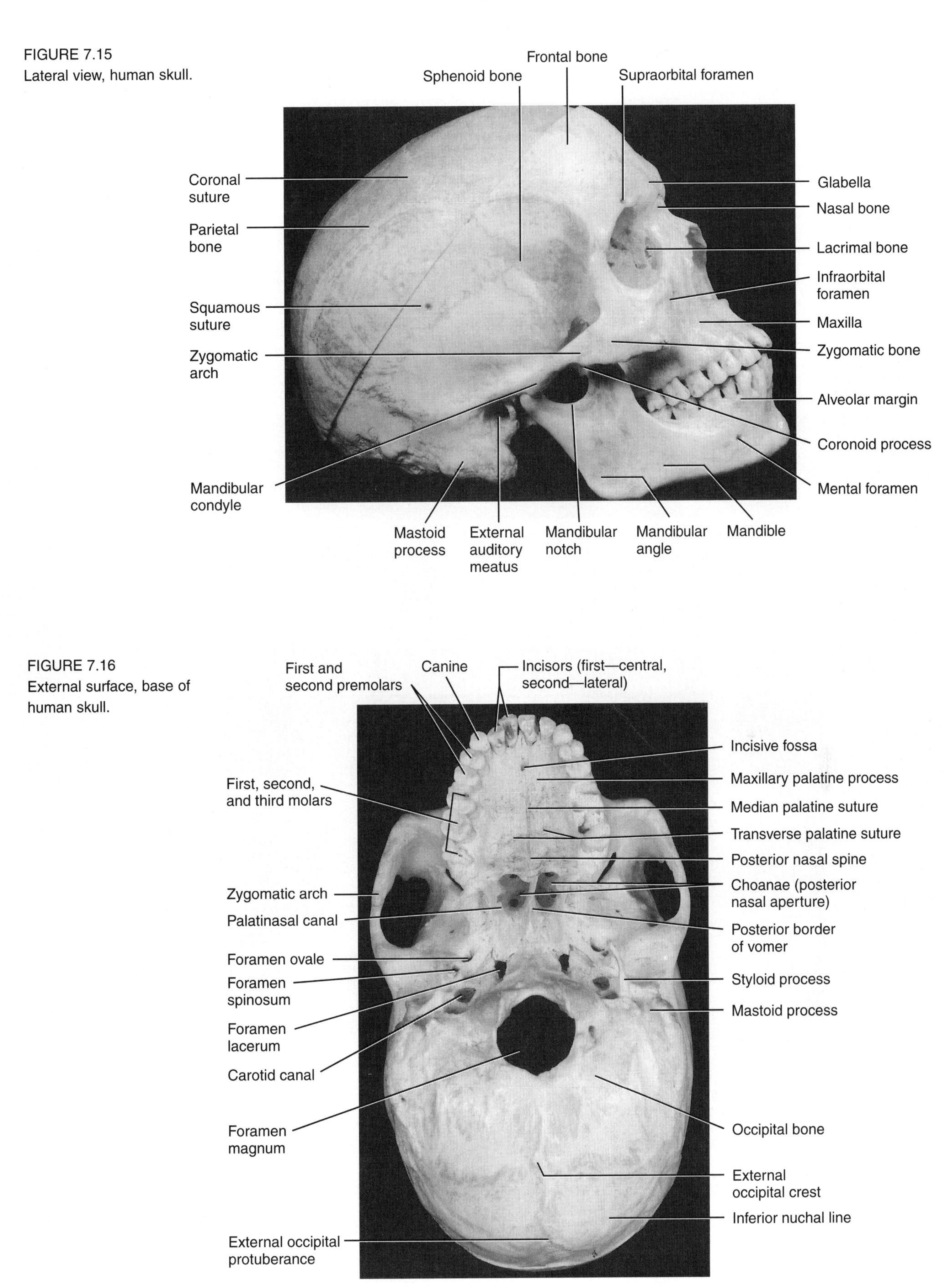

FIGURE 7.15
Lateral view, human skull.

FIGURE 7.16
External surface, base of human skull.

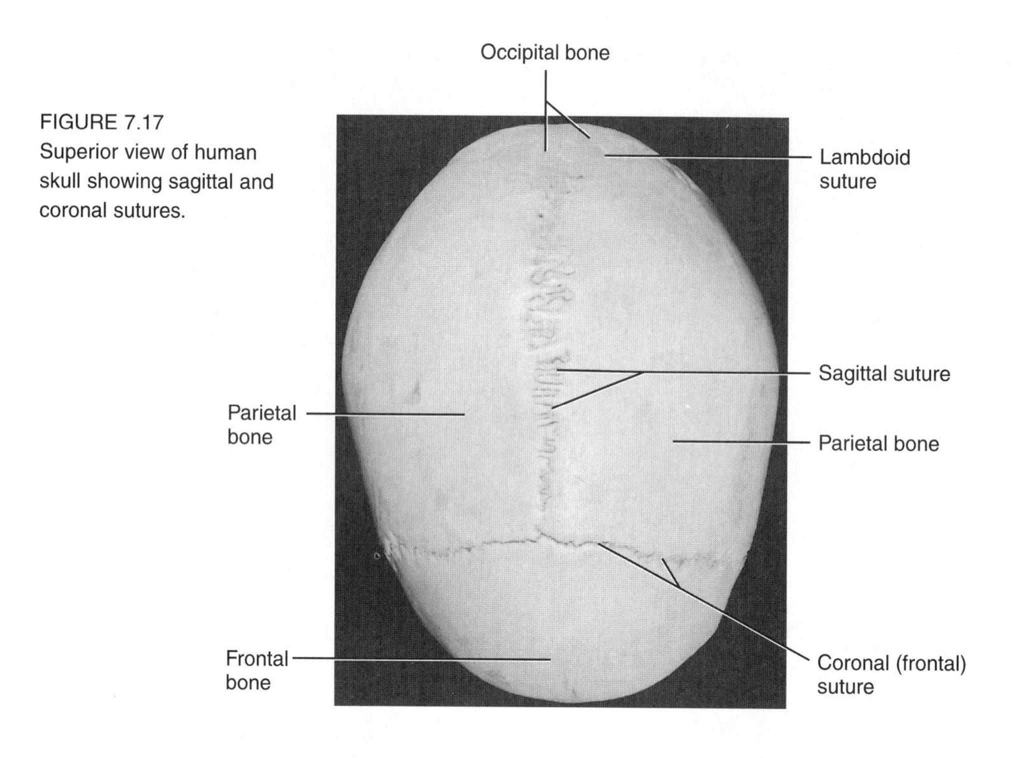

FIGURE 7.17
Superior view of human skull showing sagittal and coronal sutures.

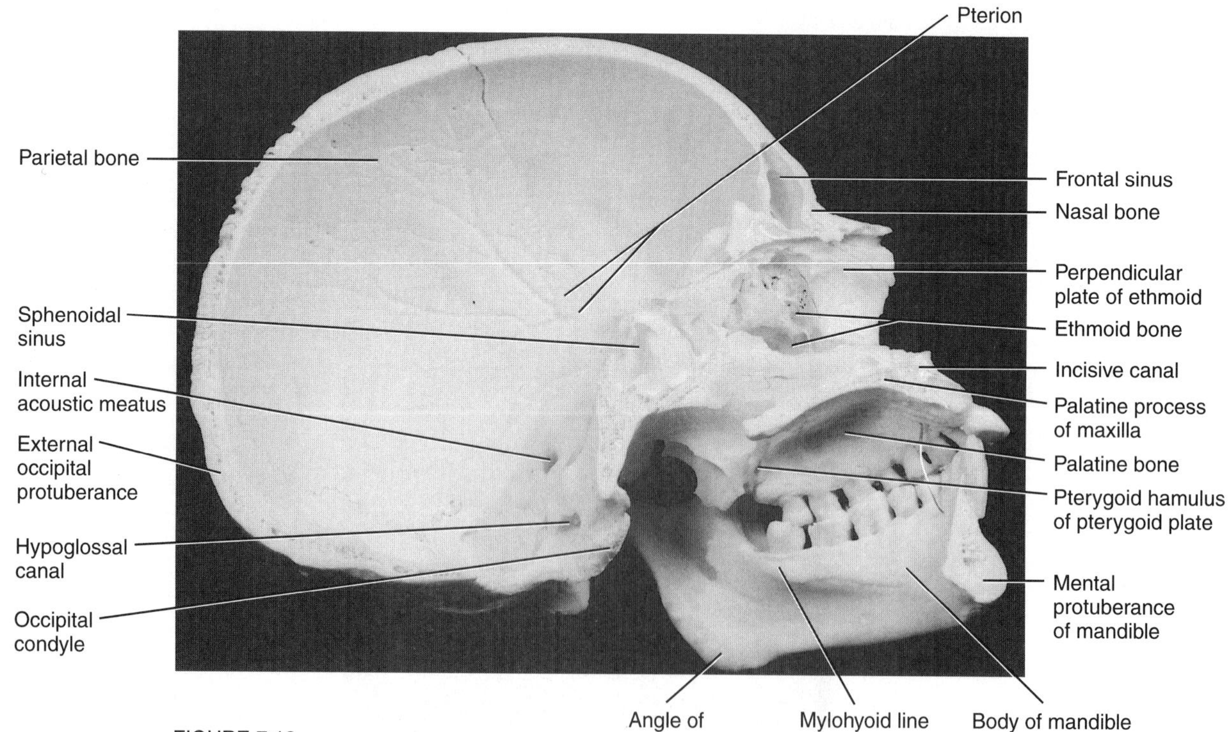

FIGURE 7.18
Sagittal section of human skull showing internal structures.

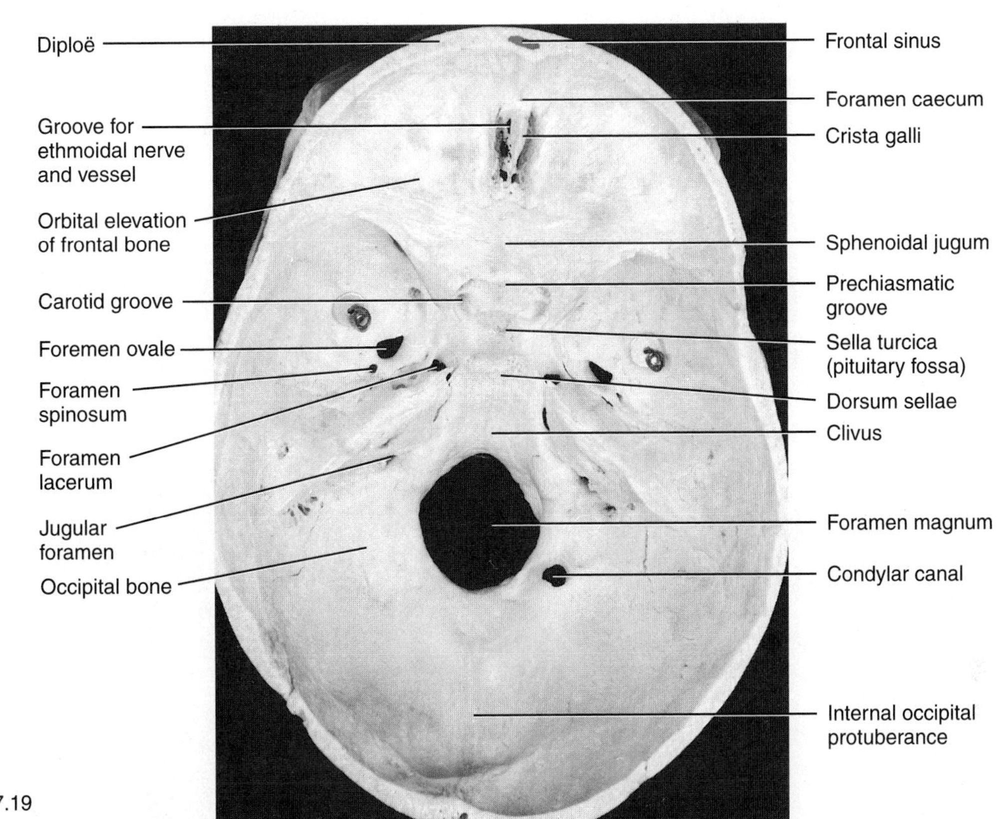

FIGURE 7.19
Superior view of cranial vault.

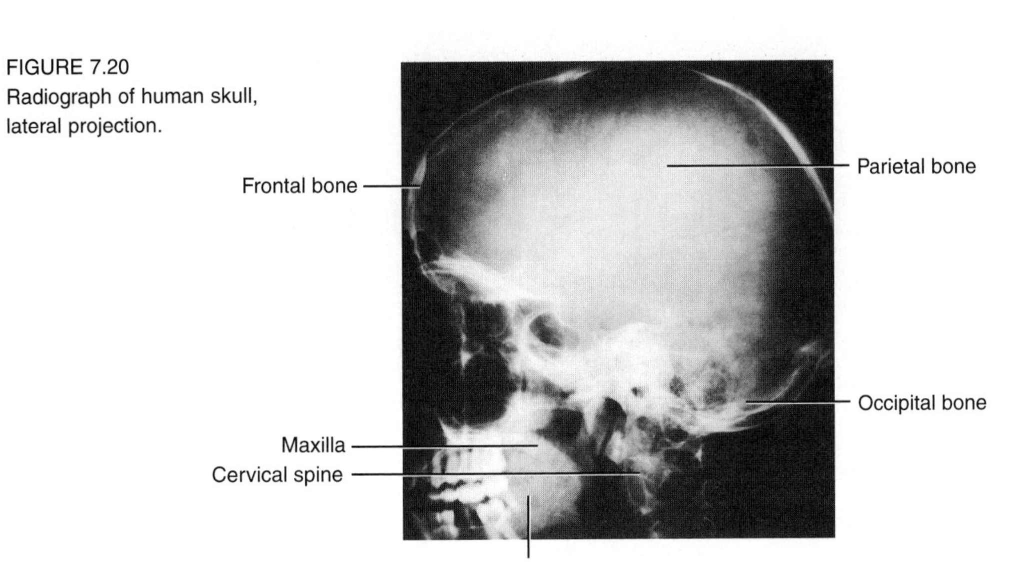

FIGURE 7.20
Radiograph of human skull, lateral projection.

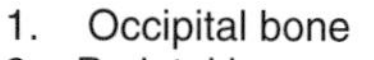

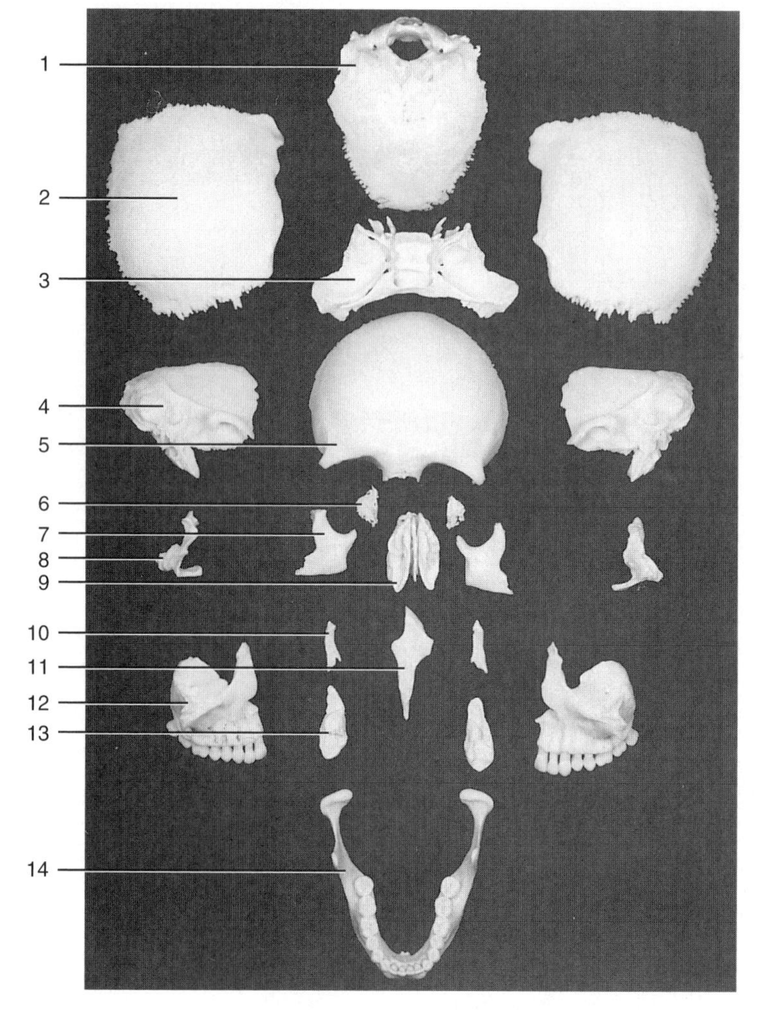

FIGURE 7.21
Disarticulated skull bones.

1. Occipital bone
2. Parietal bone
3. Sphenoid bone
4. Temporal bone
5. Frontal bone
6. Lacrimal bone
7. Zygomatic bone
8. Palatine bone
9. Ethmoid bone
10. Nasal bone
11. Vomer
12. Maxilla
13. Inferior nasal concha
14. Mandible

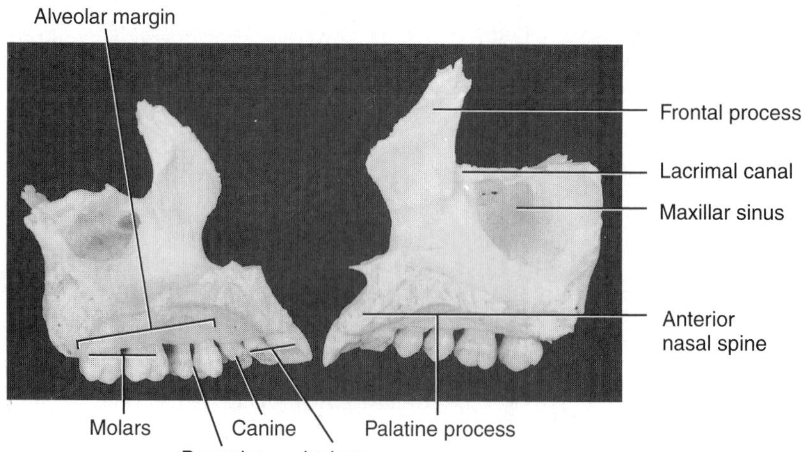

FIGURE 7.22
Medial view of left and right maxillae.

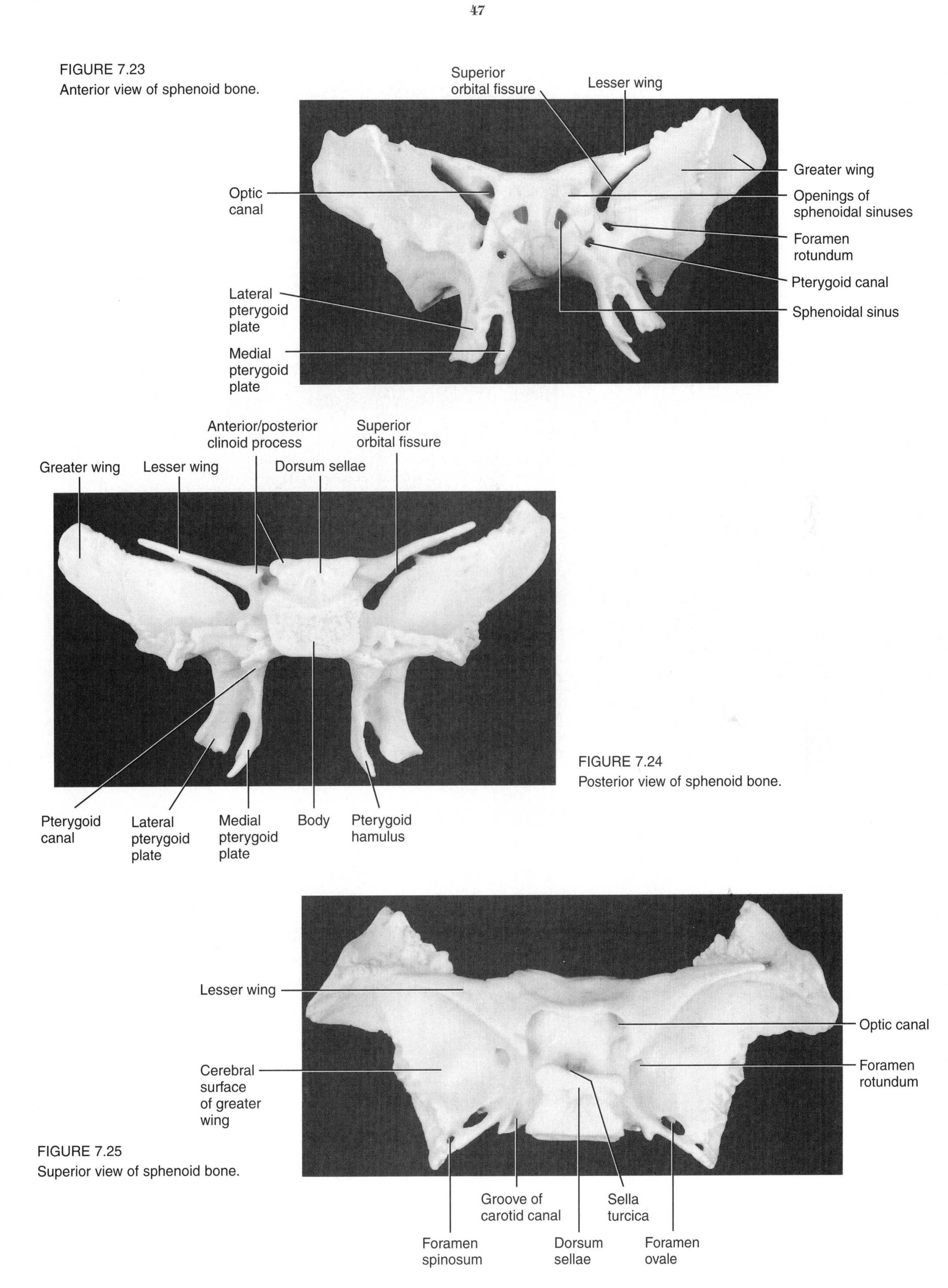

FIGURE 7.23
Anterior view of sphenoid bone.

FIGURE 7.24
Posterior view of sphenoid bone.

FIGURE 7.25
Superior view of sphenoid bone.

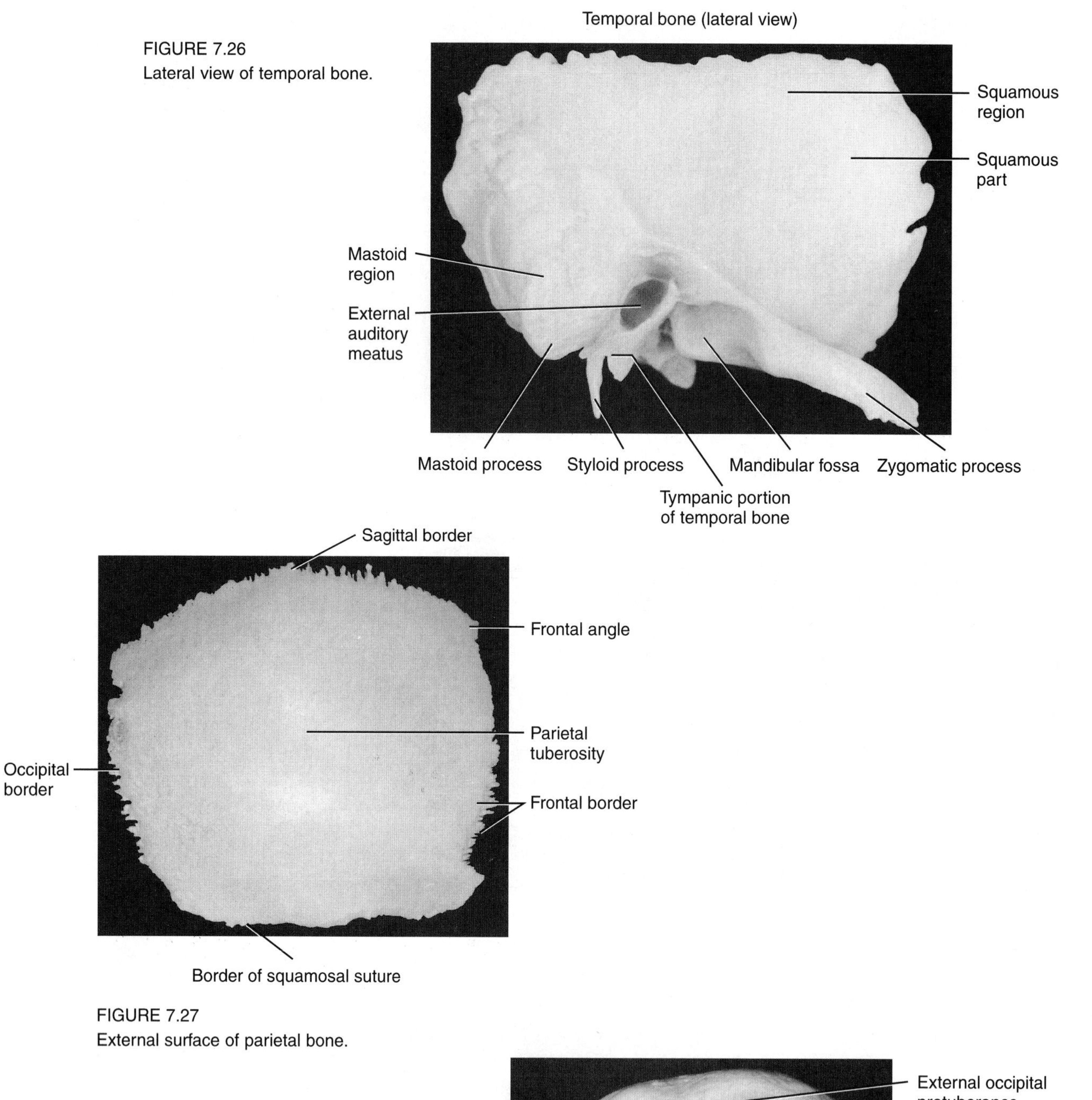

FIGURE 7.26
Lateral view of temporal bone.

FIGURE 7.27
External surface of parietal bone.

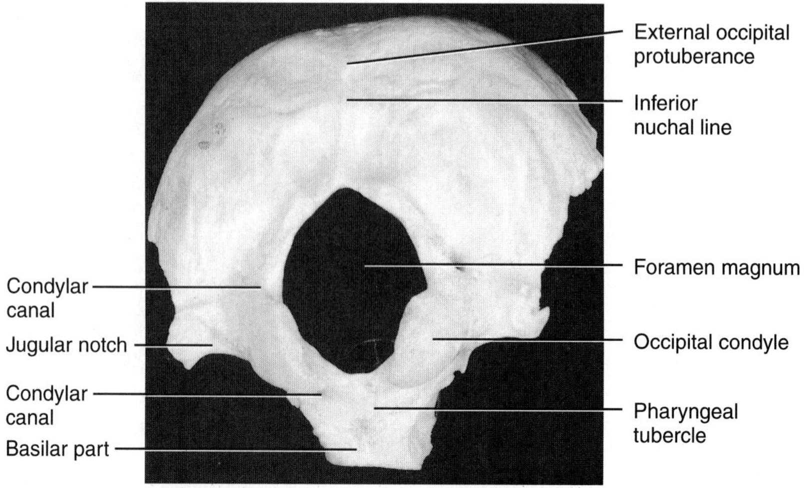

FIGURE 7.28
External surface of occipital bone as seen from below.

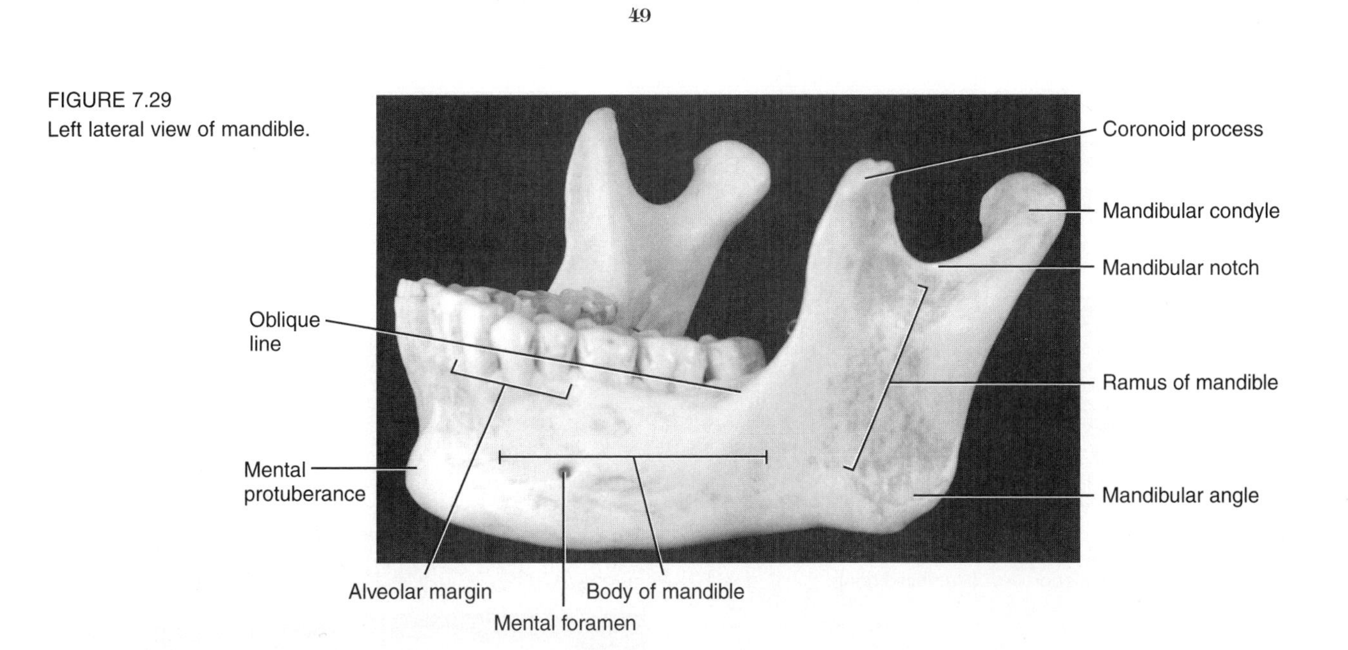

FIGURE 7.29
Left lateral view of mandible.

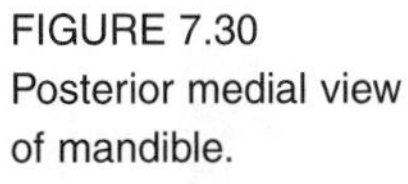

FIGURE 7.30
Posterior medial view of mandible.

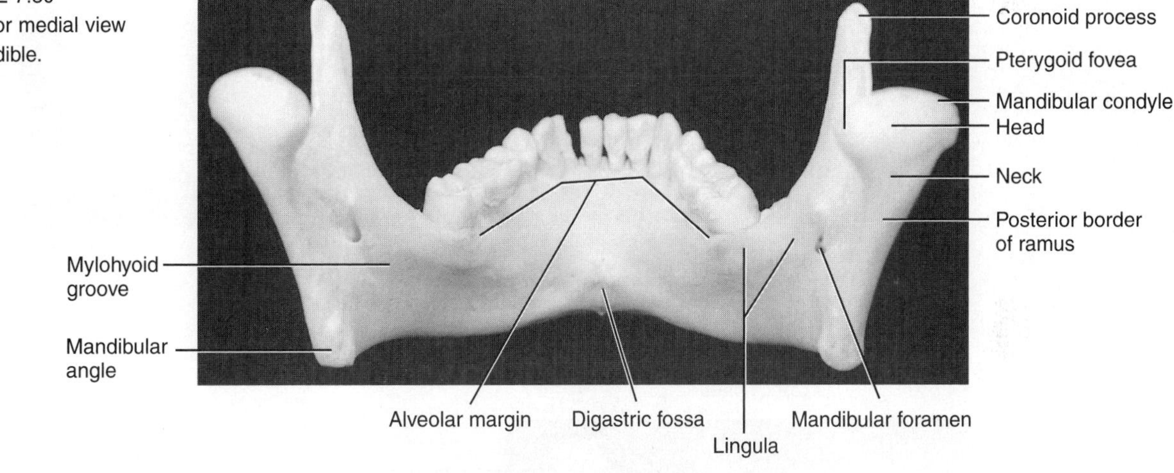

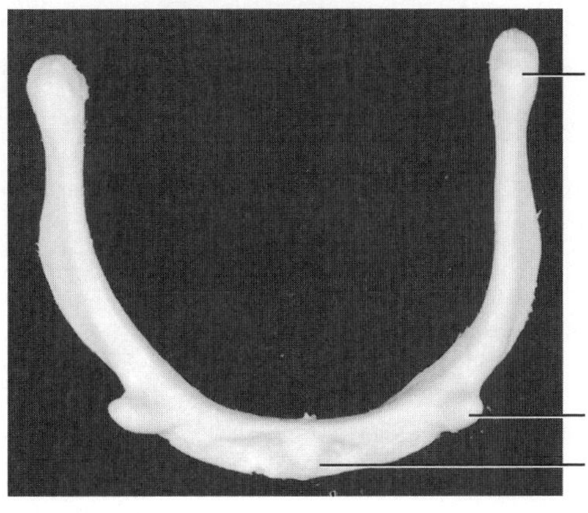

FIGURE 7.31
Superior view of hyoid bone in a vertical position displaying the cornu.

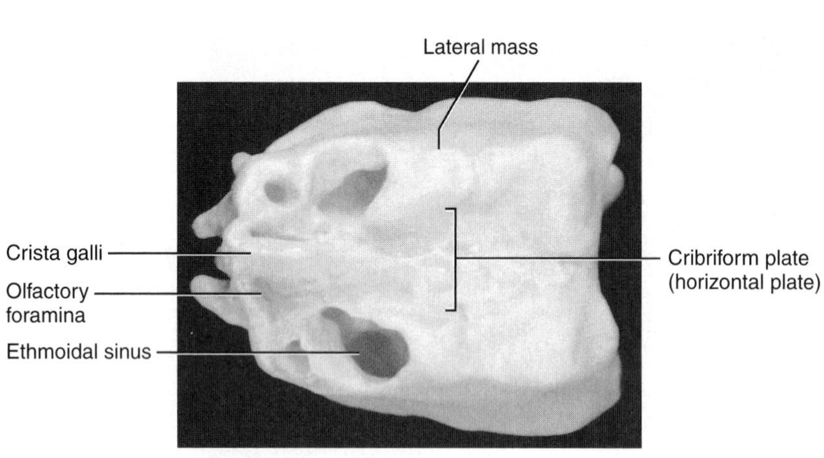

FIGURE 7.32
Superior view of ethmoid bone.

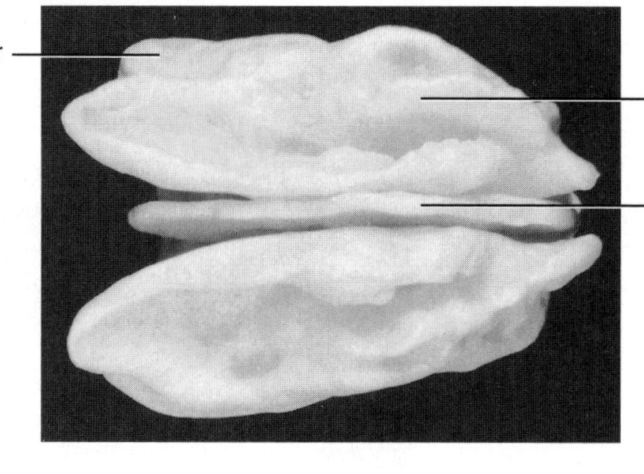

FIGURE 7.33
Inferior view of ethmoid bone.

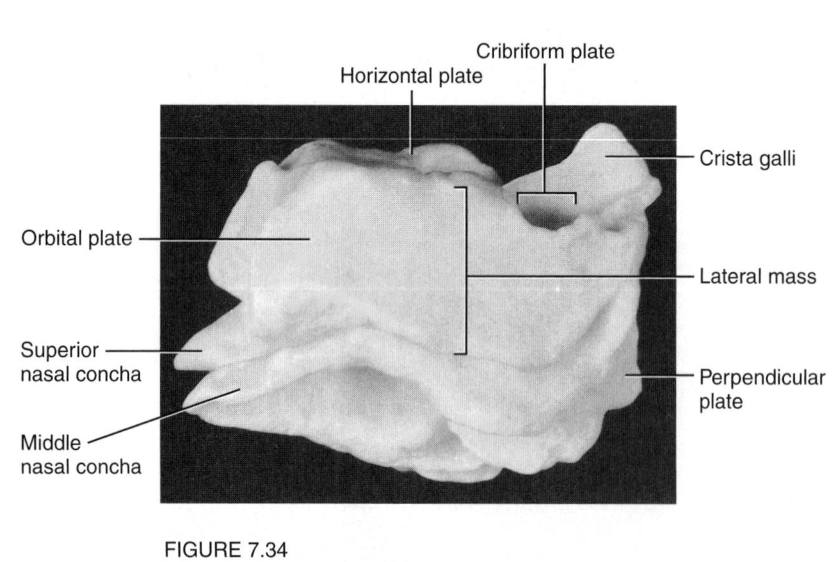

FIGURE 7.34
Lateral view of ethmoid bone.

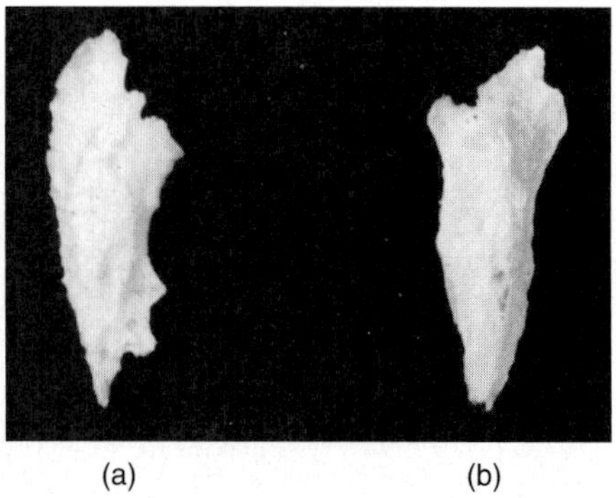

(a) (b)

FIGURE 7.35
Superior (a) and inferior (b) views of inferior nasal concha.

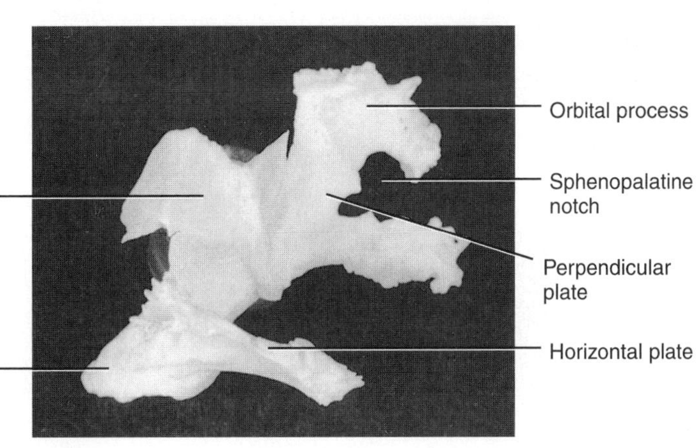

FIGURE 7.36
Medial view of right palatine bone.

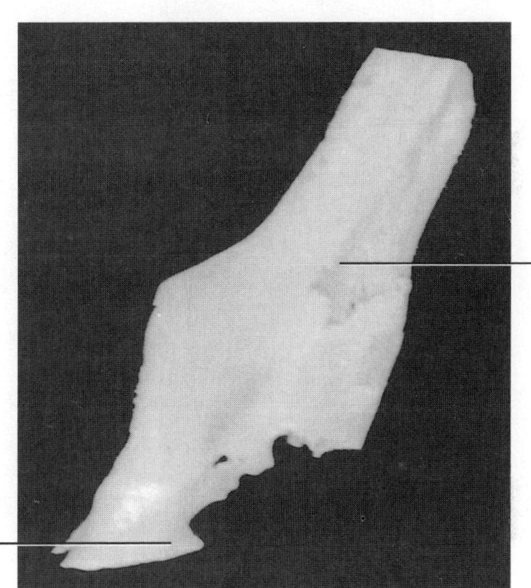

FIGURE 7.37
Lateral view of vomer bone.

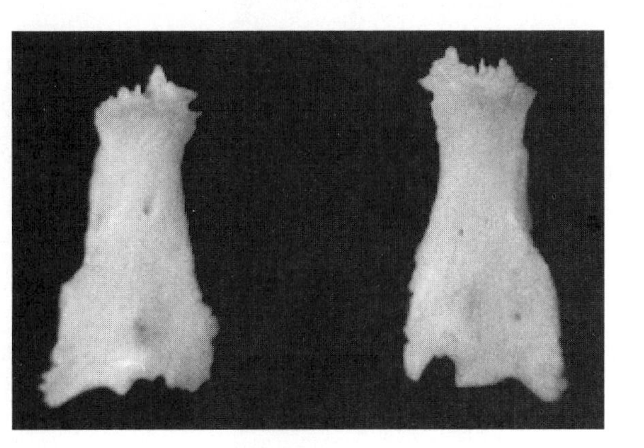

FIGURE 7.38
Superior view of nasal bones.

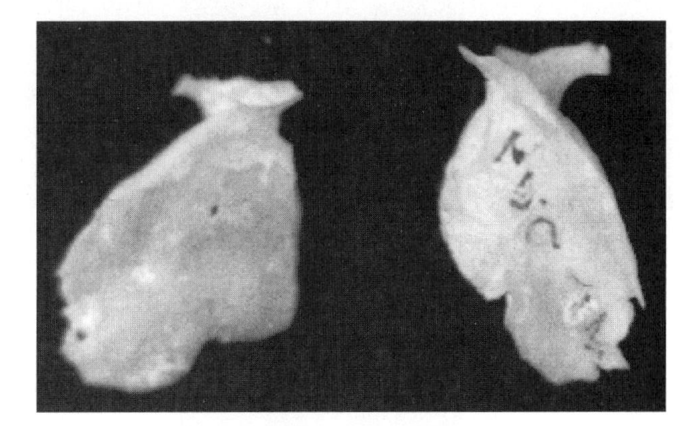

FIGURE 7.39
Lateral (a) and medial (b) surfaces of lacrimal bone.

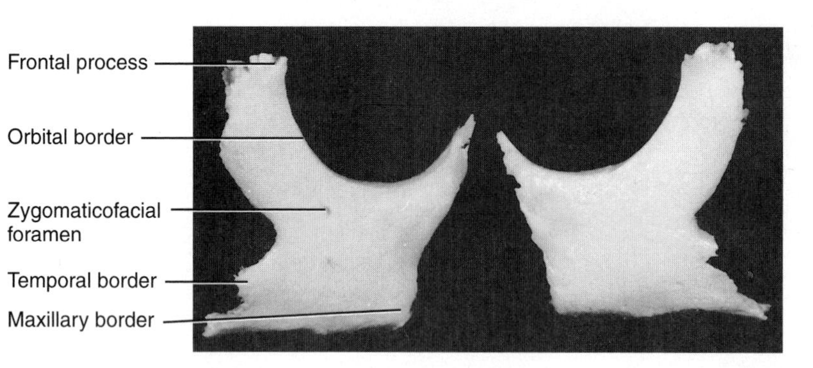

FIGURE 7.40
Lateral view of left and right zygomatic bone.

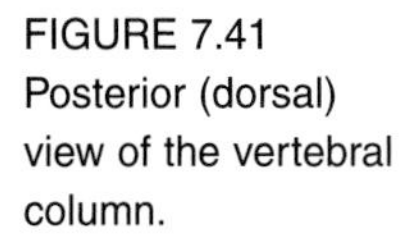

FIGURE 7.41
Posterior (dorsal) view of the vertebral column.

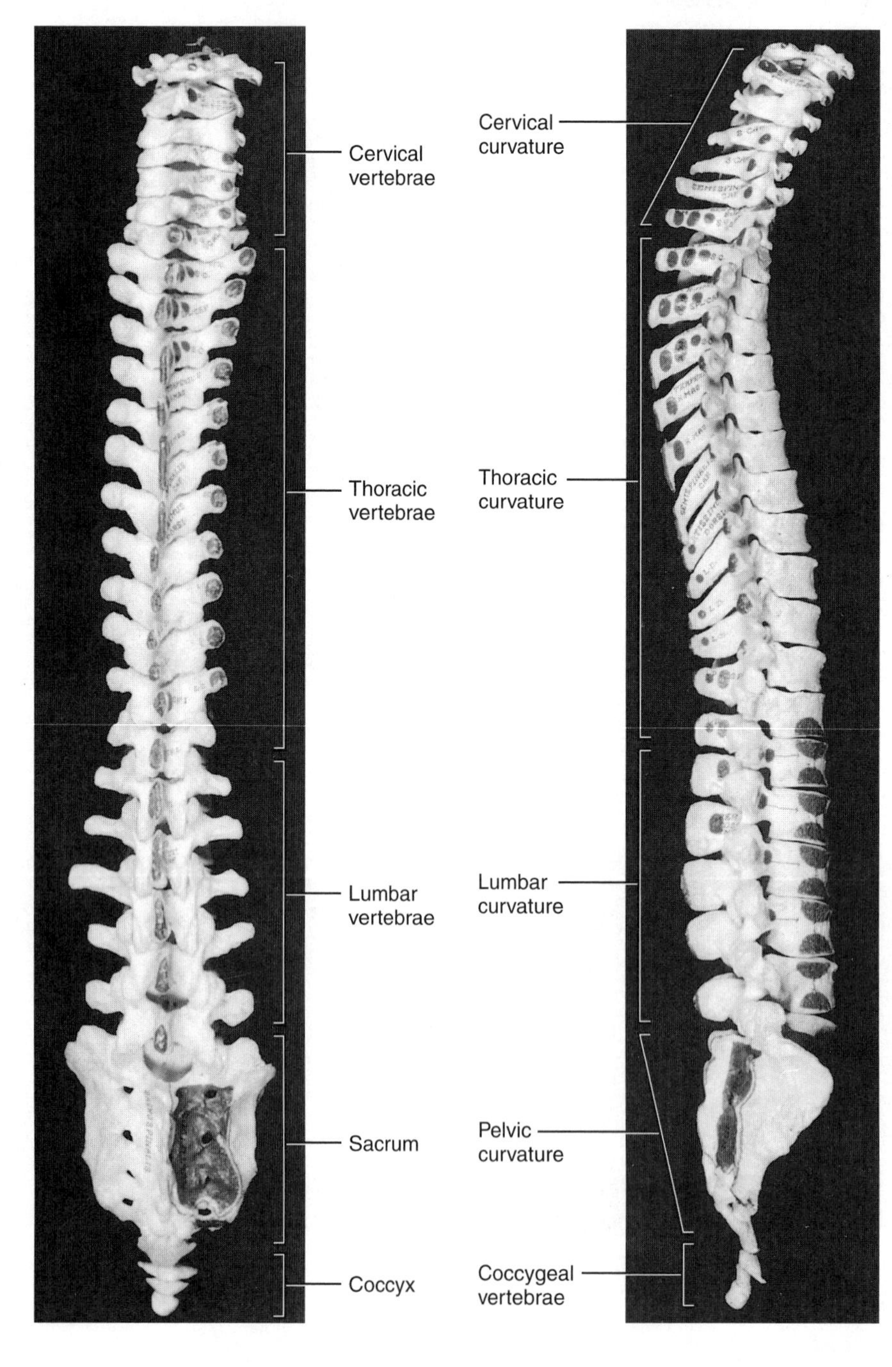

FIGURE 7.42
Lateral view of the vertebral column displaying cervical, thoracic, lumbar, and pelvic curvatures.

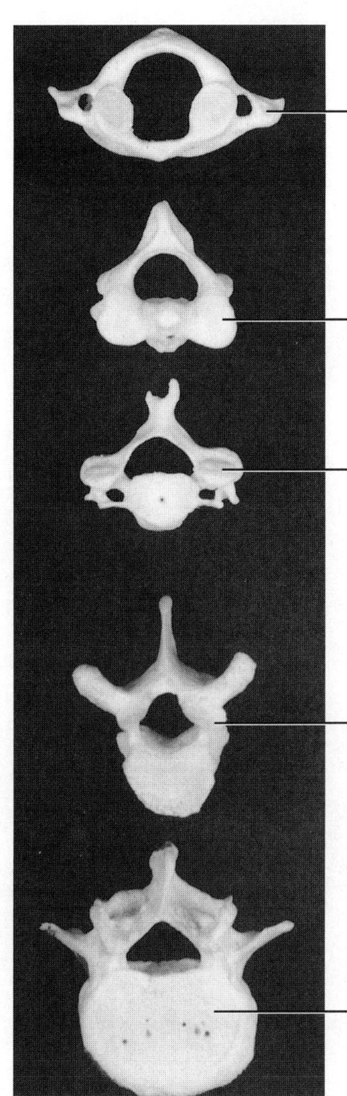

FIGURE 7.43
Superior view of atlas, axis, typical cervical, thoracic, and lumbar vertebrae.

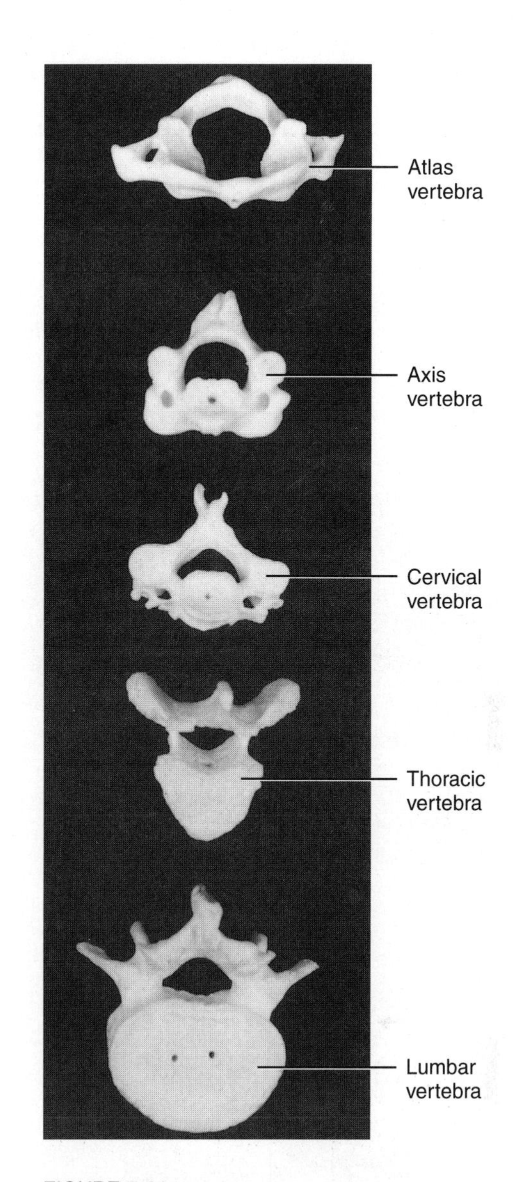

FIGURE 7.44
Inferior view of atlas, axis, cervical, thoracic, and lumbar vertebrae.

Posterior tubercle
Posterior arch
Vertebral foramen
Transverse process
Transverse foramen
Superior articular facet
Anterior tubercle
Facet for dens of axis

FIGURE 7.45
Superior view of first (atlas) cervical vertebra.

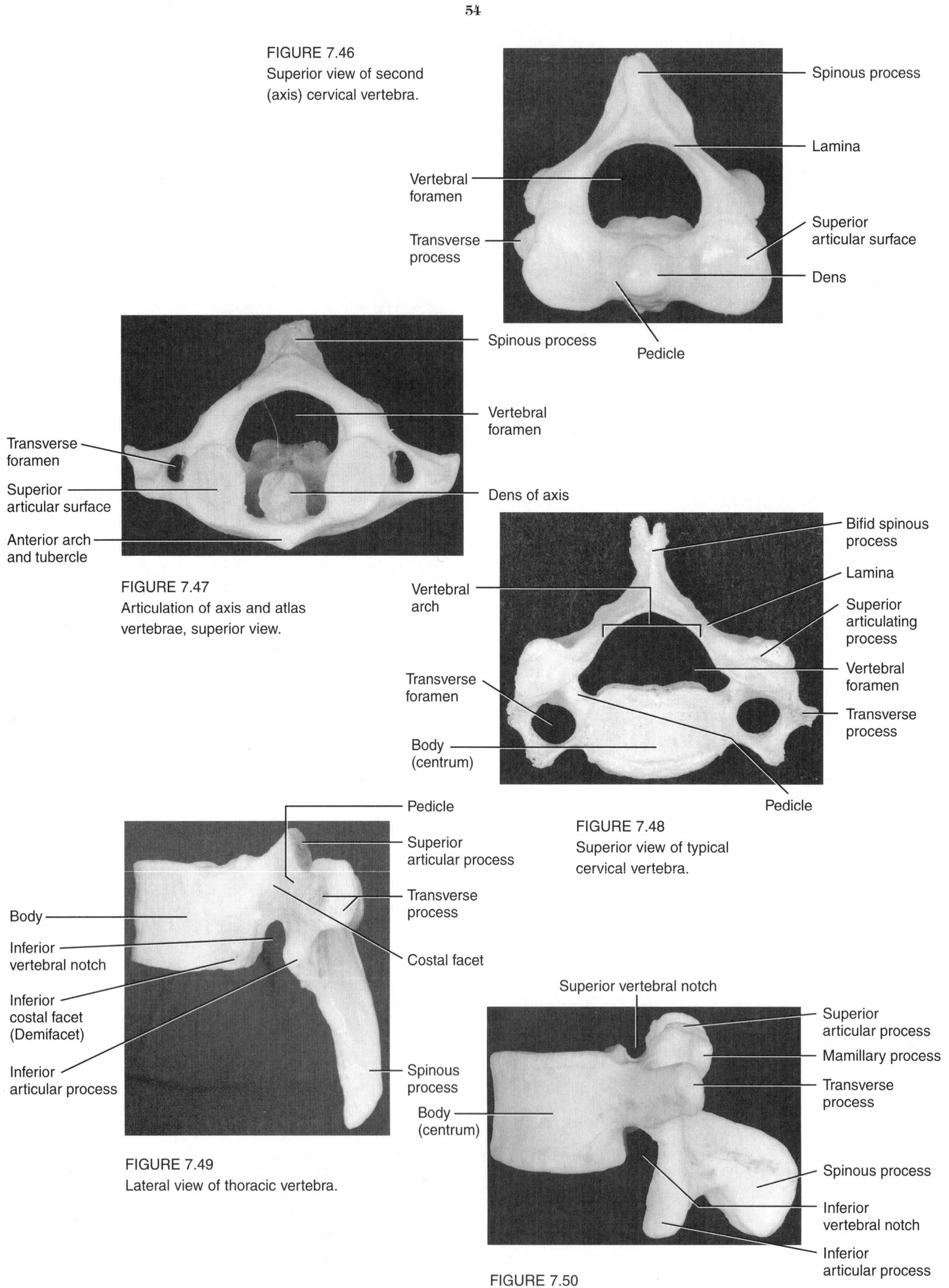

FIGURE 7.46
Superior view of second (axis) cervical vertebra.

FIGURE 7.47
Articulation of axis and atlas vertebrae, superior view.

FIGURE 7.48
Superior view of typical cervical vertebra.

FIGURE 7.49
Lateral view of thoracic vertebra.

FIGURE 7.50
Lateral view, fifth lumbar vertebra.

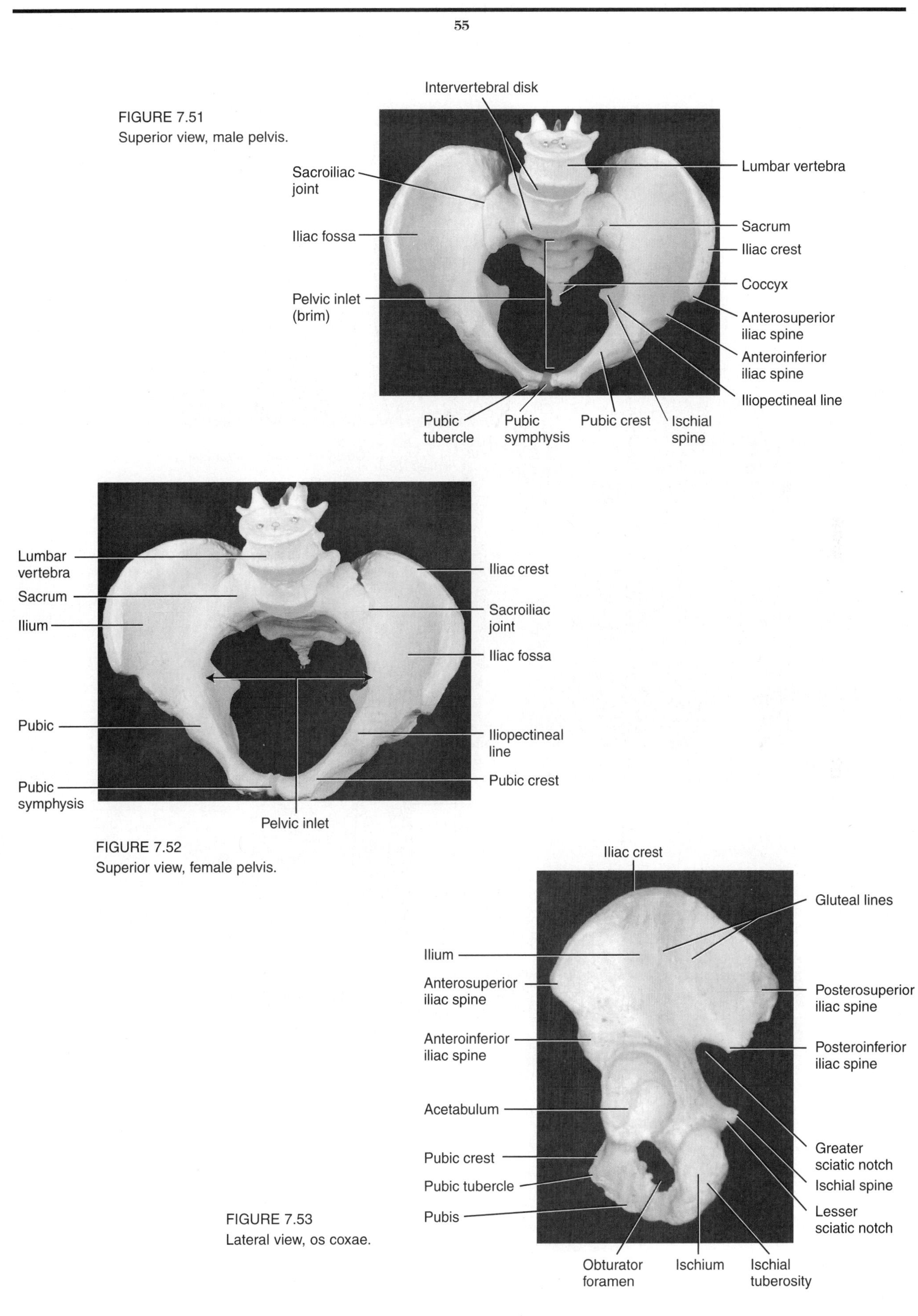

FIGURE 7.51
Superior view, male pelvis.

FIGURE 7.52
Superior view, female pelvis.

FIGURE 7.53
Lateral view, os coxae.

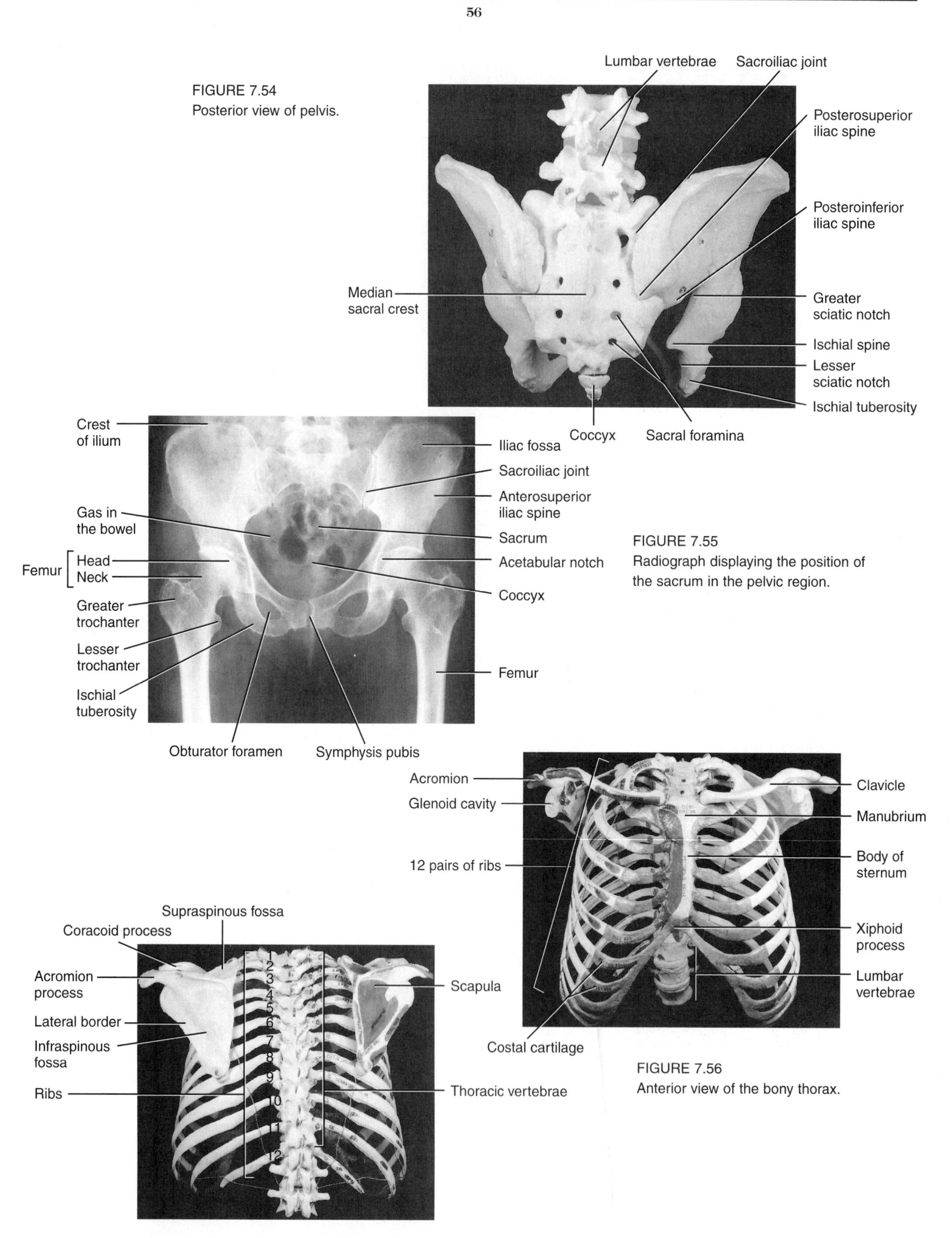

FIGURE 7.54
Posterior view of pelvis.

FIGURE 7.55
Radiograph displaying the position of the sacrum in the pelvic region.

FIGURE 7.56
Anterior view of the bony thorax.

FIGURE 7.57
Posterior view of the bony thorax.

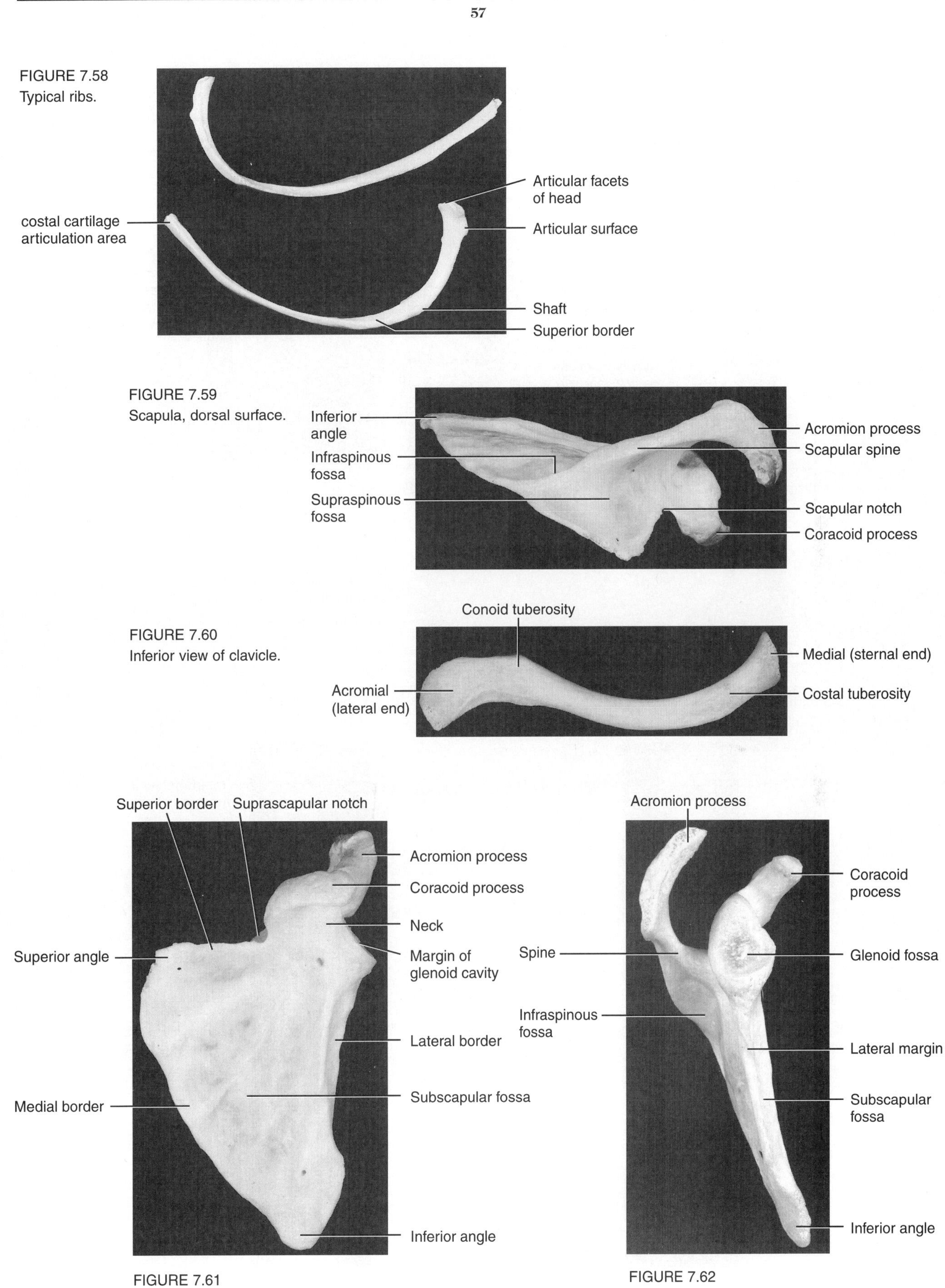

FIGURE 7.58
Typical ribs.

FIGURE 7.59
Scapula, dorsal surface.

FIGURE 7.60
Inferior view of clavicle.

FIGURE 7.61
Anterior view of scapula.

FIGURE 7.62
Lateral view of scapula.

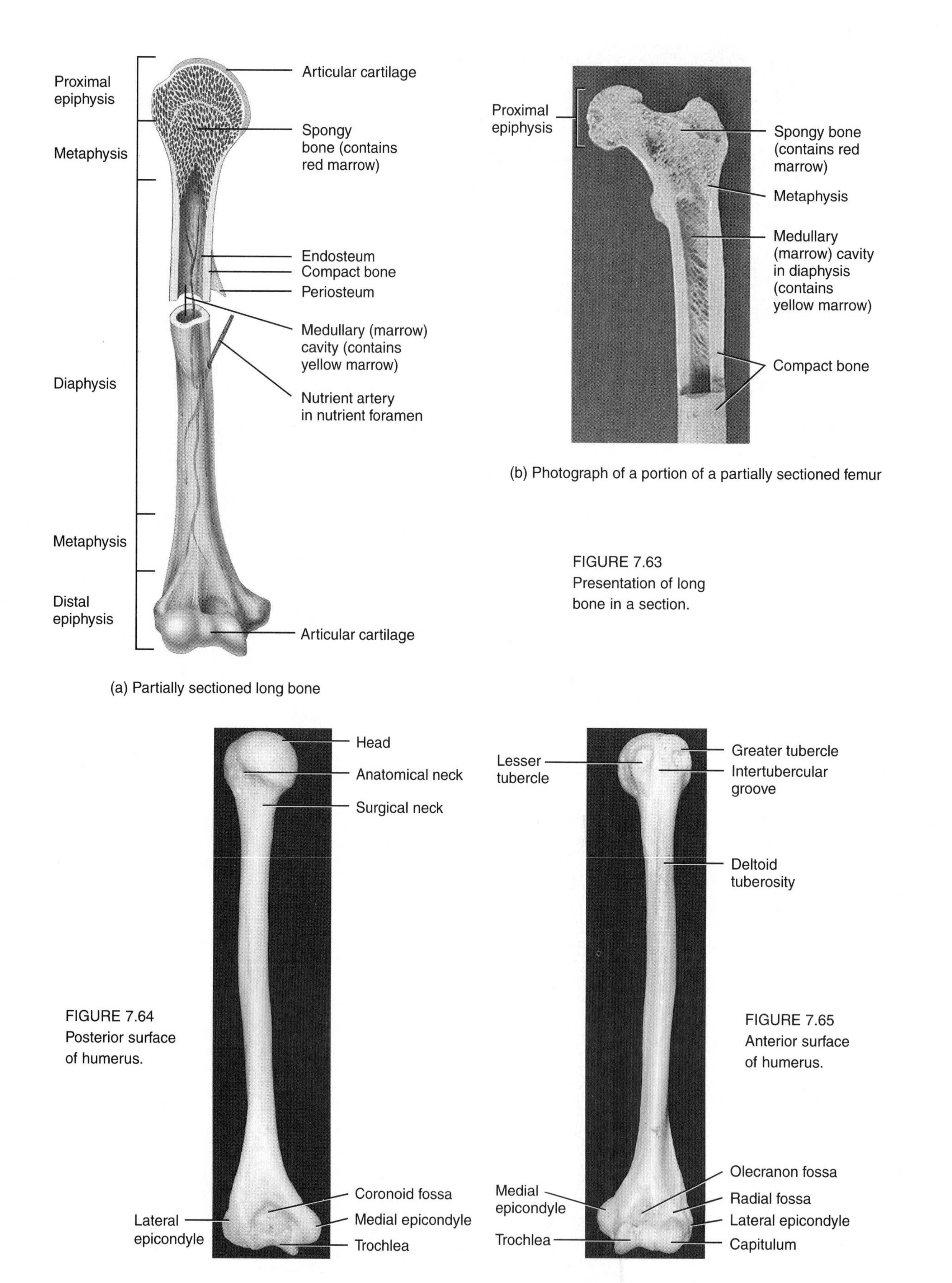

(a) Partially sectioned long bone

(b) Photograph of a portion of a partially sectioned femur

FIGURE 7.63
Presentation of long bone in a section.

FIGURE 7.64
Posterior surface of humerus.

FIGURE 7.65
Anterior surface of humerus.

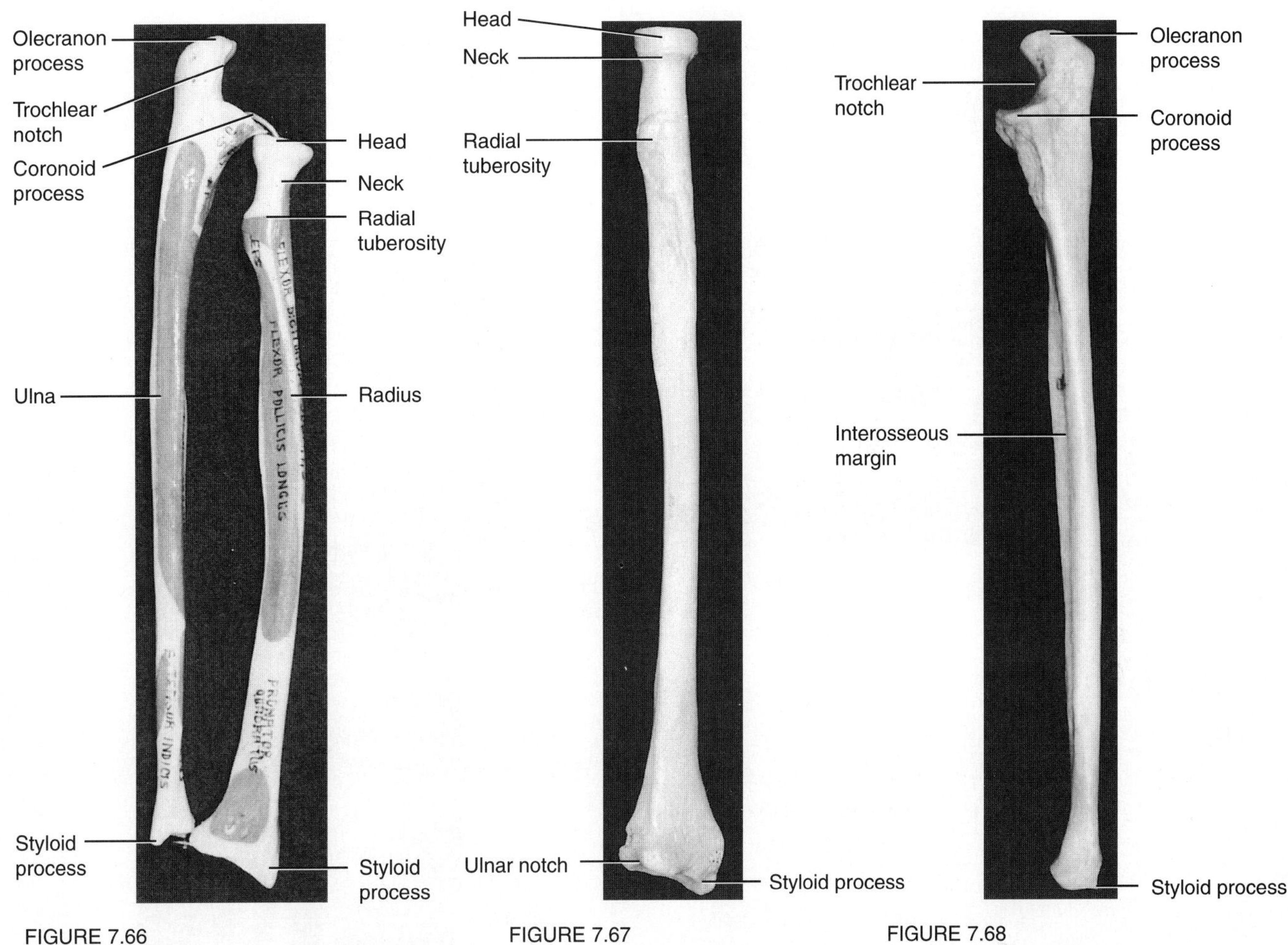

FIGURE 7.66
Radius and ulna bones of the forearm.

FIGURE 7.67
Medial view of the radius bone.

FIGURE 7.68
Medial view of the ulna bone.

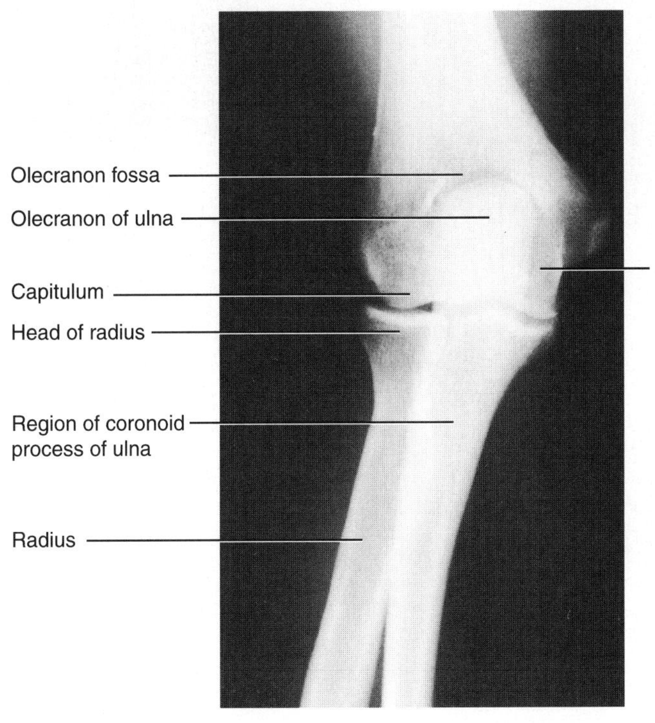

FIGURE 7.69
Radiograph of the elbow joint displaying the articulation between the humerus, radius, and ulna bones.

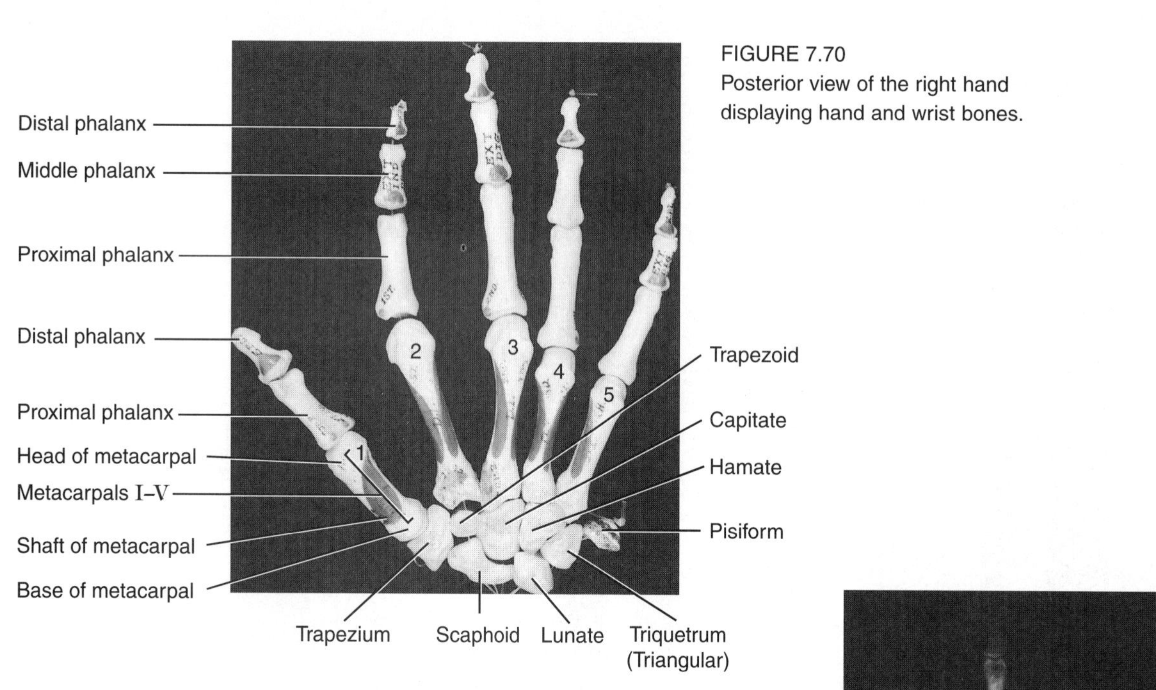

FIGURE 7.70
Posterior view of the right hand displaying hand and wrist bones.

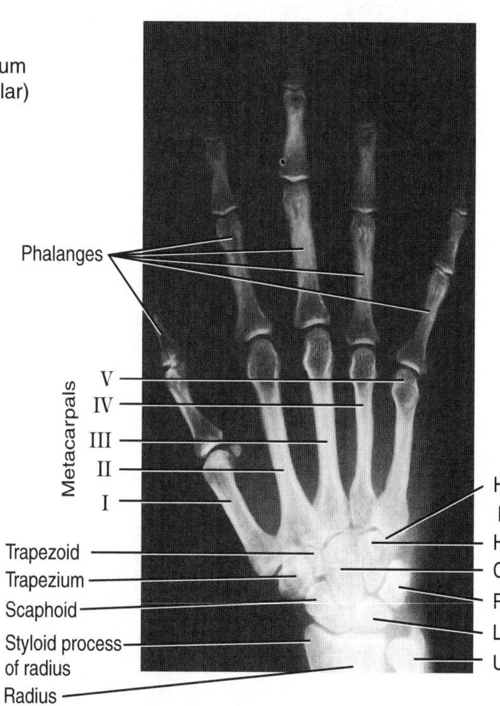

FIGURE 7.71
Radiograph of posterior surface of right hand displaying wrist and hand bones.

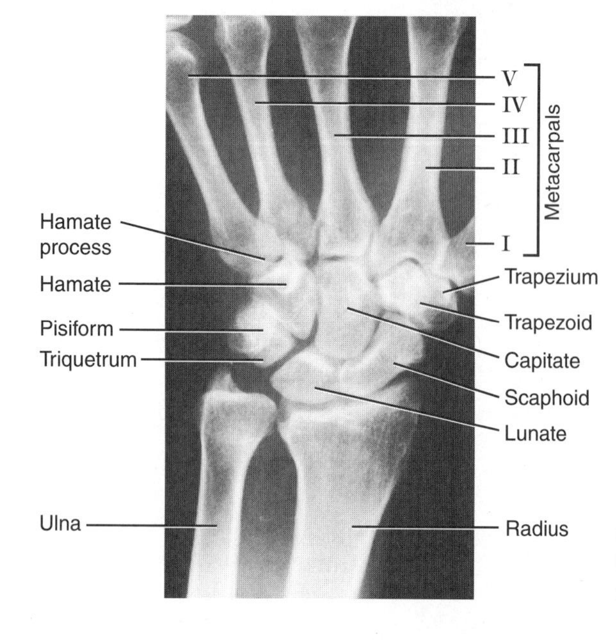

FIGURE 7.72
Radiograph of anterior view of right hand showing the carpals.

FIGURE 7.73
Anterior view of the femur.

FIGURE 7.74
Posterior view of the femur.

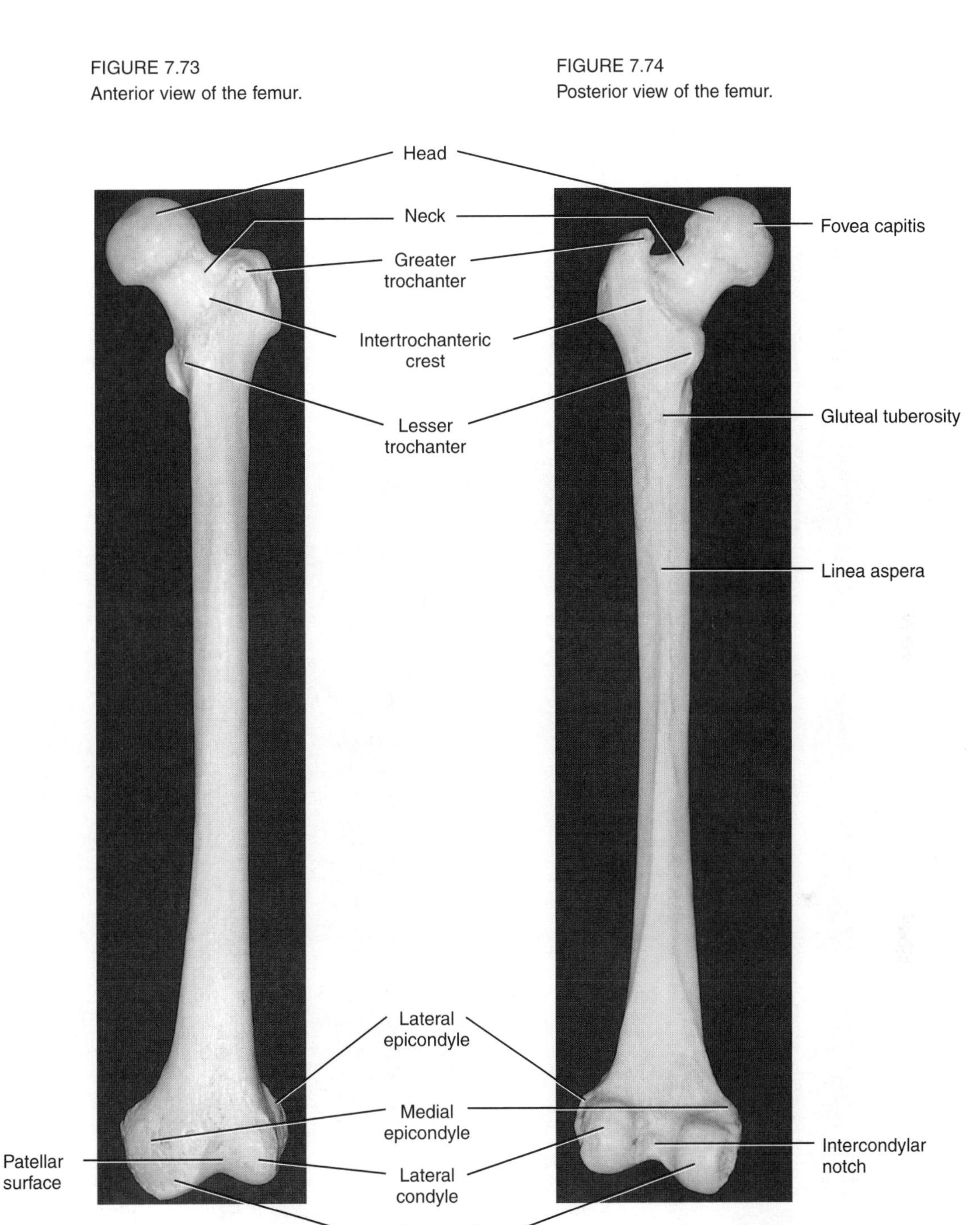

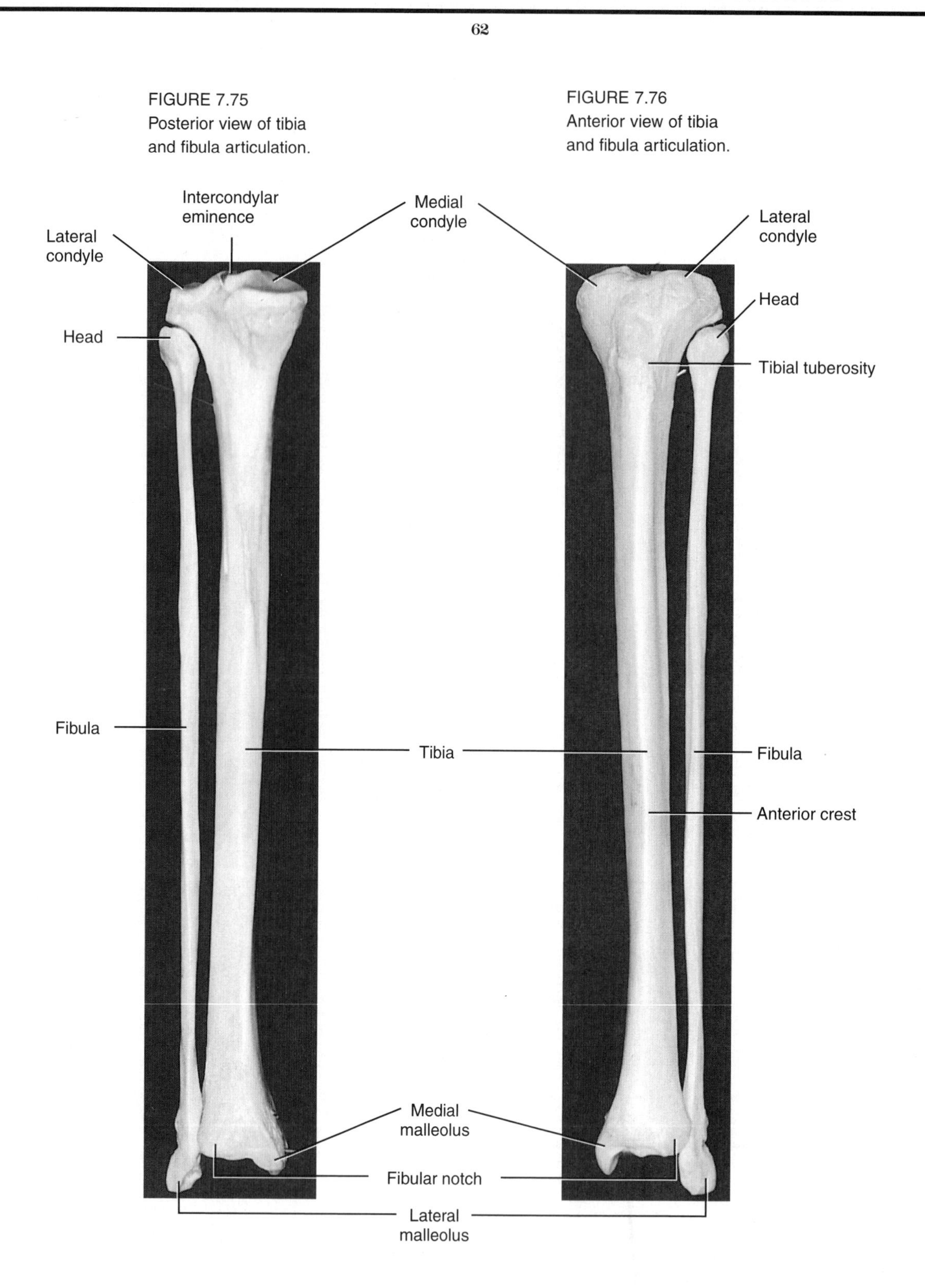

FIGURE 7.75
Posterior view of tibia and fibula articulation.

FIGURE 7.76
Anterior view of tibia and fibula articulation.

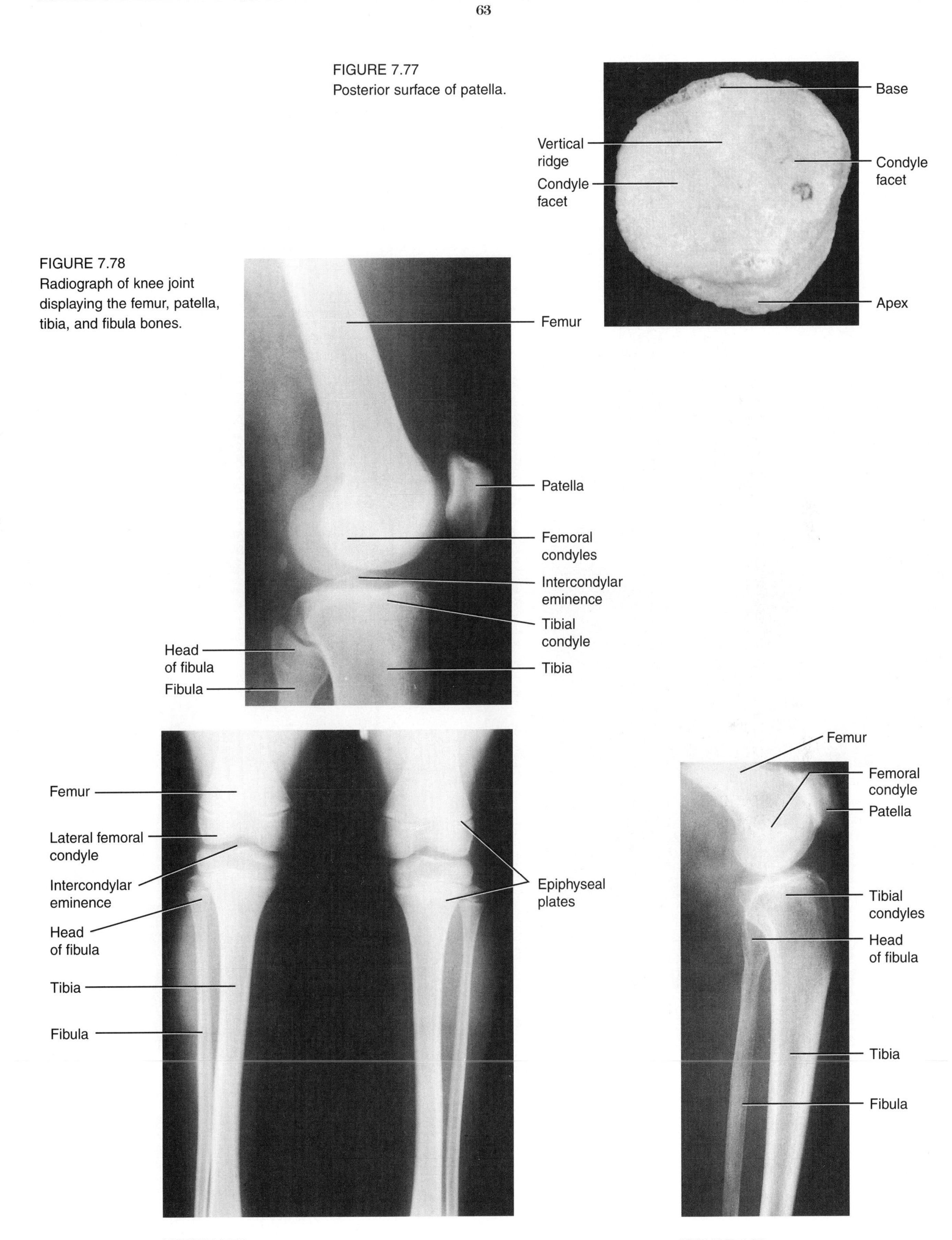

FIGURE 7.77
Posterior surface of patella.

FIGURE 7.78
Radiograph of knee joint displaying the femur, patella, tibia, and fibula bones.

FIGURE 7.79
Radiograph of left and right tibia, fibula, and femur forming knee joints.

FIGURE 7.80
Radiograph of knee joint, lateral view.

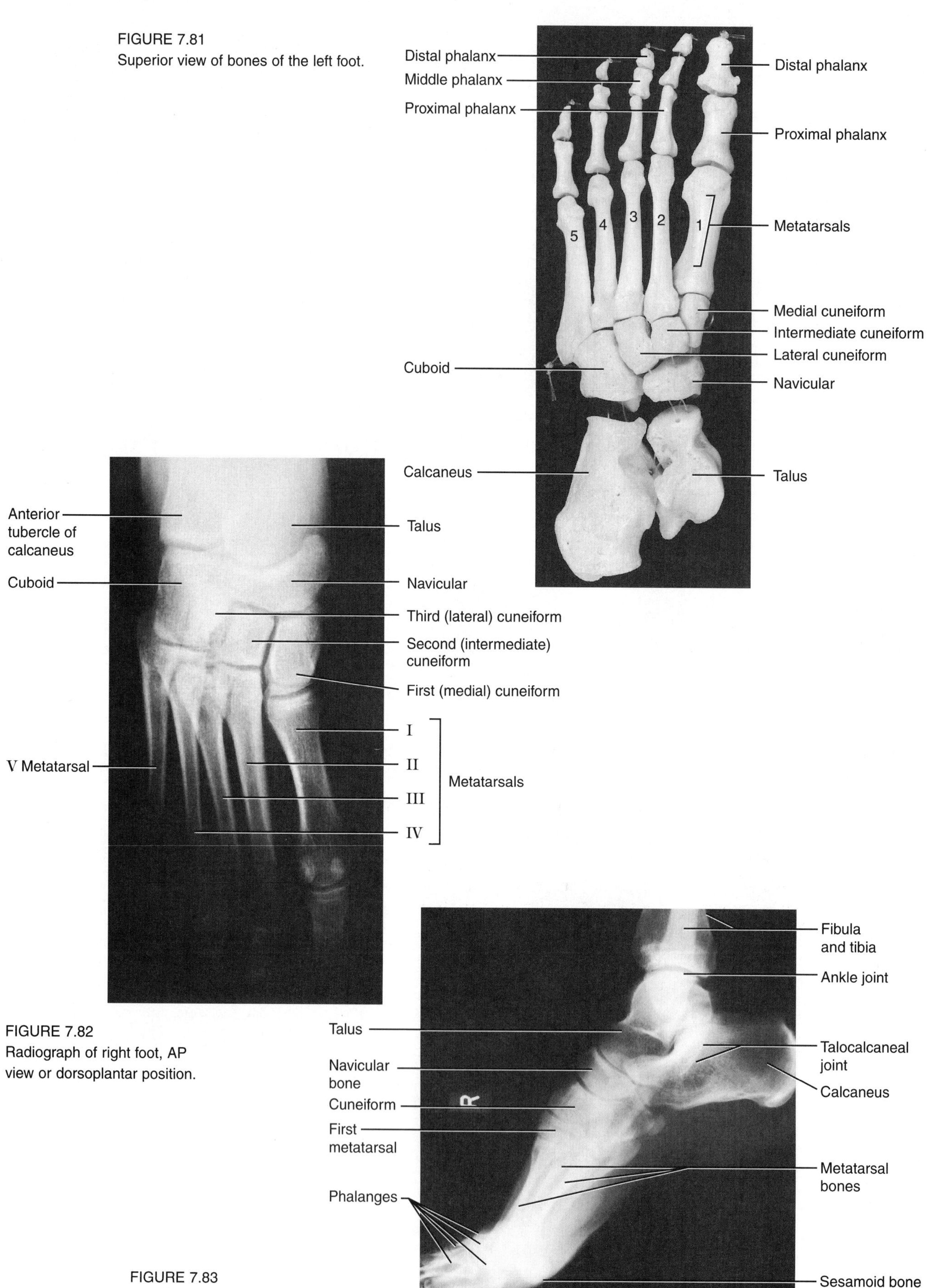

FIGURE 7.81
Superior view of bones of the left foot.

FIGURE 7.82
Radiograph of right foot, AP view or dorsoplantar position.

FIGURE 7.83
Radiograph of right foot, lateral view.

Articulation

The manner in which the bones **articulate** with each other determines the movements of the joints and body parts. Without the mobility of joints, our muscles will **atrophy.** However, not all joints in our body are movable joints. Diversity is essential for the normal functioning of the human body.

Joints can be classified in different ways. The general approach is to classify the joints on the basis of their morphology and movements. Thus, **synarthroses** display no movement, **amphiarthroses** show slight movement, and **diarthroses** are freely movable **synovial** joints.

The synarthroses, amphiarthroses, and diarthroses are further classified on the basis of their morphology. Therefore, synarthroses can be (1) **fibrous** joints, as seen in **sutures** associated with skull bones; (2) **gomphoses,** as seen in the articulation of teeth and the **alveolar bone;** and (3) **cartilaginous joints,** or **synchondroses,** as seen in the formation of the **epiphyseal plates** of long bones.

Amphiarthroses may be of two types: (1) **syndesmosis,** as in the articulation of the tibia and fibula; and (2) **symphysis,** as in the **symphysis pubis.**

Diarthroses (freely movable joints) include the synovial joints. The synovial joints can be of several types: (1) **gliding** or **arthrodial** joints as seen between **carpals,** between **tarsal** bones, and between the **clavicle** and the **sternum;** (2) **hinge** or **ginglymus joints,** for example, in the knee and elbow joints; (3) **pivot** or **trochoid joints,** for example, in the articulation of the **atlas** and **axis** vertebrae; (4) **condyloid** or **ellipsoidal joints,** as seen in the wrist and tarsal bones; (5) **saddle** or **sellaris** joints, as seen in the articulation of the **trapezium** and the first metacarpal bone; and (6) **ball and socket** joints, for example, the hip and the shoulder joints.

There are several types of movements associated with the synovial joints, such as (1) **gliding movement,** as seen between the carpals and the tarsal bones; (2) **angular movement,** which may include **flexion, extension, adduction,** and **abduction;** (3) **rotation movement,** which could be medial or lateral; and (4) **circumduction movement,** where the distal end of the joint moves but the proximal end of the joint remains stable (e.g., pitching a ball).

Several types of special movements are associated with joints. These movements may include (1) **elevation** and **depression,** (2) **inversion** and **eversion,** (3) **dorsiflexion** and **planter flexion,** (4) **supination** and **pronation,** and (5) **protrusion** and **retraction.**

The following chart followed by **photographs** and **radiographs** will give the students a general understanding of the joints, their classification, **morphology,** and movements.

Articulations

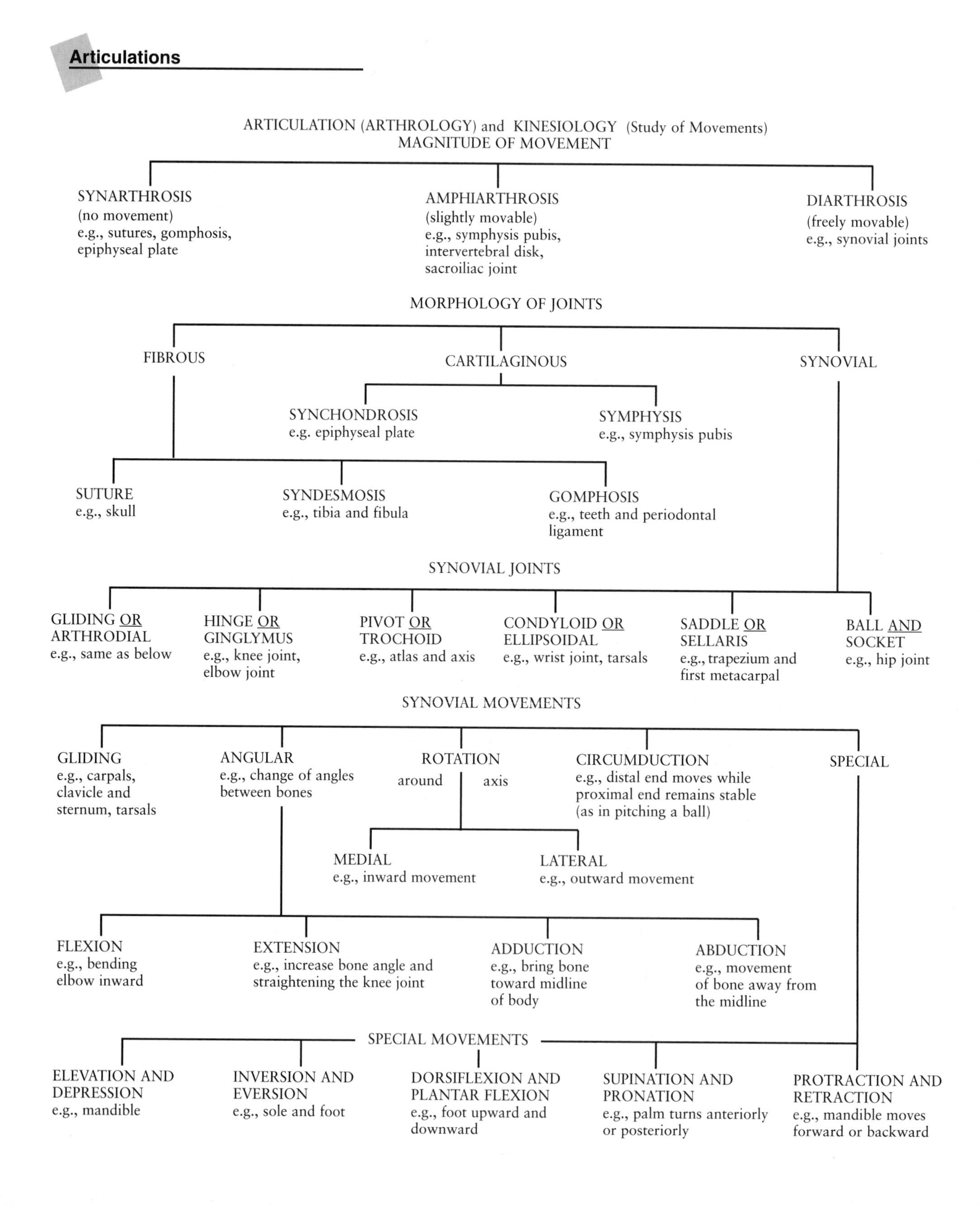

The Fibrous Joints

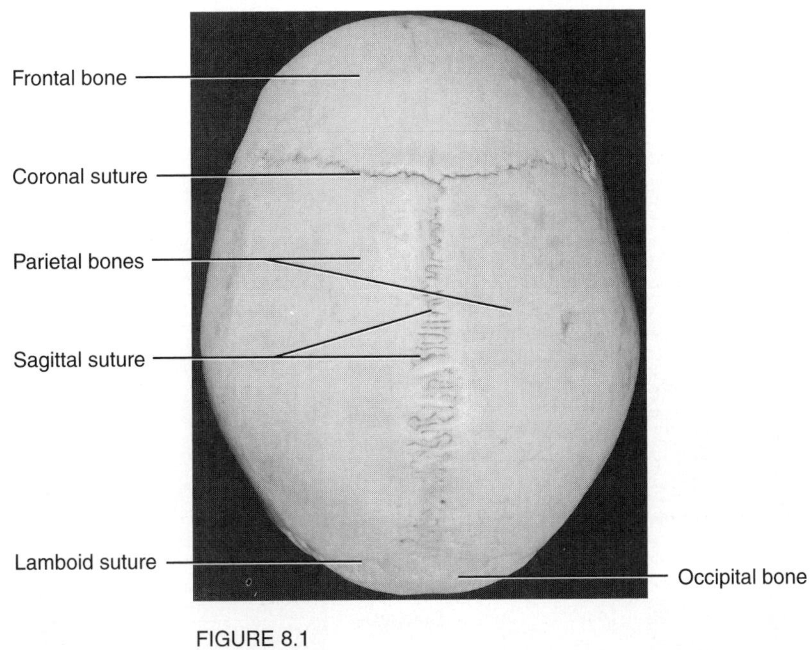

FIGURE 8.1
Coronal and sagittal sutures, examples of fibrous joint.

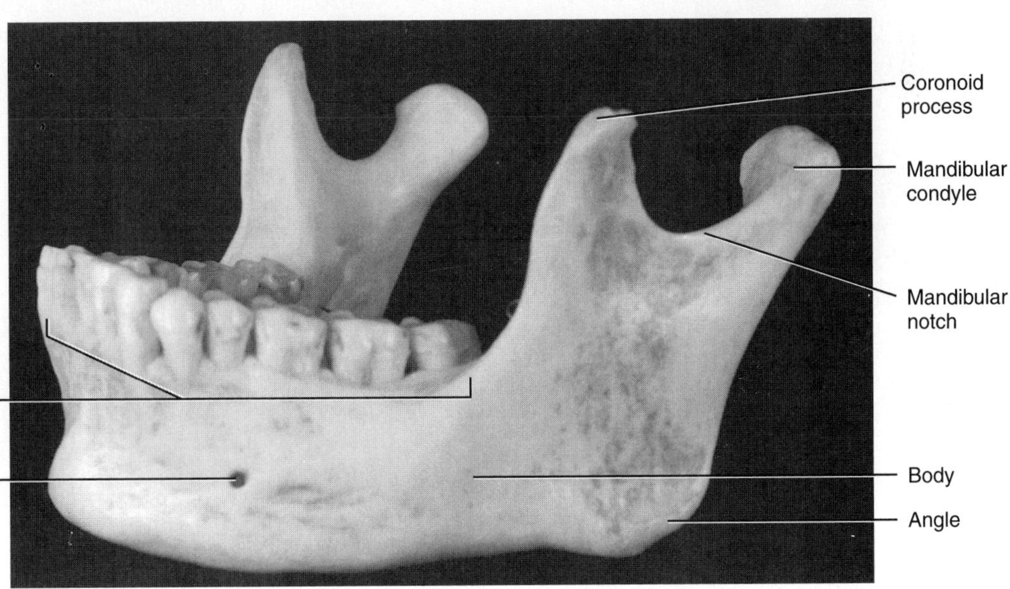

FIGURE 8.2
Gomphosis, a fibrous joint as seen between the teeth and alveolar bone.

FIGURE 8.3
Syndesmosis, an amphiarthrodial (slightly movable) fibrous joint, as seen in the articulation of the tibia and fibula bones at the distal end.

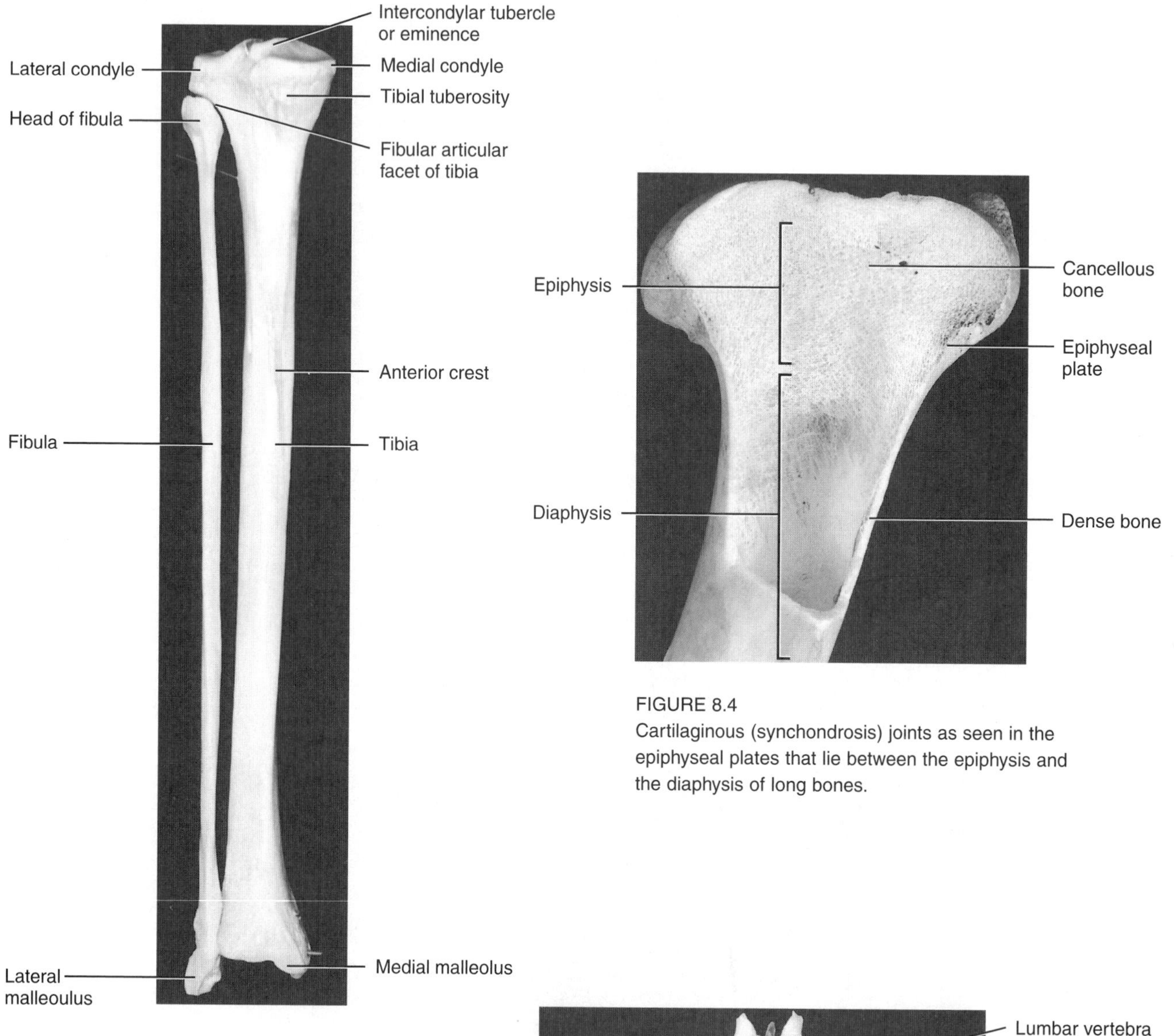

FIGURE 8.4
Cartilaginous (synchondrosis) joints as seen in the epiphyseal plates that lie between the epiphysis and the diaphysis of long bones.

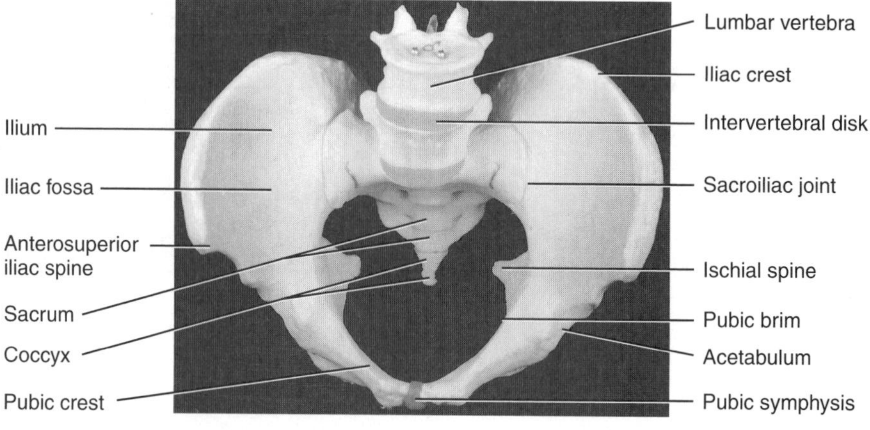

FIGURE 8.5
Cartilaginous (symphysis) joints as seen in the pubic symphysis bones.

The Synovial Joints

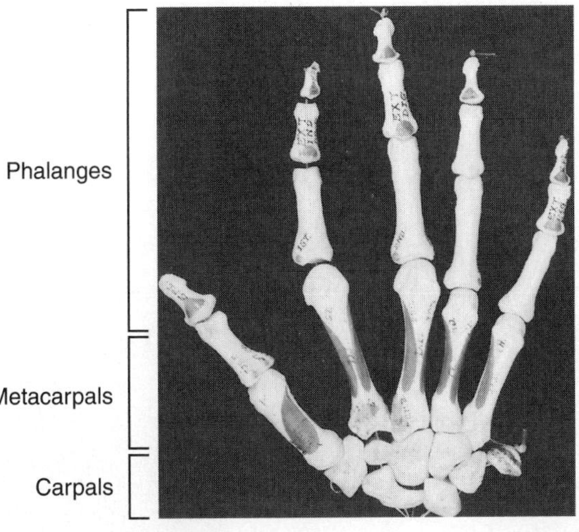

FIGURE 8.6
Gliding or arthrodial joints between opposing surfaces of bones that are slightly curved or flattened, as seen in carpals or wrist bones.

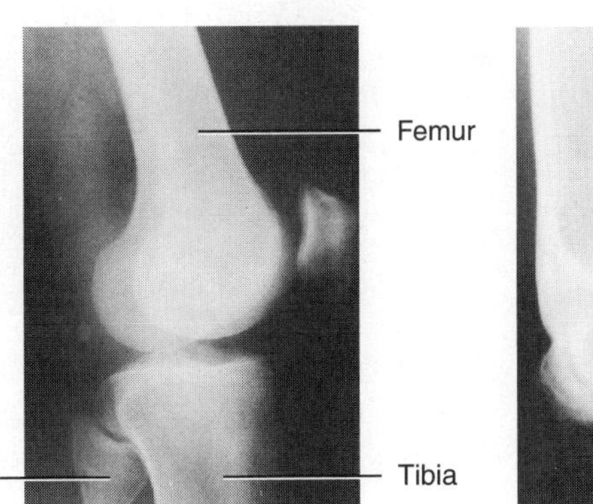

FIGURE 8.7
Radiograph of a synovial diarthrodial (freely movable), hinge or ginglymus, tibiofemoral (knee) joint.

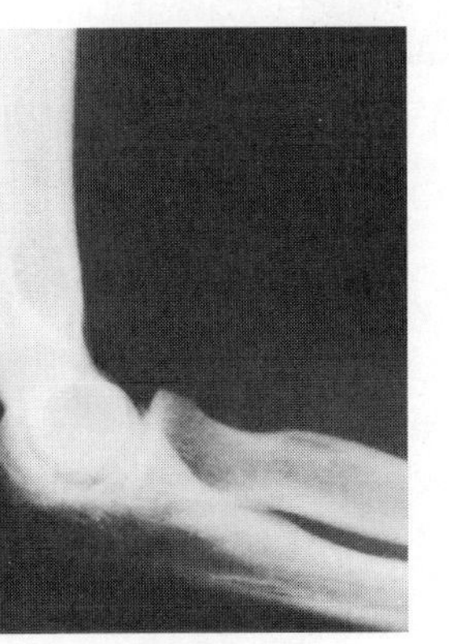

FIGURE 8.8
Radiograph of a synovial, diarthrodial, hinge, or ginglymus joint as seen in humeroulnar (elbow) articulation.

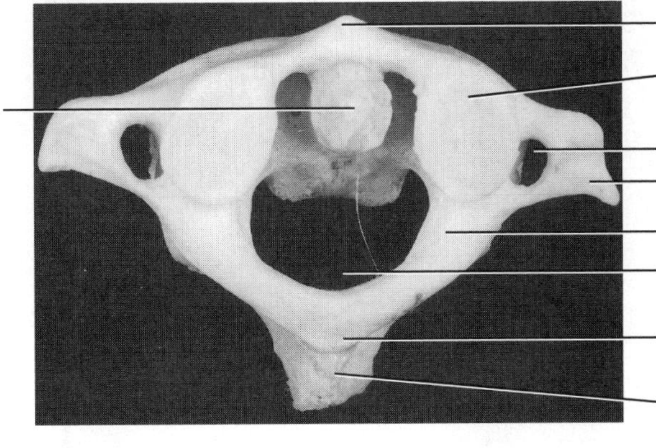

FIGURE 8.9
A synovial pivot or trochoid joint, where a depression of one bone articulates with the conical surface of another bone, thus bringing about a central axis rotation.

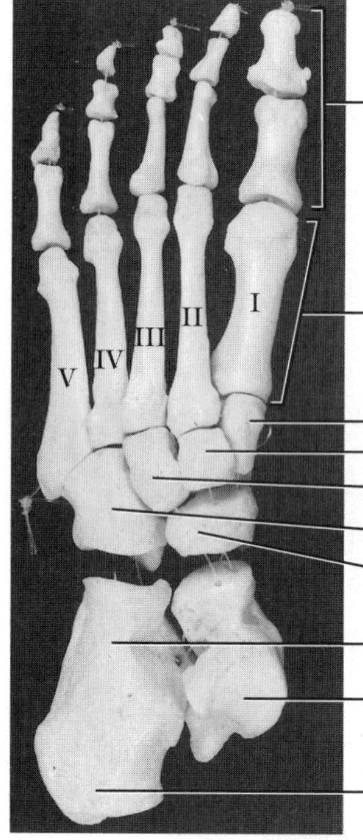

FIGURE 8.10
A synovial condyloid or ellipsoidal joint, where an elliptical cavity of one bone articulates with the oval condyle of another bone, as seen in the tarsal bones of the foot.

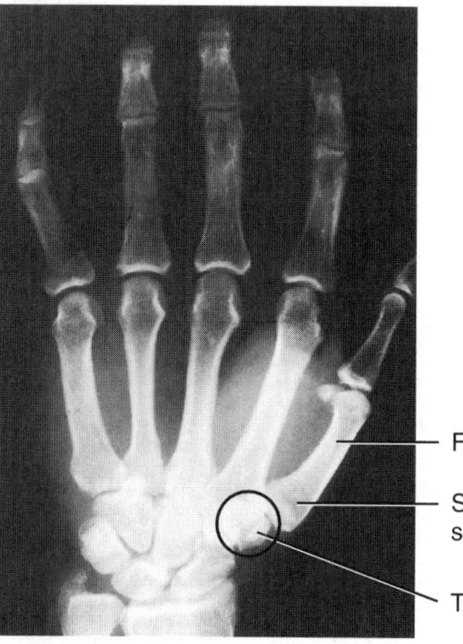

FIGURE 8.11
Radiograph of a saddle or sellaris synovial joint as seen in the articulation of the trapezium and the first metacarpal bones.

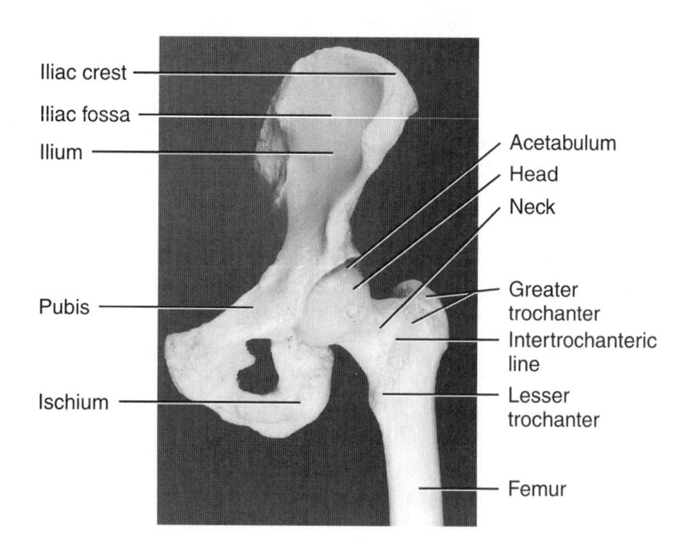

FIGURE 8.12
A ball-and-socket synovial joint where a cup-shaped socket articulates with a rounded convex ball-like structure, as seen in the hip joint.

Synovial Movements

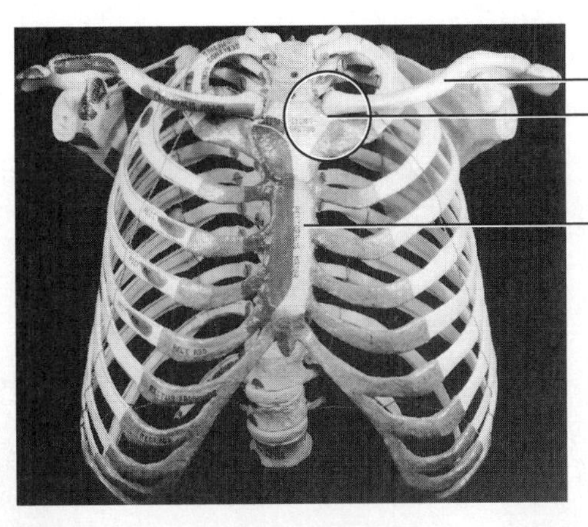

FIGURE 8.13
Gliding movement as seen between the sternum and clavicle bone.

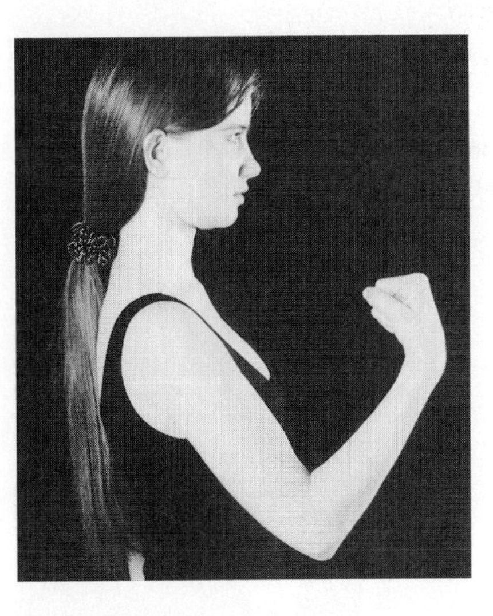

FIGURE 8.14
Angular flexion movement, where the angle between two bones is decreased.

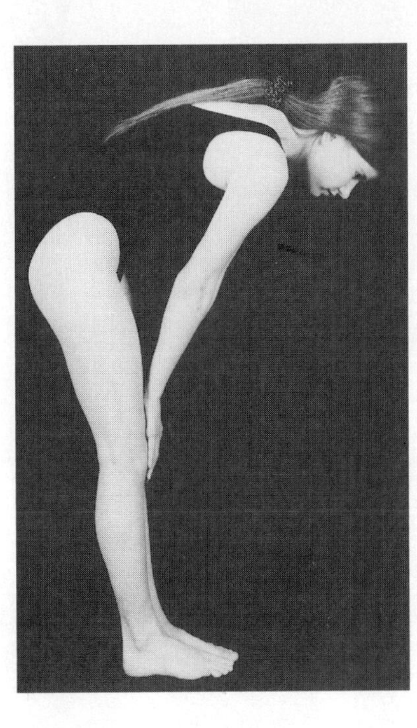

FIGURE 8.15
Angular flexion movement in the vertebral column.

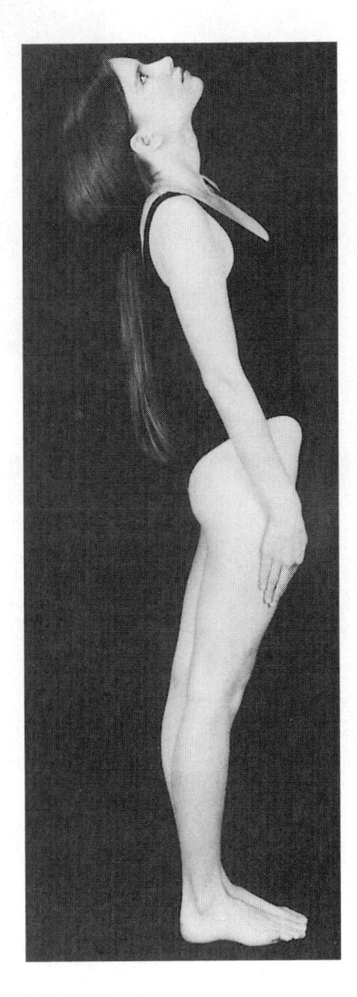

FIGURE 8.16
Angular extension movement at the vertebral column.

FIGURE 8.17
Angular extension movement at the hip and shoulder joints.

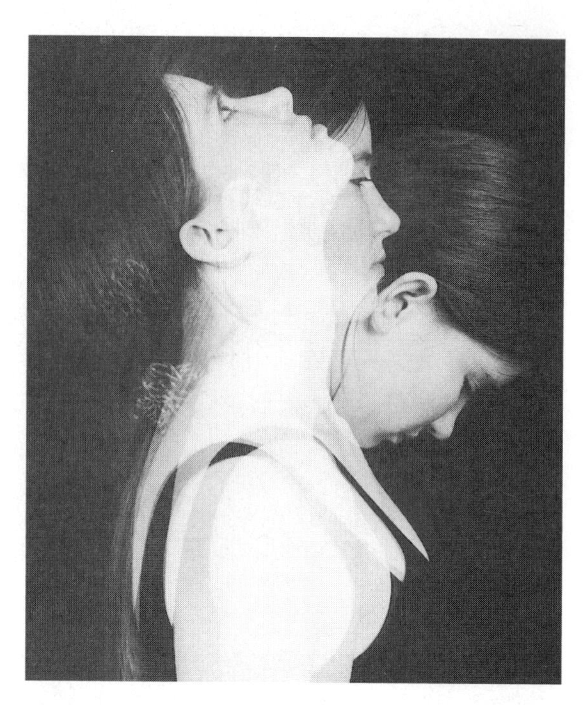

FIGURE 8.18
Angular flexion and extension movements of the head at the cervical articulation.

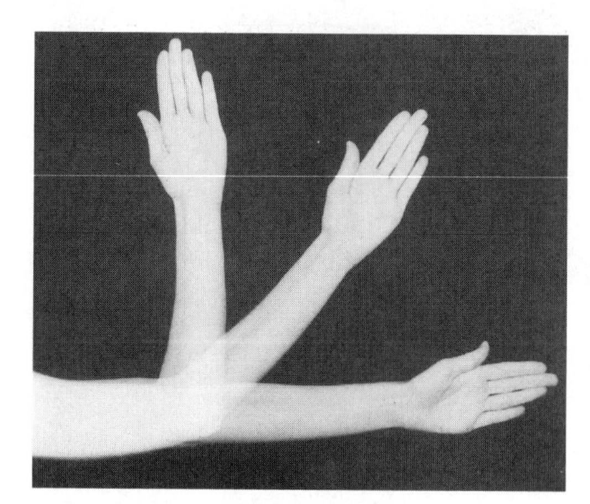

FIGURE 8.19
Angular flexion and extension movements at the elbow joint.

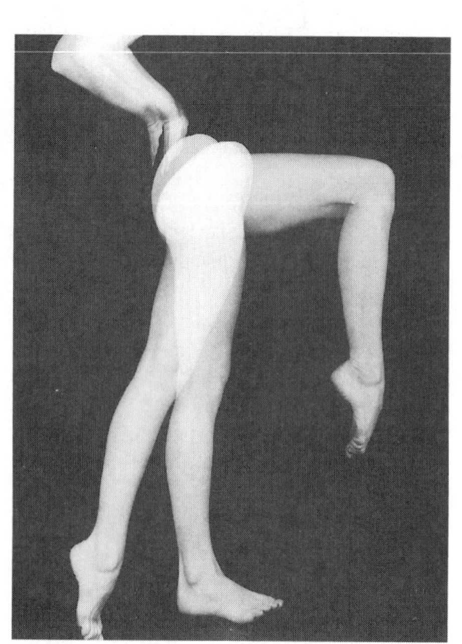

FIGURE 8.20
Angular flexion and extension movements at the knee and hip joints.

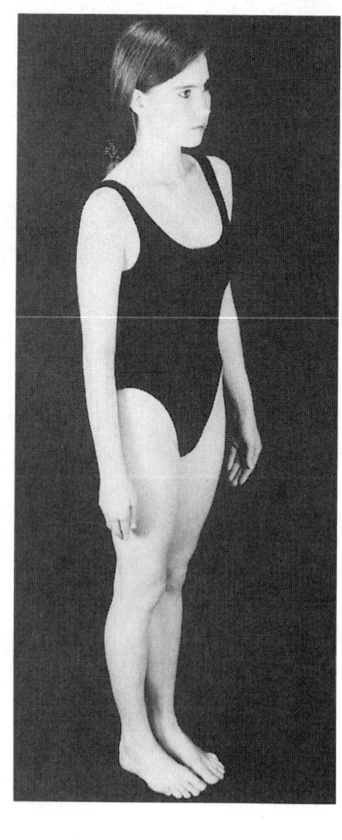

FIGURE 8.21
Adduction movement that brings the body parts toward the axis of the body.

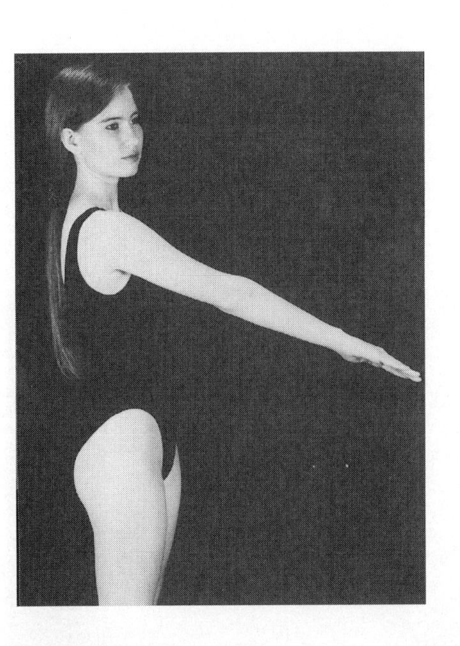

FIGURE 8.22
Abduction movements that take the body part away from the axis of the body.

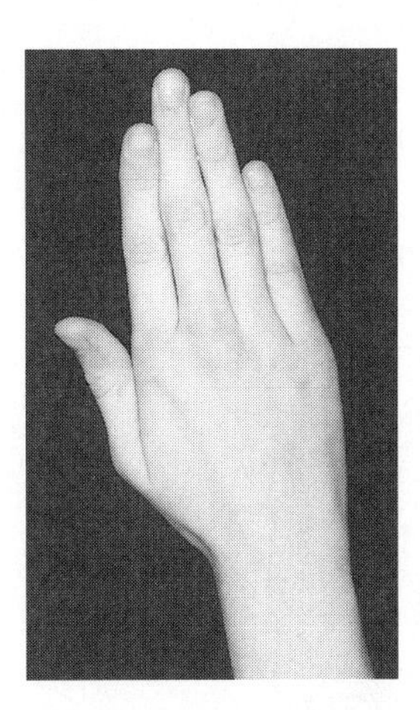

FIGURE 8.23
Adduction of the fingers at the metacarpophalangeal joints.

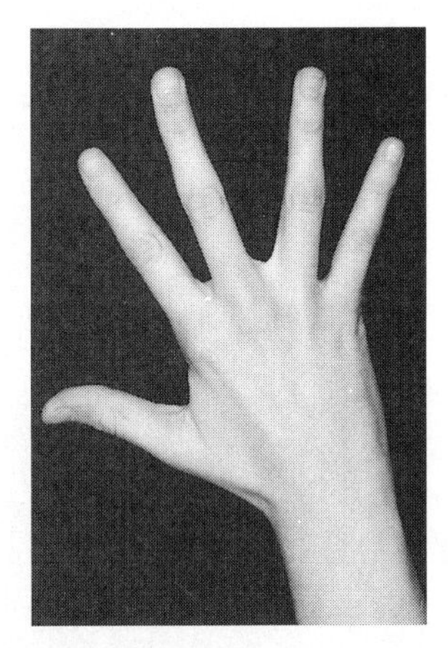

FIGURE 8.24
Abduction of the fingers at the metacarpophalangeal joints.

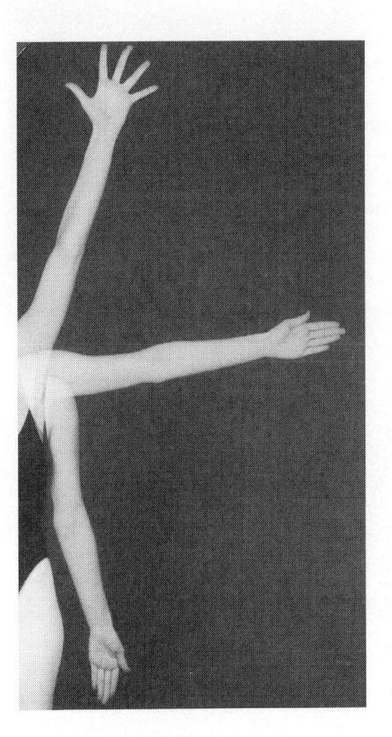

FIGURE 8.25
Abduction and adduction of the left arm and the phalanges at their respective joints.

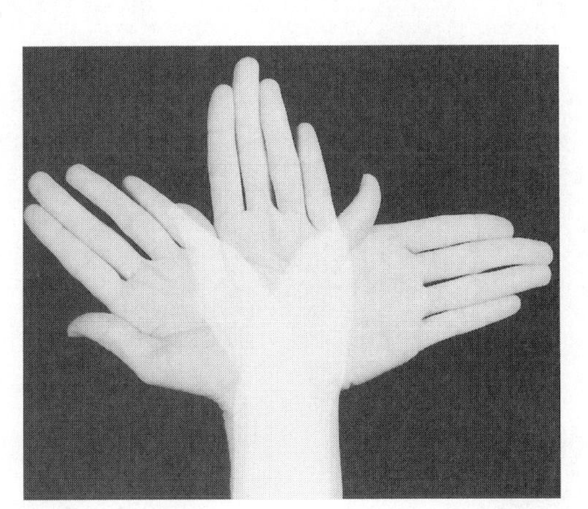

FIGURE 8.26
Adduction and abduction of the hand at the carpometacarpal joints.

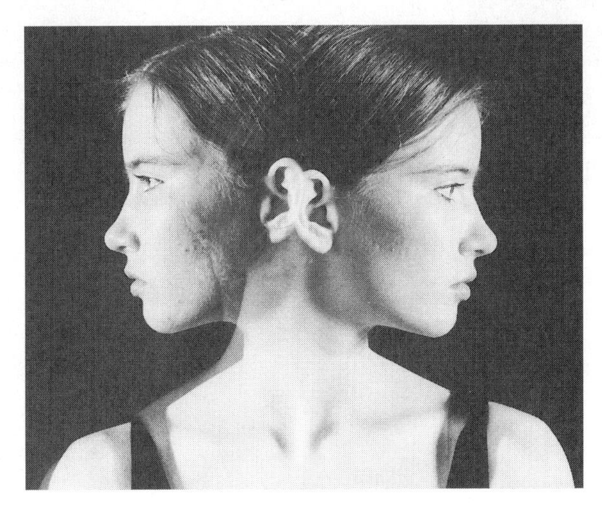

FIGURE 8.27
Rotation movement of the head at the atlantoaxial cervical joint.

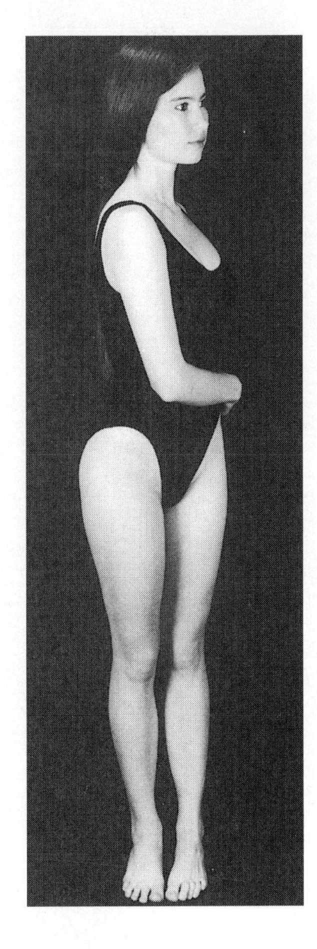

FIGURE 8.28
Rotation movement at the vertebral column.

Special Movements

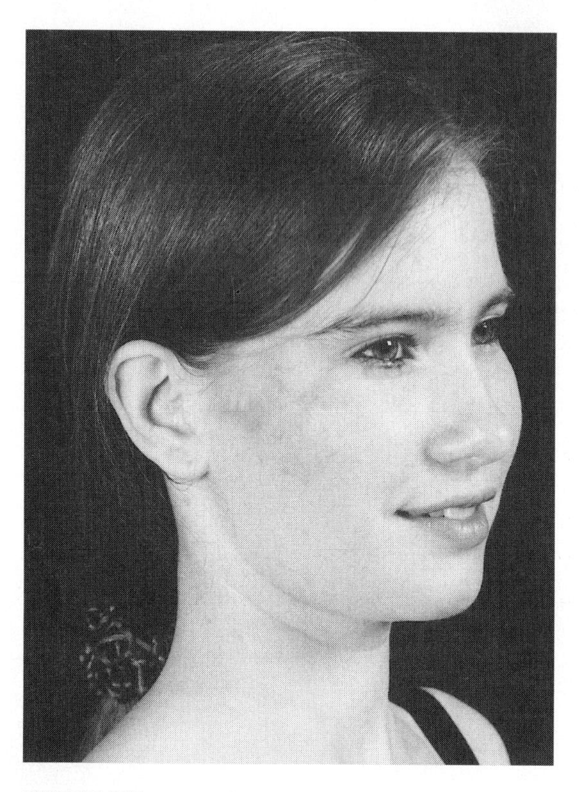

FIGURE 8.29
Elevation of the mandible.

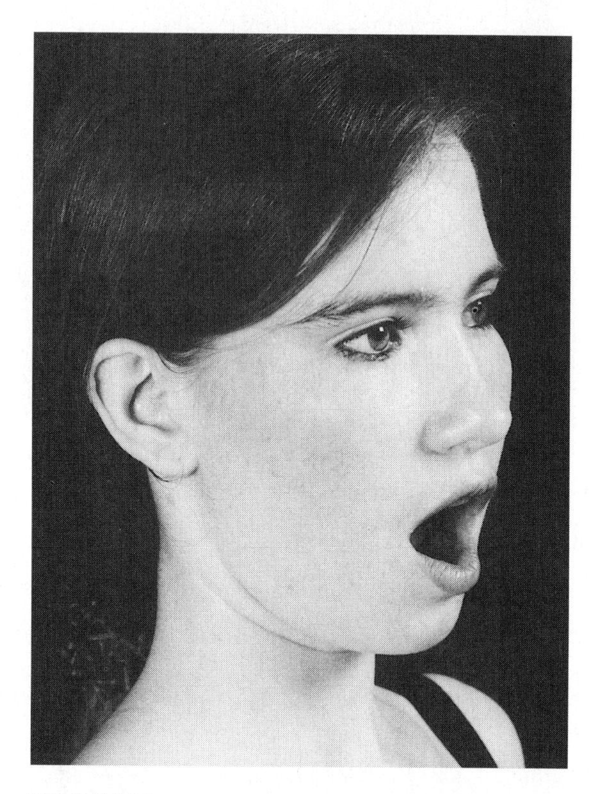

FIGURE 8.30
Depression of the mandible.

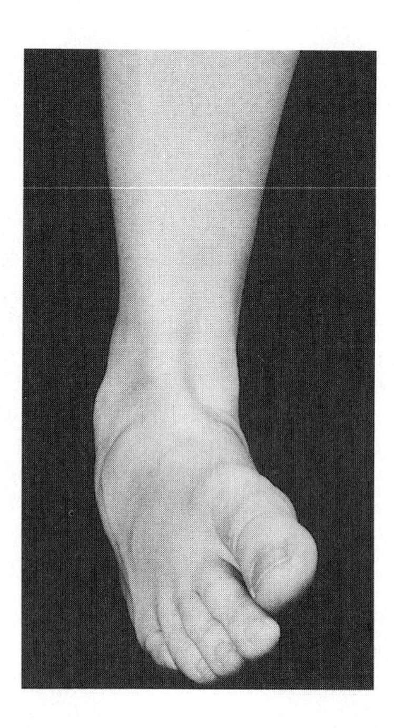

FIGURE 8.31
Inversion of the sole of the foot.

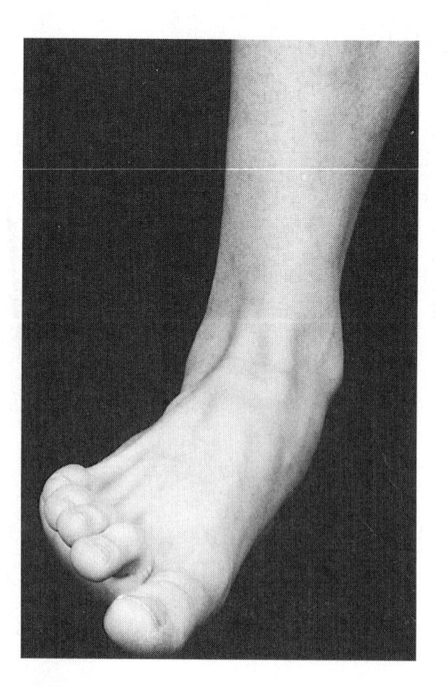

FIGURE 8.32
Eversion of the sole of the foot.

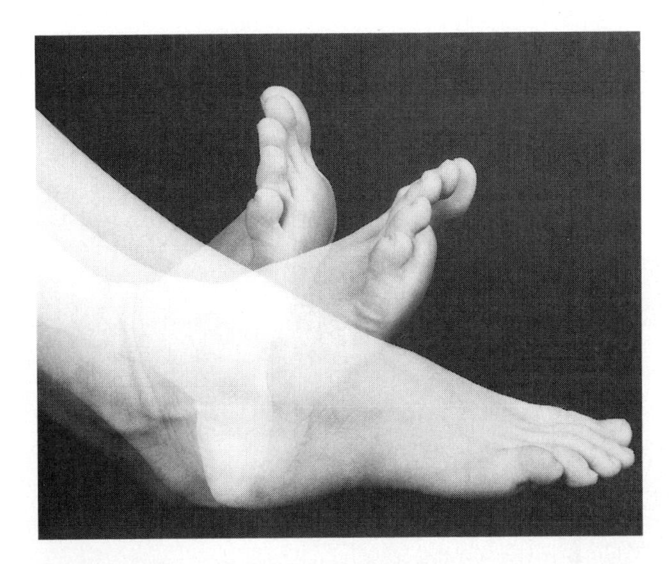

FIGURE 8.33
Dorsiflexion (movement upward) and plantar flexion (movement downward) of the foot.

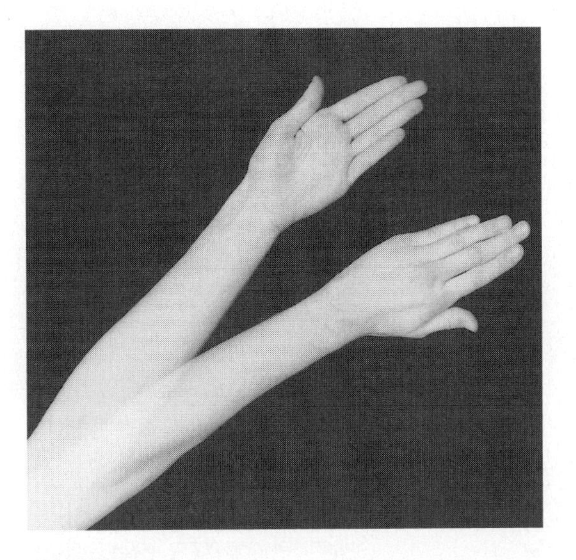

FIGURE 8.34
Supination (palm facing anteriorly) and pronation (palm facing posteriorly).

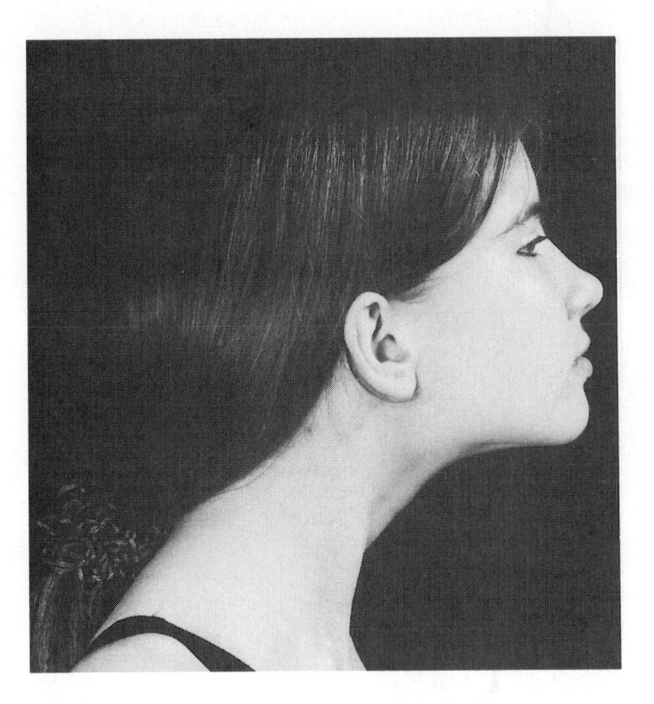

FIGURE 8.35
Temporomandibular joint displaying protraction.

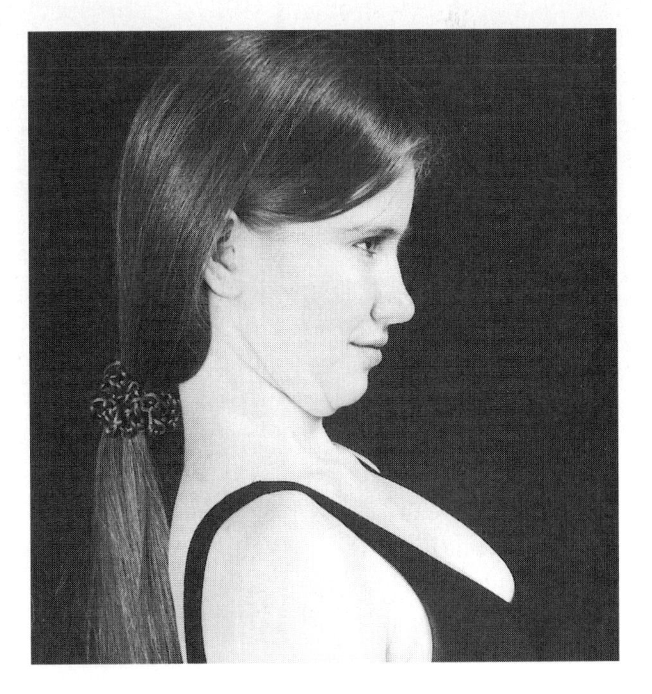

FIGURE 8.36
Temporomandibular joint displaying retraction.

Joints and Ligaments

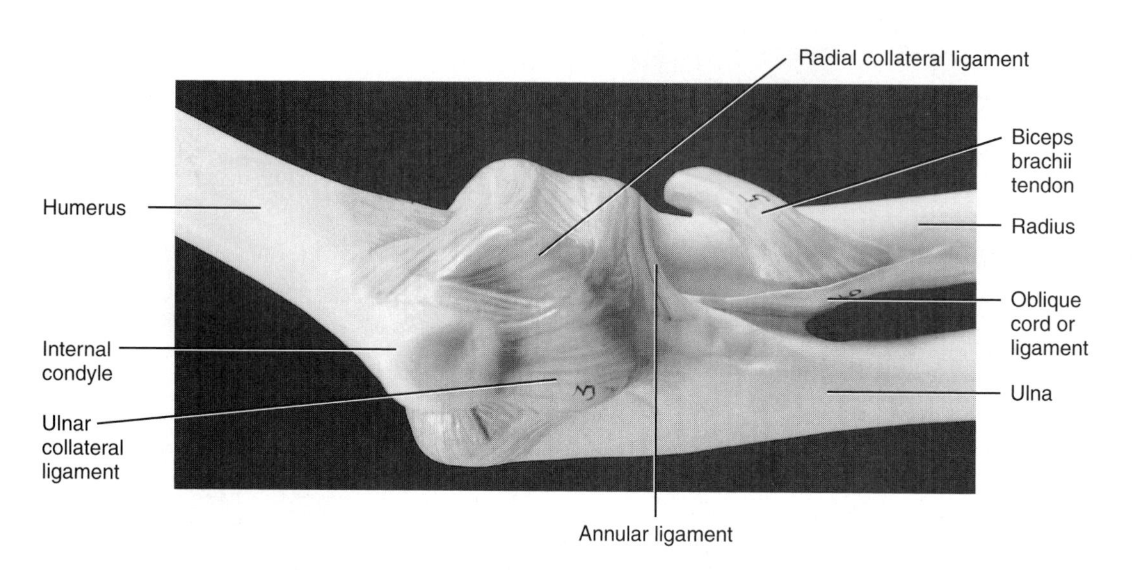

FIGURE 8.37
Superficial view of the elbow joint.

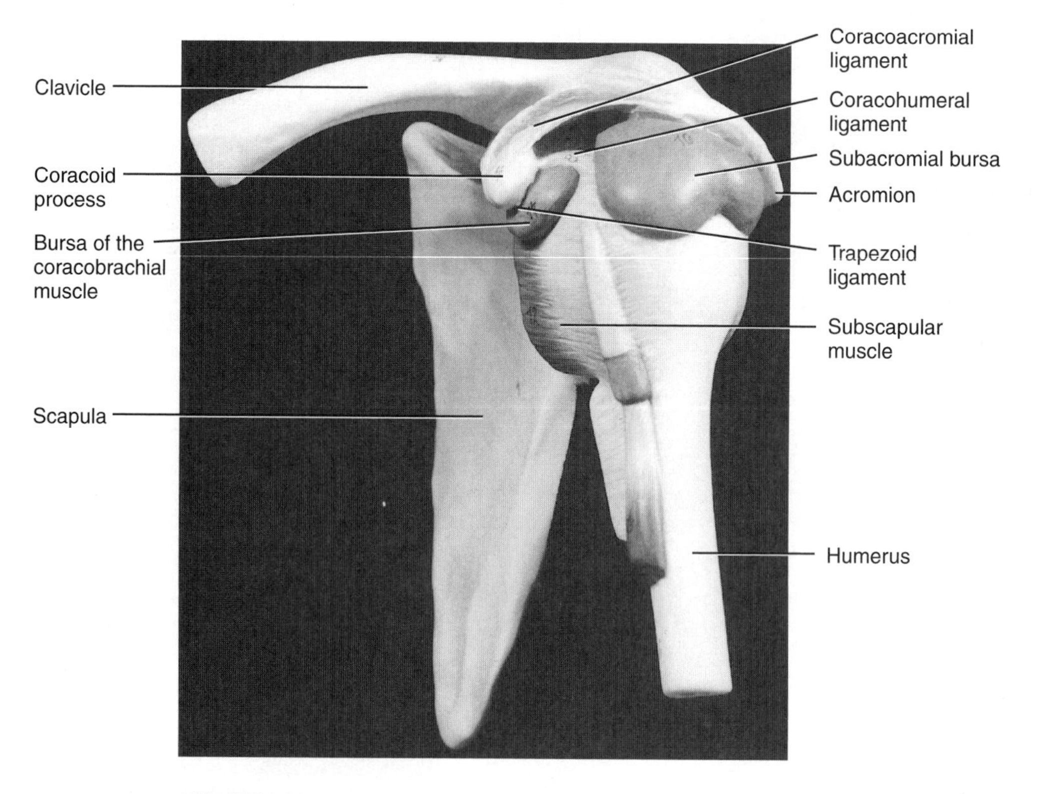

FIGURE 8.38
Anterior view of the shoulder joint showing muscle, bursa, and ligaments.

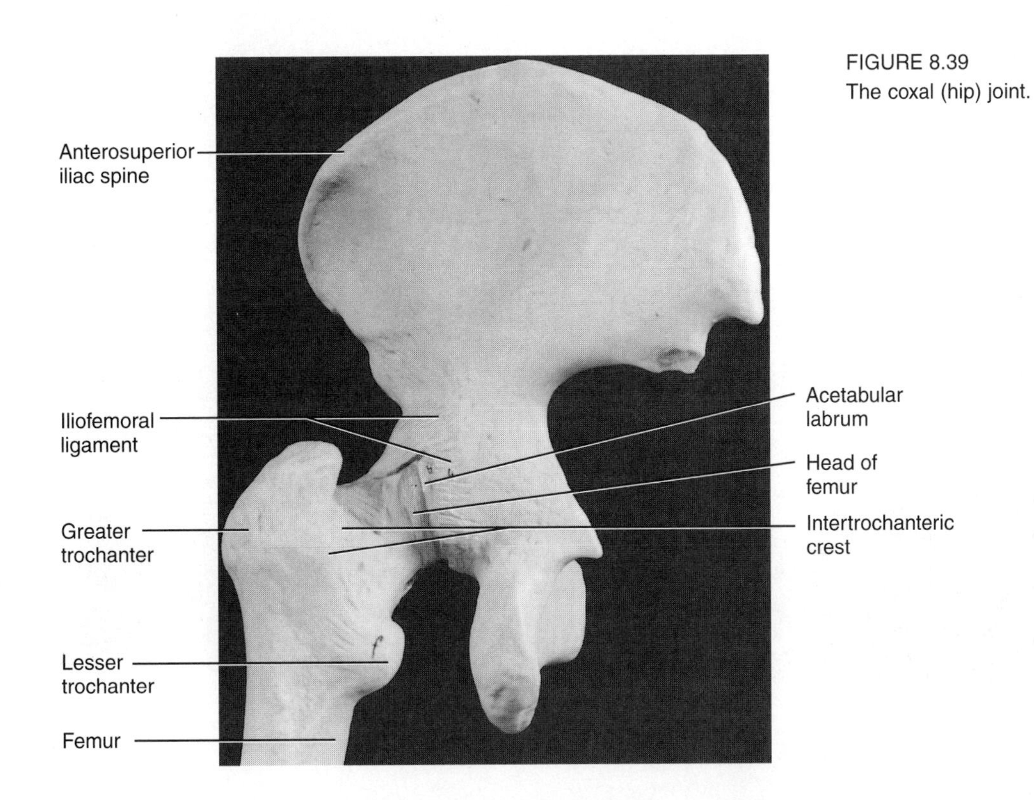

FIGURE 8.39
The coxal (hip) joint.

FIGURE 8.40
The knee joint displaying ligaments, medial and lateral menisci, and the patella.

Femur
Patellar surface
Articular cartilage
Tibial collateral ligament (medial ligament)
Base of patella
Ligamentum patellae
Fibular collateral ligament
Lateral meniscus
Medial meniscus
Medial facets of patella
Medial vertical facet
Fibula
Tibia

Specialized Connective Tissue: Blood and Bone Marrow

The **blood** is a specialized connective tissue in which millions of cells are suspended in a fluid medium called plasma. Blood has several diverse functions, including the transport of nutrients, gases, hormones, proteins, cellular waste products, lipids, and, of course, cells.

Blood plasma is an aqueous solution composed of inorganic salts, proteins (**prothrombin, globulins, fibrinogen**), and **albumins.** These proteins and electrolytes collectively create a colloidal osmotic pressure within the cardiovascular system that facilitates the regulation of aqueous solutions between extracellular (interstitial) fluid and plasma. In the plasma, albumin proteins bind to fatty acids, which are then transported to different parts of the body. In this way, albumin proteins serve as transport proteins for large molecules.

Globulins (antibodies) are complex proteins that are part of the lymphatic system and are involved in the immune process. Specific antibodies bind to specific antigens, forming an antigen–antibody complex that destroys the antigen.

Fibrinogen and **prothrombin** (proteins found in the blood) are involved in a series of chemical reactions that lead to the formation of a blood clot at the site of rupture in a blood vessel.

Blood cells are produced in the bone marrow by **hematopoiesis.** An exception to this rule is the white blood cells (WBCs or leukocytes), which are formed in the bone marrow and the lymphatic system by extramedullary hematopoiesis. However, extramedullary hematopoiesis in a developing embryo can also occur in such places as the yolk sac, liver, spleen, and lymph nodes. After birth, hematopoiesis is confined to the red marrow found in the long and flat bones of the body. In adults, hematopoiesis is confined to the flat bones.

Various types of blood cells can be identified in a sample of peripheral blood smear under a microscope. **Erythrocytes,** or red blood cells (RBCs), make up over 99% of the blood cells. The remaining (less than 1% of total) blood cells are comprised of **leukocytes,** or white blood cells (WBCs), and fragments of megakaryocyte cells called **platelets** or **thrombocytes.** These three groups of blood cells are classified, according to morphology and function, as follows:

1. **Erythrocytes** in mature form do not have nuclei. They stain with acid dyes because of the basic nature of hemoglobin. The cells are uniform in diameter, generally 7 to 8 micrometers (μm), and are responsible for the transport of oxygen (O_2) and small amounts of carbon dioxide (CO_2).

2. **Leukocytes,** or white blood cells (WBCs), are complete cells, each containing a nucleus and cell organelles. The WBCs can be divided into two distinct groups based on the presence or absence of granules: granulocytes and nongranulocytes (also called agranulocytes). The granulocytes include **eosinophils, neutrophils,** and **basophils.** The agranular leukocytes include **monocytes** and **lymphocytes.** The agranulocytes contain some nonspecific granules, whereas the granulocytes always contain specific granules that are larger than the nonspecific granules. The three types of granular cells and two types of agranular cells have different functions.

3. **Platelets,** or **thrombocytes,** are fragments of megakaryocytes found in the marrow. These small disk-shaped fragments are approximately 2 to 4 μm in diameter and number from 200,000 to 350,000 per cubic millimeter of blood. Their specific function is associated with the clotting of blood.

Bone Marrow

In adult humans, hematopoiesis is confined to the bone marrow and the lymphoid tissue. However, in the embryonic and fetal stages, hematopoiesis occurs in the yolk sac, spleen, bone marrow, and liver. Under certain pathological conditions, the adult human spleen and liver may be stimulated to resume a role in hematopoiesis.

Bone marrow fills the medullary cavities of the cancellous, long, and intramembranous bones. Bone marrow cells are densely packed between blood vessels and reticular fiber stroma. Connective tissue cells—**osteoblasts, osteoclasts, plasma cells, macrophages, mast cells,** and bone marrow cells, which are multinucleated **megakaryocytes** and **adipocytes**—are also found in the hematopoietic tissue. Different sizes of blood vessels are also present in the marrow cavity.

The bone marrow has several functions: (1) the creation of blood cells and their migration into the peripheral blood, (2) the phagocytosis of defective cells, (3) the genesis of B and T lymphocytes, and (4) the formation of bone (**osteogenesis**) by the osteoblasts lining the endosteum.

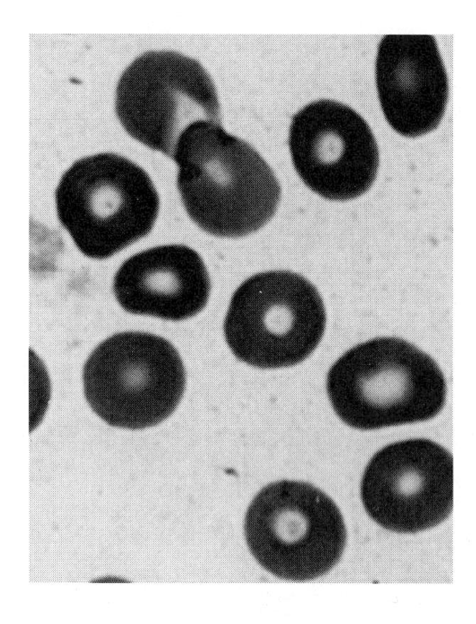

FIGURE 9.1
Light micrograph (LM) of erythrocytes. (1000×)

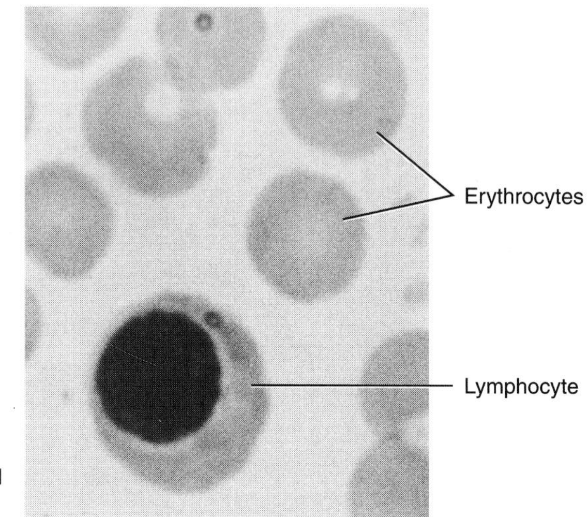

FIGURE 9.2
LM of a lymphocyte and erythrocytes. (1000×)

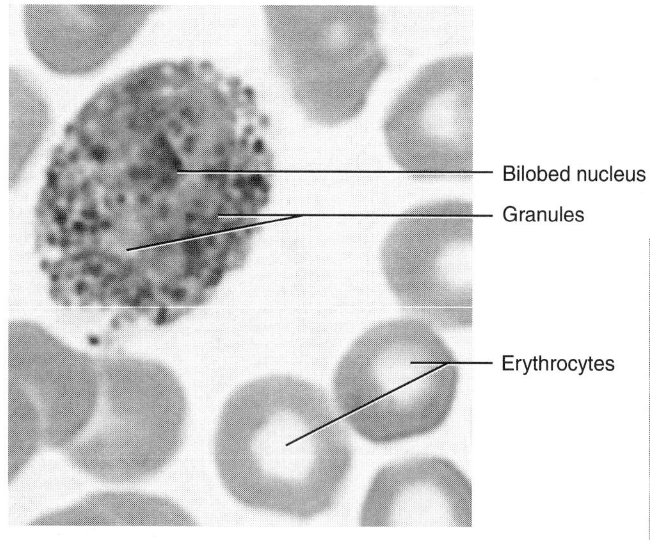

FIGURE 9.3
LM of a basophil.

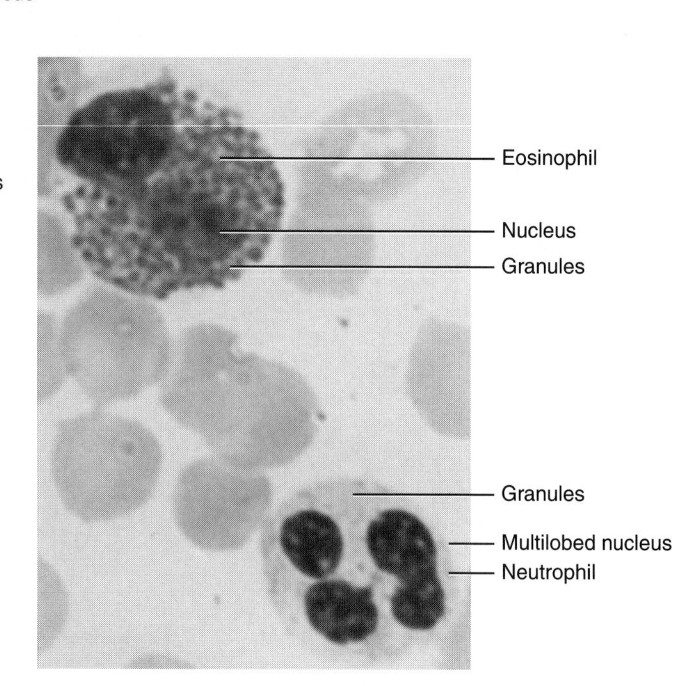

FIGURE 9.4
LM of an eosinophil and a neutrophil. (1000×)

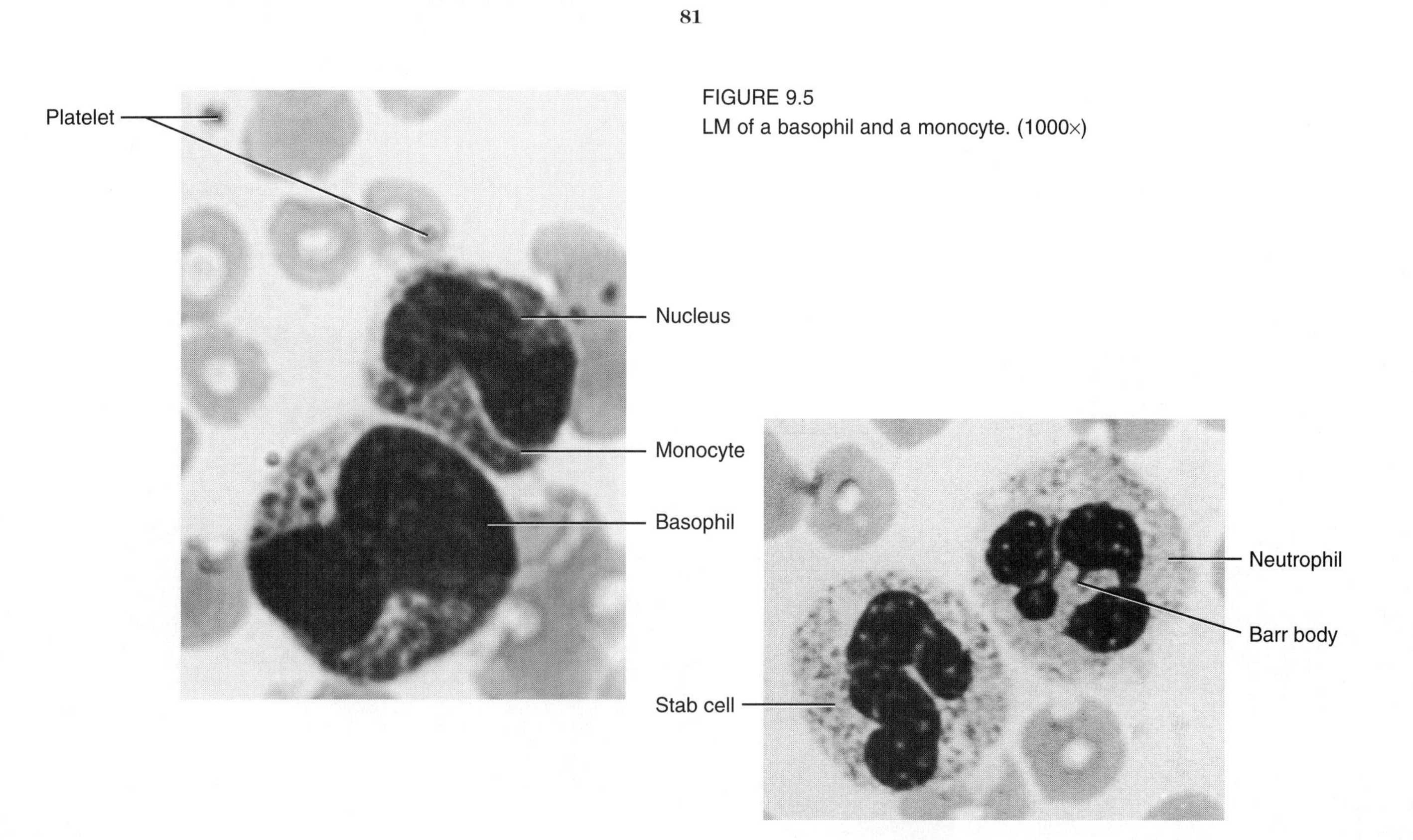

FIGURE 9.5
LM of a basophil and a monocyte. (1000×)

FIGURE 9.6
LM of a neutrophil with a Barr body and a stab cell. (1000×)

Bone Marrow

Orthochromatic erythroblast
Lymphocyte
Basophilic erythroblast
Metamyelocyte
Adipose cells (fat cell)
Megakaryoctye
Megakaryocyte
Metamyelocyte
Hematopoietic cords
Neutrophil myelocyte
Adipose cell (fat cell)

FIGURE 9.7
LM of red bone marrow. (1000×)

Muscle

Muscle fibers can be classified into three groups based on location, morphology, and function: smooth muscle, skeletal muscle, and cardiac muscle fibers. All muscle fibers have distinctive functional similarities. However, there are slight differences in the morphologies of muscle fibers. The histological terminology of muscle tissue is slightly different from the terminology used in defining other tissues: for example, the muscle plasma membrane is called sarcolemma, endoplasmic reticulum is called sarcoplasmic reticulum, and cytoplasm is called sarcoplasm.

Skeletal Muscle

The **skeletal muscle** is composed of voluntary fibers associated with the movement of the skeleton and certain organs, such as the tongue and eye. The cross-banding and alternating striation of actin and myosin are characteristic features of skeletal muscle. Under polarized light, anisotropic dark bands, called **A bands,** alternating with isotropic **I bands,** are visible. Running transversely in the center of the I band is a narrow dense band called the **Z band.** A well-prepared muscle section under an electron microscope shows a narrow band, called the **H band,** passing through the middle of the A band. A dark **M** line traverses the H band. Skeletal muscle fibers are multinucleate elongated cells surrounded by a membrane called the **sarcolemma.** The cells have a high concentration of Golgi bodies, mitochondria, sarcoplasmic reticulum, and myofibrils. The myofibrils run parallel to the sarcoplasmic reticulum and form the myosin proteins.

Cardiac Muscle

The cardiac muscle fibers are composed of branching elongated cells, which indicate periodic contours where cells have formed junctions with adjoining cells. A characteristic feature of cardiac muscle is the presence of 0.5- to 1-nm thick intercalated disks between cells. The disks are highly refractive in fresh muscle tissue and stain deeply in a fixed tissue. The single large oval nucleus is centrally located in each cell. Occasionally, there may be two nuclei in a given cell. However, this is quite rare.

Smooth Muscle

Like striated and cardiac muscle, **smooth muscle** is also derived from mesodermal mesenchymal cells. Exceptions are the iridic smooth muscle of the eye and the modified smooth muscle associated with the walls of sweat glands. These muscles are derived from the ectoderm.

Mature smooth muscle cells are **fusiform** or spindle-shaped and have flattened oval or elongated nuclei that are centrally located in abundant cytoplasm.

Visceral smooth muscle cells are densely packed and lie roughly parallel to each other in such a way that the wide portion of one cell nestles next to the narrow portion of the adjoining cell. Smooth muscles may differ in shape depending on their location: (1) short and thick in the walls of the small arteries; (2) twisted and folded, as a result of the contraction of elastic fibers, in the walls of large arteries; or (3) slender and very long in the body walls of the gastrointestinal tract.

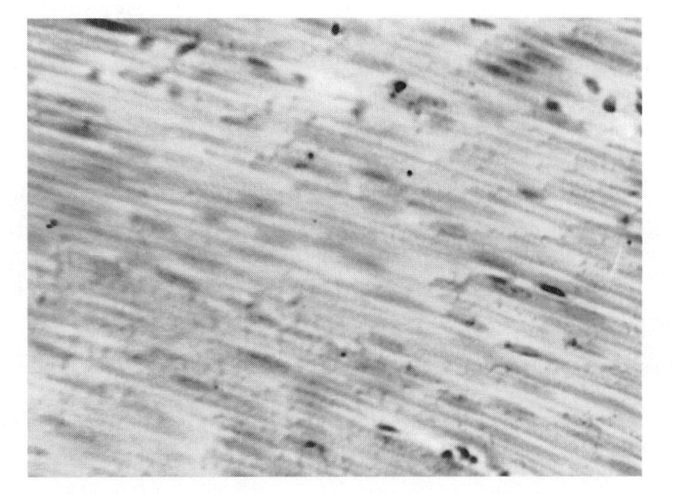

FIGURE 10.1
Light micrograph (LM) of visceral smooth muscle. (200×)

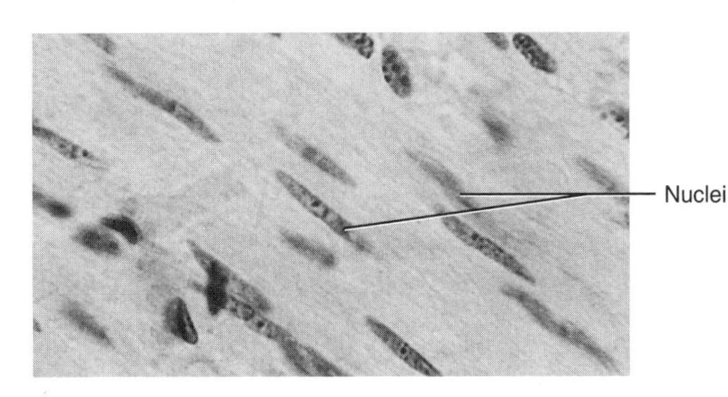

FIGURE 10.2
LM of visceral smooth muscle at a higher magnification. (400×)

FIGURE 10.3
LM of skeletal muscle. (400×)

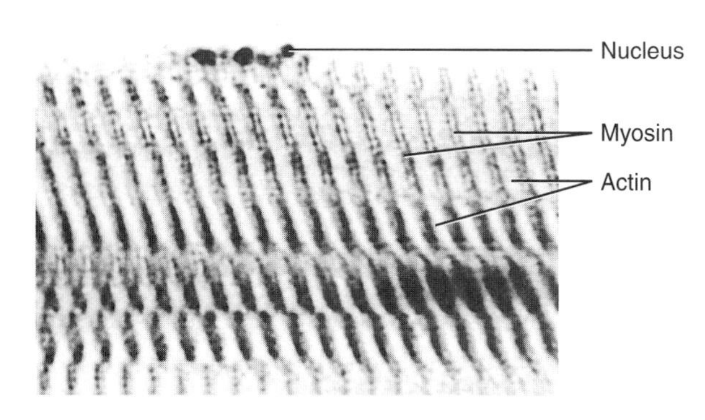

FIGURE 10.4
LM of skeletal muscle at a higher magnification. (1000×)

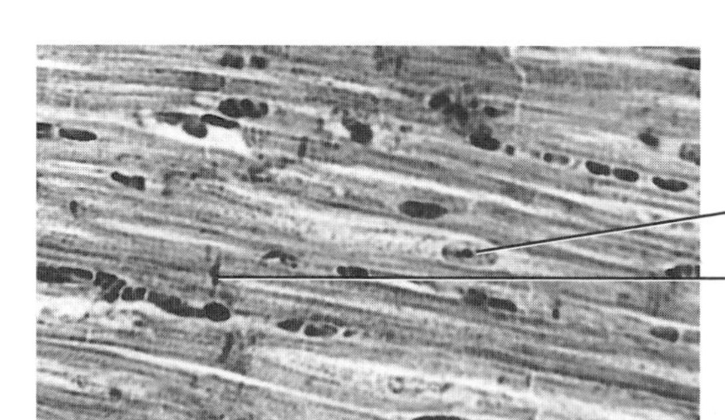

FIGURE 10.5
LM of cardiac muscle. (1000×)

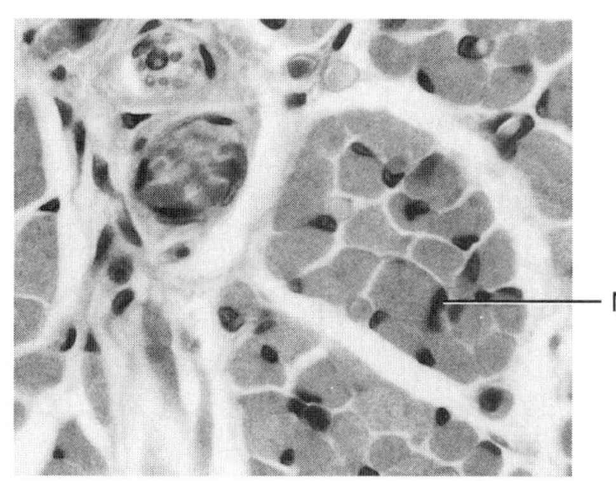

FIGURE 10.6
LM of skeletal muscle in cross section. (400×)

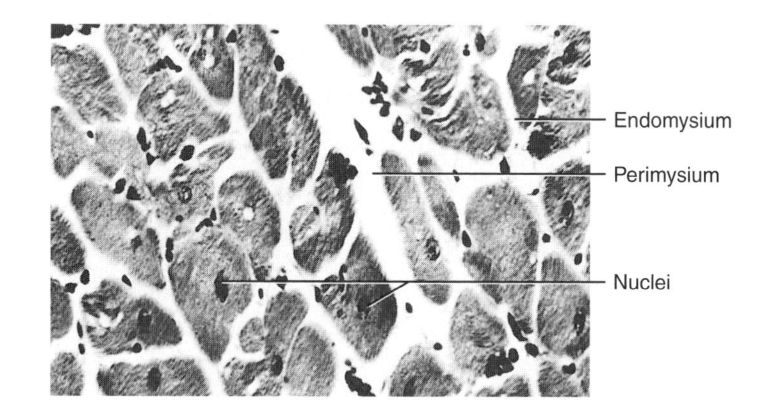

FIGURE 10.7
LM of cardiac muscle in cross section. (400×)

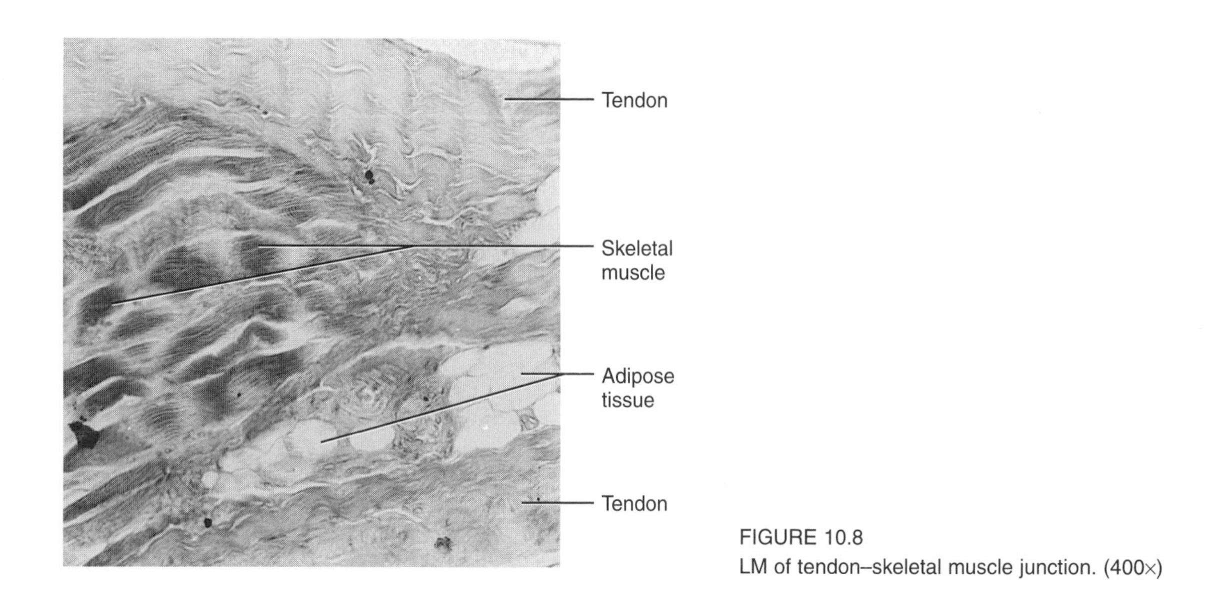

FIGURE 10.8
LM of tendon–skeletal muscle junction. (400×)

FIGURE 10.11
Anterior view of facial muscles.

Galea aponeurotica (epicranial aponeurosis)
Frontalis
Orbicularis oculi
Levator labii superioris
Zygomaticus minor
Zygomaticus major
Risorius
Platysma
Thyroid cartilage (Adam's apple)
Frontal bone
Corrugator supercilii
Levator palpebra superioris
Lacrimal gland
Tarsal plates
Zygomatic bone
Nasalis
Nasal cartilage
Maxilla
Masseter
Buccinator
Orbicularis oris
Mandible
Depressor labii inferioris
Mentalis
Sternocleidomastoid
Omohyoid
Sternohyoid
DANK

(a) Anterior superficial view (b) Anterior deep view

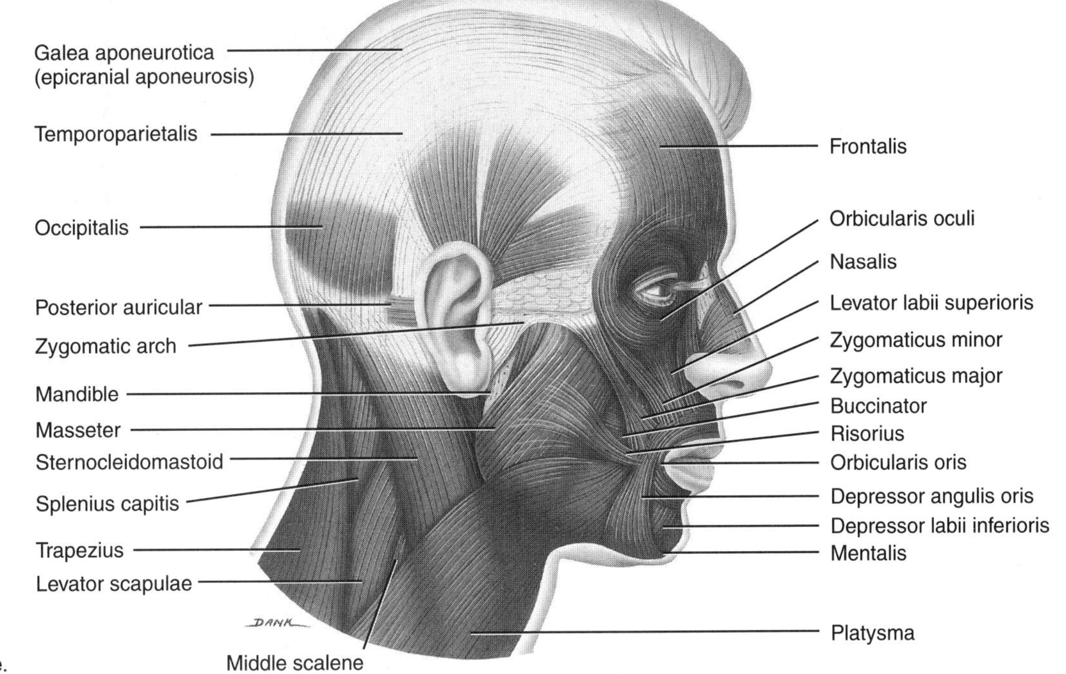

FIGURE 10.12
Lateral view of facial muscle.

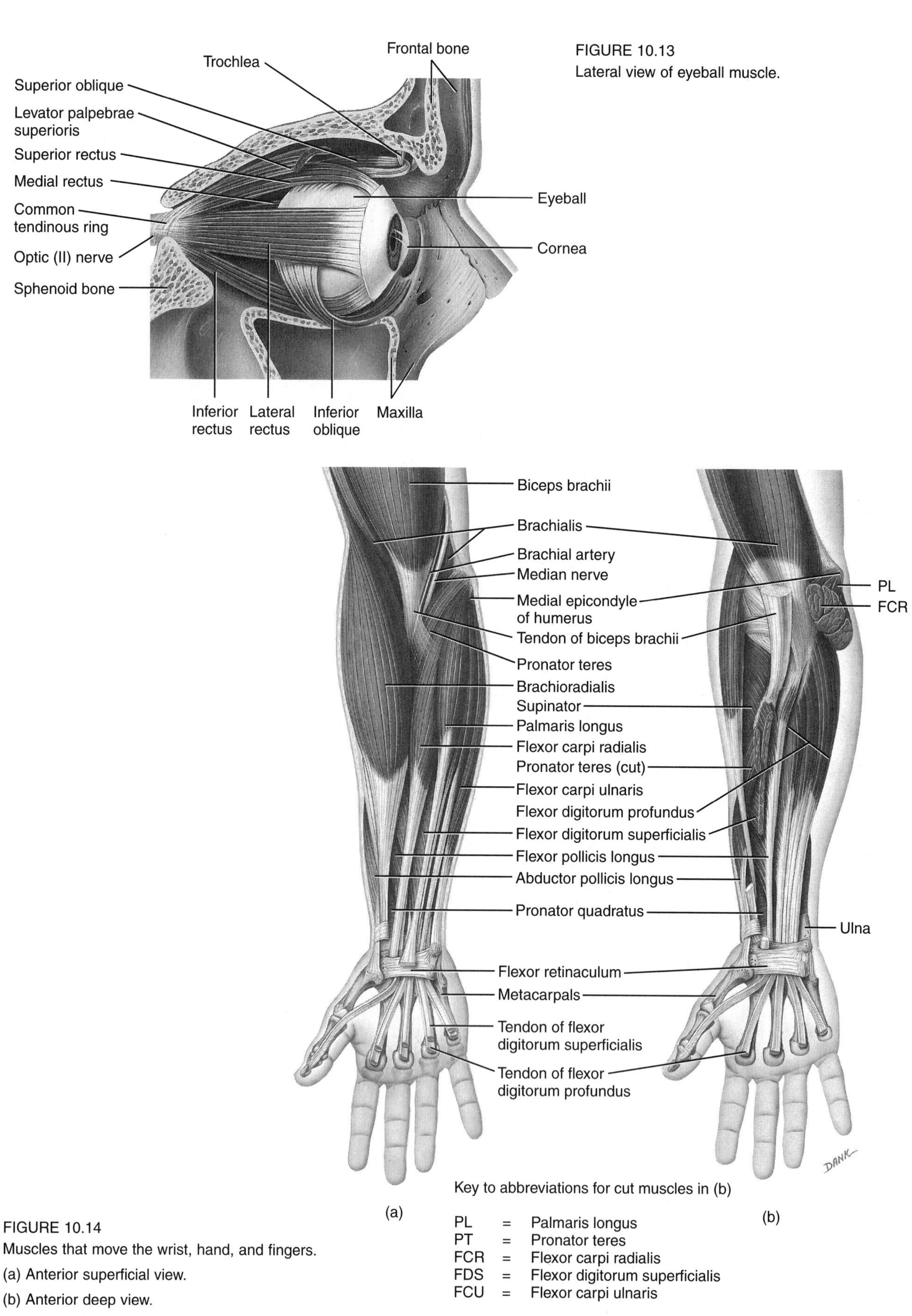

FIGURE 10.13
Lateral view of eyeball muscle.

Key to abbreviations for cut muscles in (b)

PL	=	Palmaris longus
PT	=	Pronator teres
FCR	=	Flexor carpi radialis
FDS	=	Flexor digitorum superficialis
FCU	=	Flexor carpi ulnaris

FIGURE 10.14
Muscles that move the wrist, hand, and fingers.

(a) Anterior superficial view.

(b) Anterior deep view.

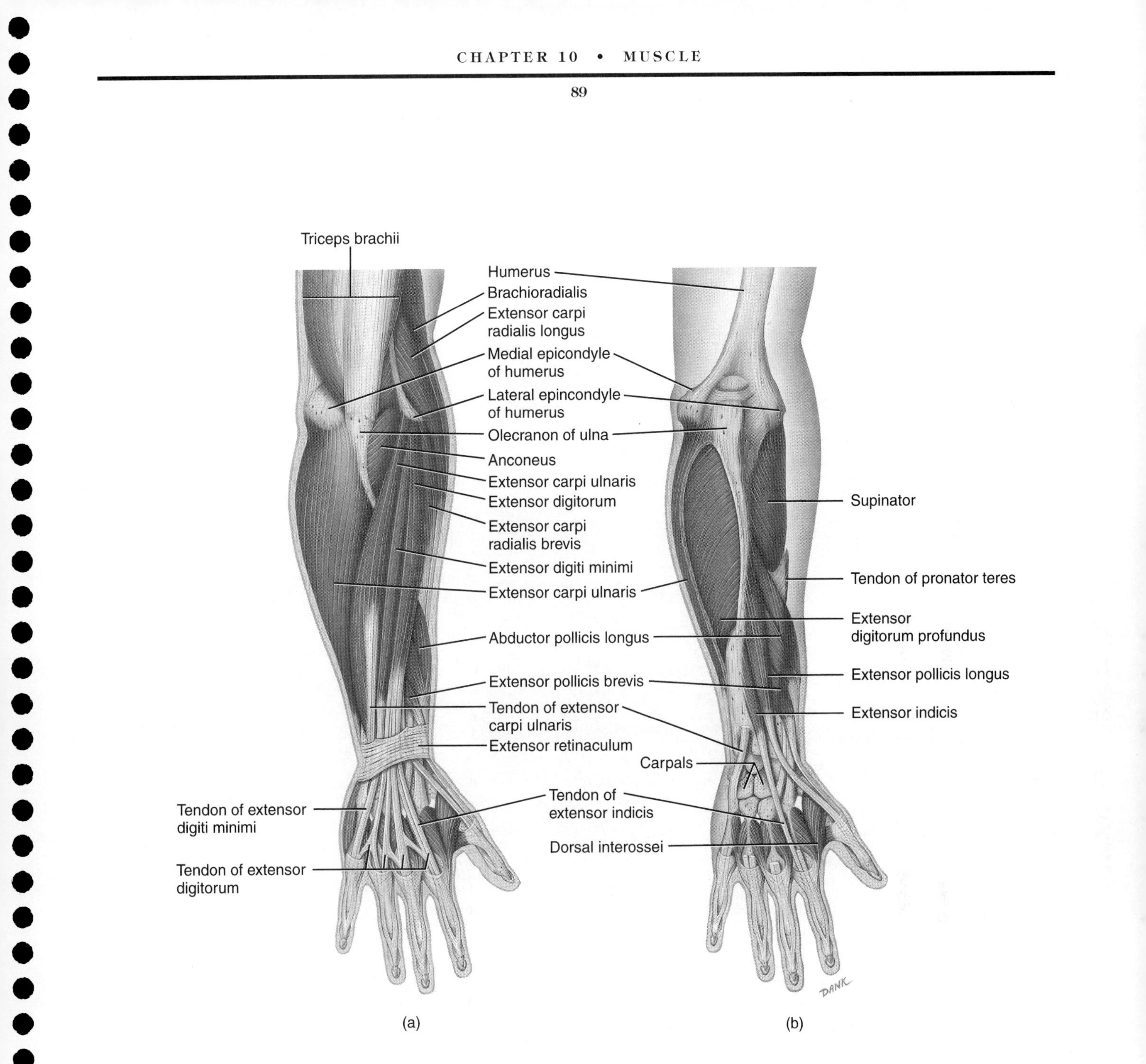

FIGURE 10.14 *(continued)*
Muscles that move the wrist, hand, and fingers.
(a) Posterior superficial view.
(b) Posterior deep view.

FIGURE 10.15
Muscles of the anterior abdominal wall and cross section of abdominal wall above the umbilicus (navel).

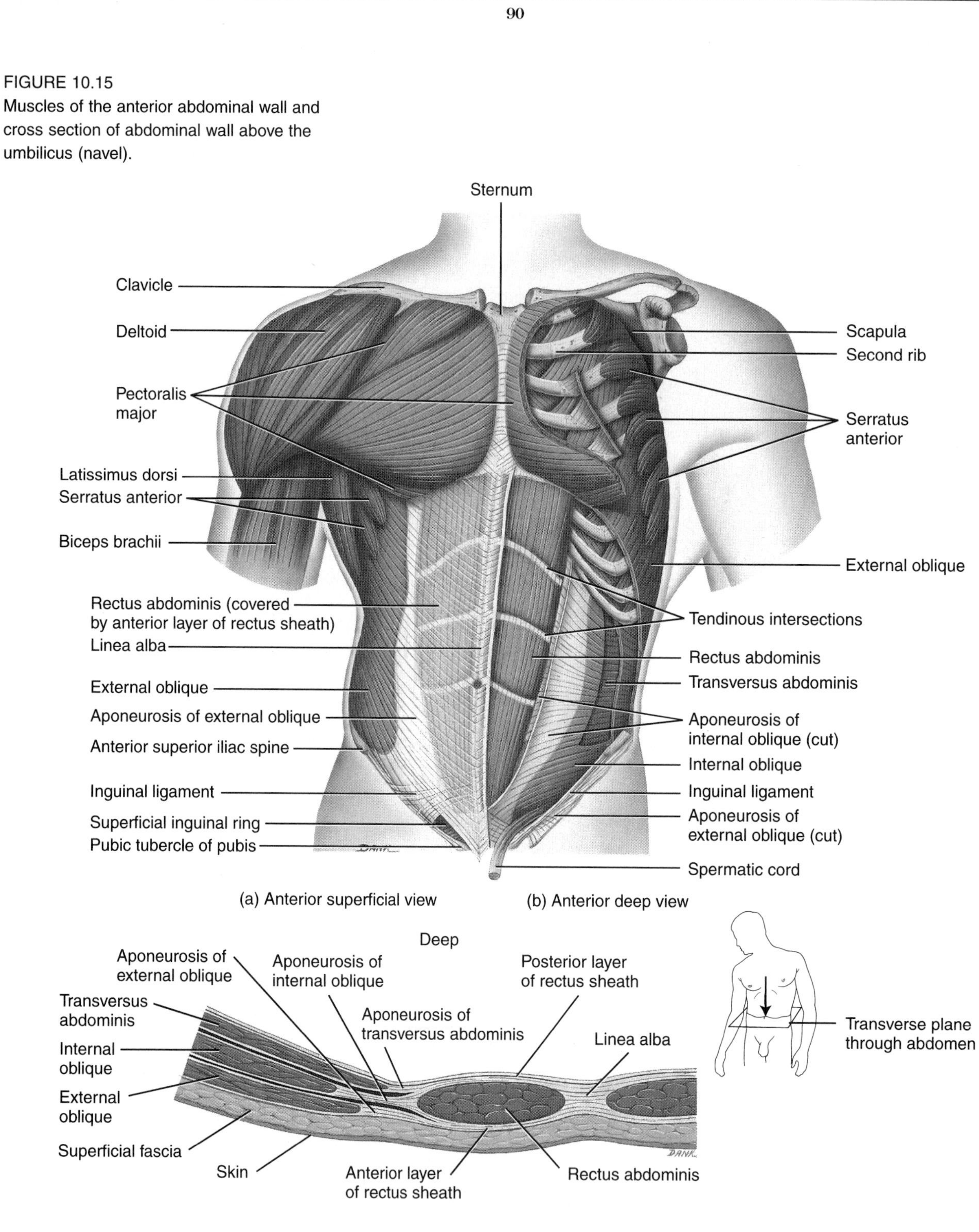

FIGURE 10.16
Muscles that move the femur (thigh), anterior superficial view.

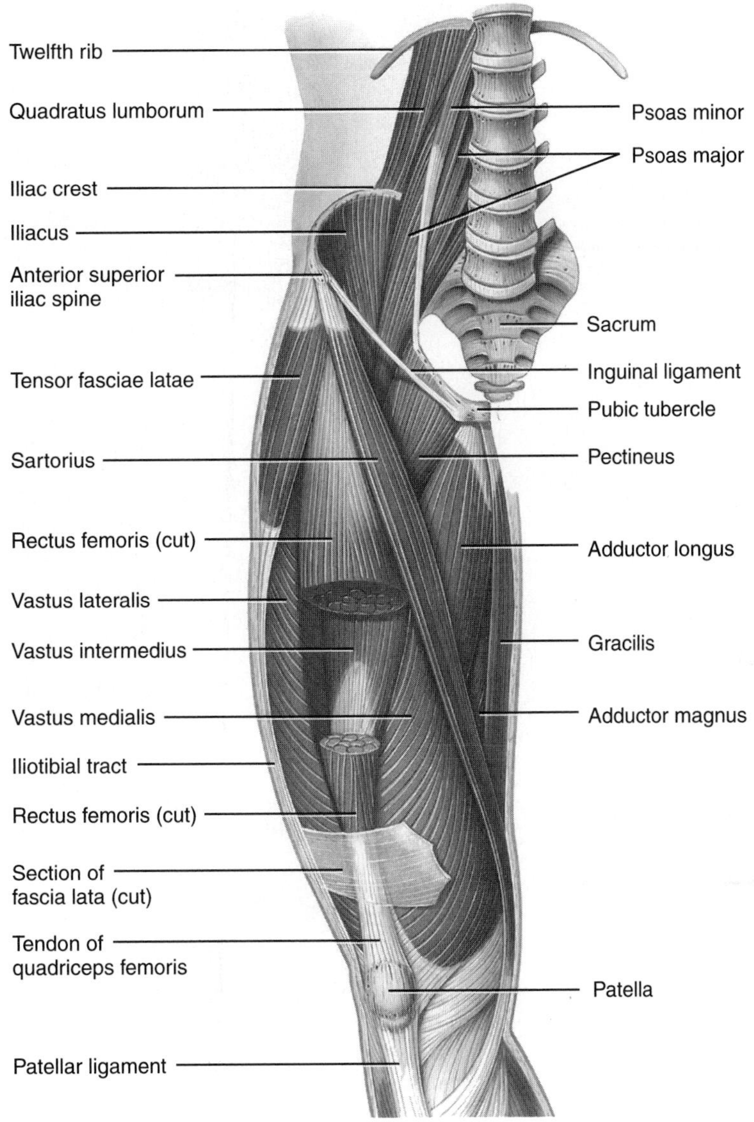

Twelfth rib
Quadratus lumborum
Iliac crest
Iliacus
Anterior superior iliac spine
Tensor fasciae latae
Sartorius
Rectus femoris (cut)
Vastus lateralis
Vastus intermedius
Vastus medialis
Iliotibial tract
Rectus femoris (cut)
Section of fascia lata (cut)
Tendon of quadriceps femoris
Patellar ligament
Psoas minor
Psoas major
Sacrum
Inguinal ligament
Pubic tubercle
Pectineus
Adductor longus
Gracilis
Adductor magnus
Patella

FIGURE 10.17
Muscles of the pelvic floor as seen in the male perineum.

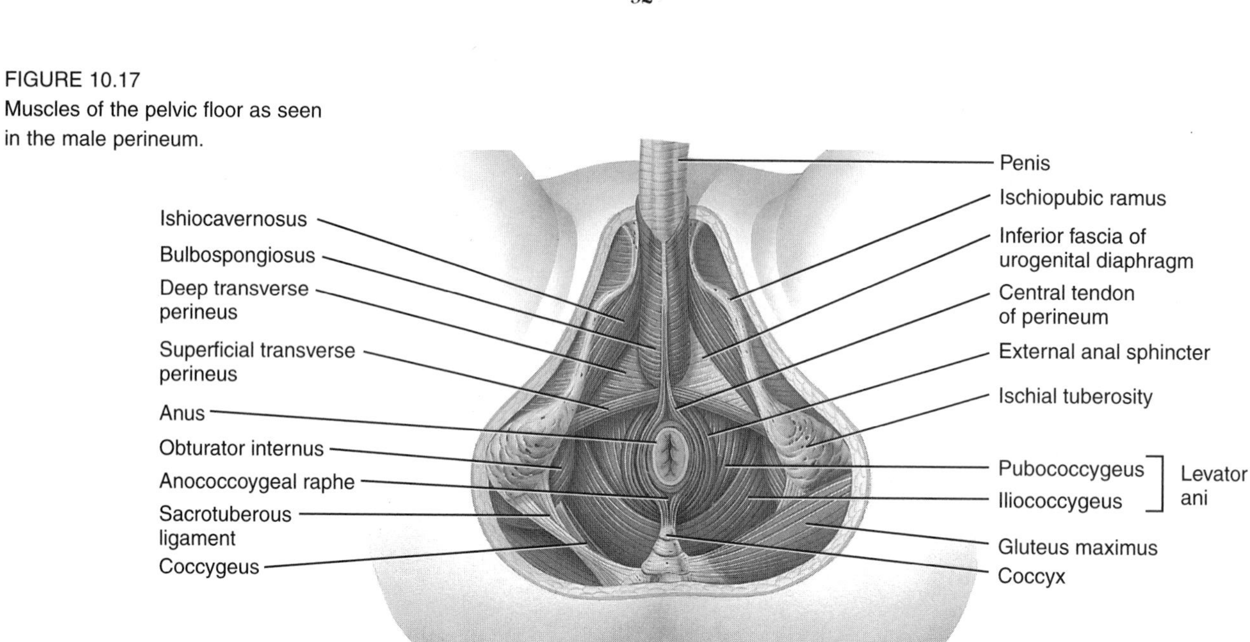

FIGURE 10.18
Muscles of the pelvic floor as seen in the female perineum.

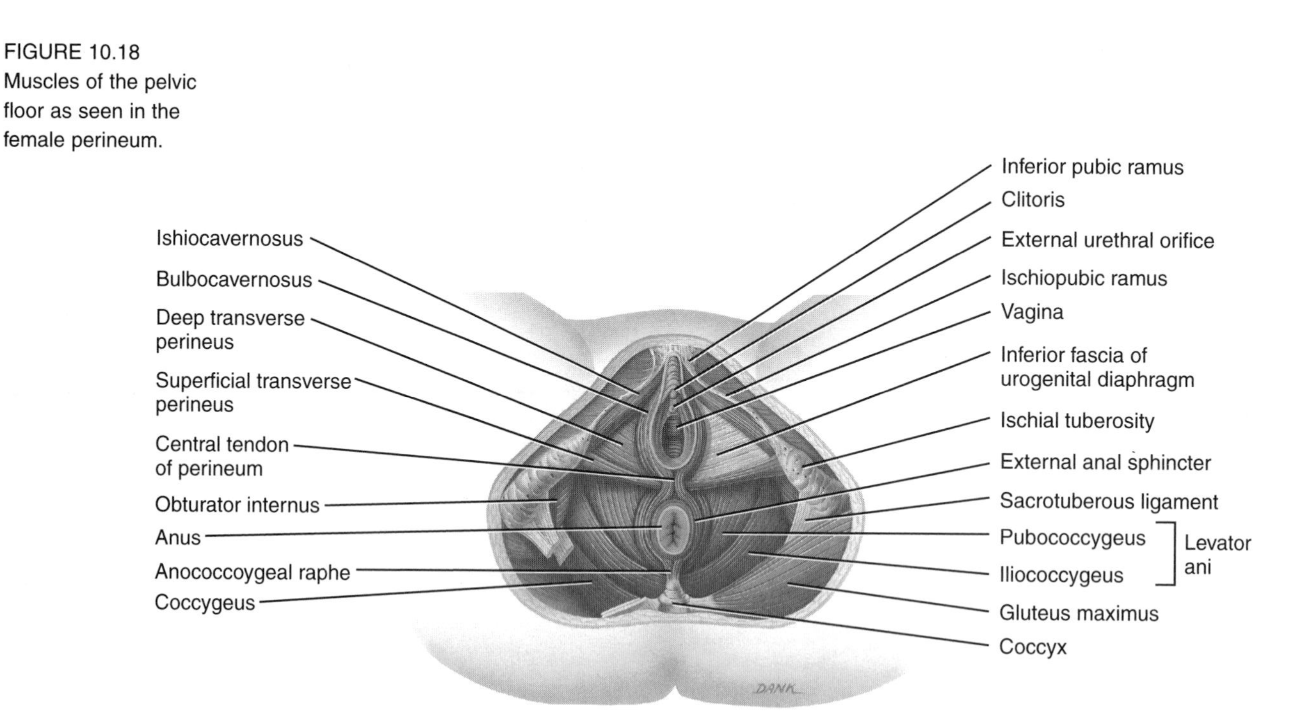

FIGURE 10.19
Superficial muscle of the lower leg:
(a) anterior view and (b) right lateral view.

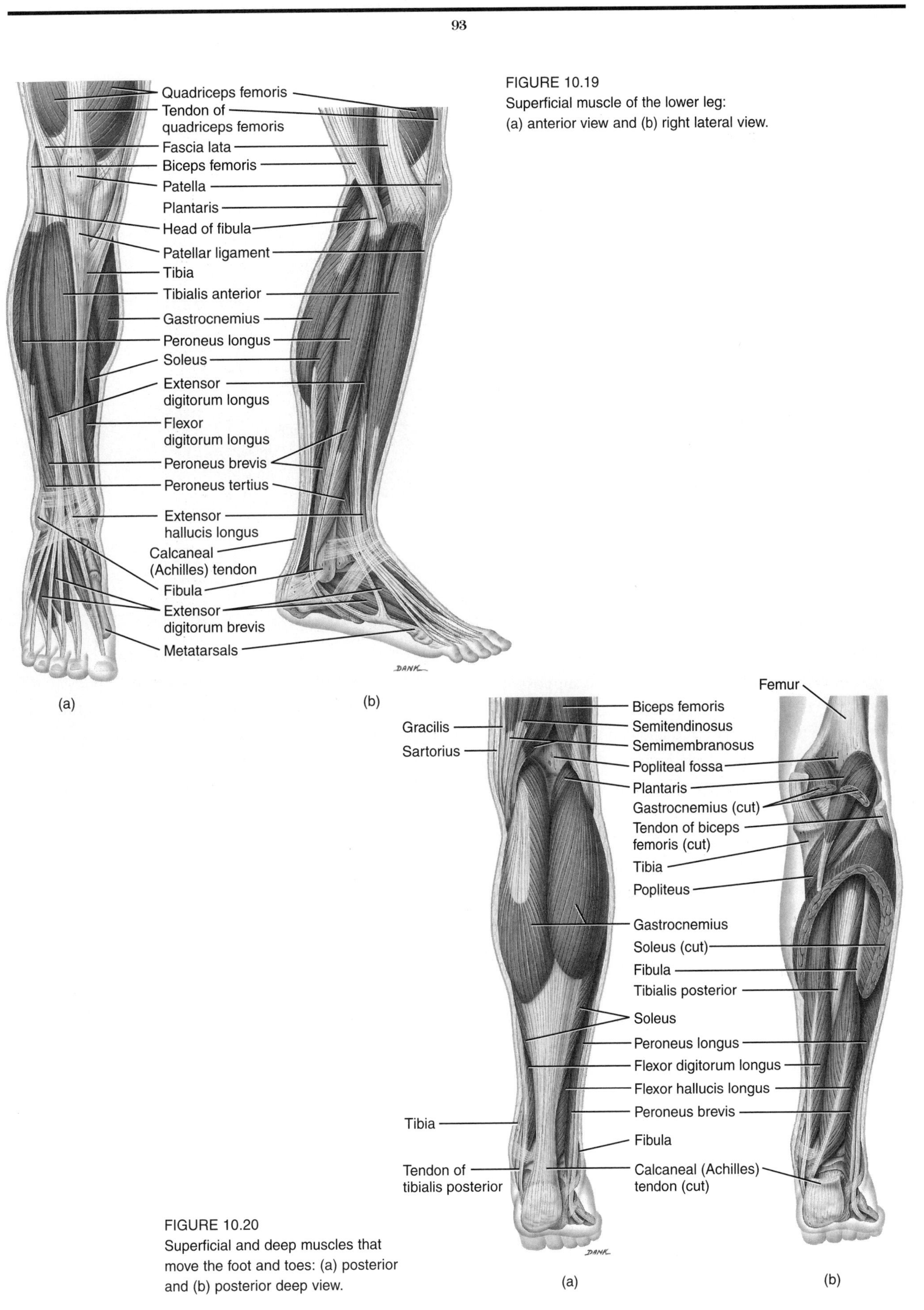

FIGURE 10.20
Superficial and deep muscles that move the foot and toes: (a) posterior and (b) posterior deep view.

Nervous Tissue and the Central Nervous System

The nervous tissue is composed of two types of cells: **neuroglia** cells (also called **supporting** cells) and **neurons** or **nerve** cells. Neurons are highly specialized and can transmit impulses from one part of the body to another. The neuroglia cells are generally smaller than neurons and are scattered throughout the nervous tissue. Neuroglia cells are a conglomerate of cells: **ependymal** cells, which line the ventricles and the neural canal; **astrocytes,** which facilitate the transfer of molecules between blood capillaries and neural cells; **oligodendrocytes,** which myelinate fibers in the central nervous system; and **microglia,** which are phagocytic cells of the nervous system. Collectively, astrocytes and oligodendrocytes are referred to as **macroglia.**

The **central nervous system** (**CNS**) consists of the **brain** and the **spinal cord.** The brain lies in a bony **cranium,** whereas the spinal cord is surrounded by a long, bony vertebral column. The CNS receives **interoceptive** nervous impulses from within the body and **exteroceptive** impulses from stimuli outside the body.

Membranes of the CNS Protection for the CNS is provided by a bony case, the cranium, that covers the brain, a vertebral column that covers the spinal cord, and a three-membrane investment called the meninges that covers both the brain and the spinal cord. The outermost investment of the meninges, which blends with the bony structure lying above it, is a tough, fibrous, relatively inelastic membrane called **dura mater** or **pachymeninx.** The middle membrane of the meninges is the **arachnoid,** composed of a reticulate fiber network, **trabeculae,** and a subarachnoid space. The innermost membrane of the meninges is the **pia mater,** a thin membrane that closely invests the brain and extends into the depths of the **cerebral sulci.** The pia mater also surrounds the entire spinal cord and extends into the depths of the **anterior median fissure.**

Spinal Cord In cross section, the spinal cord displays an H-shaped central area of gray matter, composed of nerve cells and their fibers, and a central canal lined by ependymal cells.

The **cerebrum** gray matter forms the upper surface of the cerebrum (**cerebral cortex**). The cortical surface forms convolutions called **gyri** that increase the surface area of the gray matter. Between the convoluted folds, or gyri, are depressions called **sulci.** The cortex contains nerve and neuroglia cells, fibers, and blood vessels. The cortical cells are **stellate, pyramidal, fusiform,** or **spindle-shaped** and are generally arranged in stratified layers.

The **cerebellum** coordinates movements of striated muscle and maintains equilibrium and posture. The cerebellum is divided into left and right **hemispheres.** Between the hemispheres is the **vermis,** which is segmented into lobules separated by transverse fissures. As in the case of the cerebrum, the gray matter in the cerebellum is located on the surface, forming a thin layer of **cerebellar cortex.** Below the cortex lies the white matter with myelinated fibers, and a few nerve cells lie in the middle of the cerebellum.

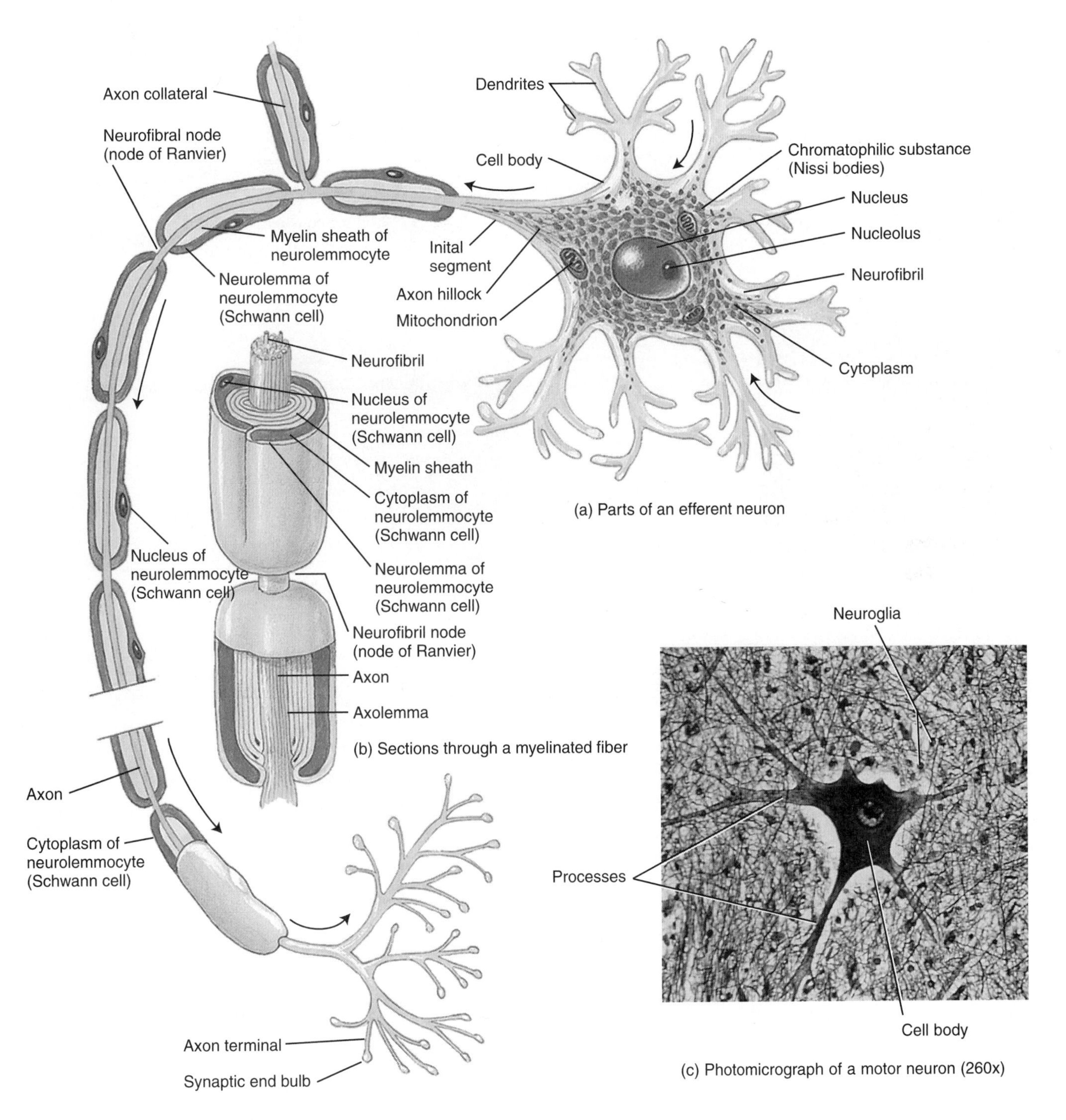

(a) Parts of an efferent neuron

(b) Sections through a myelinated fiber

(c) Photomicrograph of a motor neuron (260x)

FIGURE 11.1
A micrograph and diagrammatic presentation of a multipolar neuron.

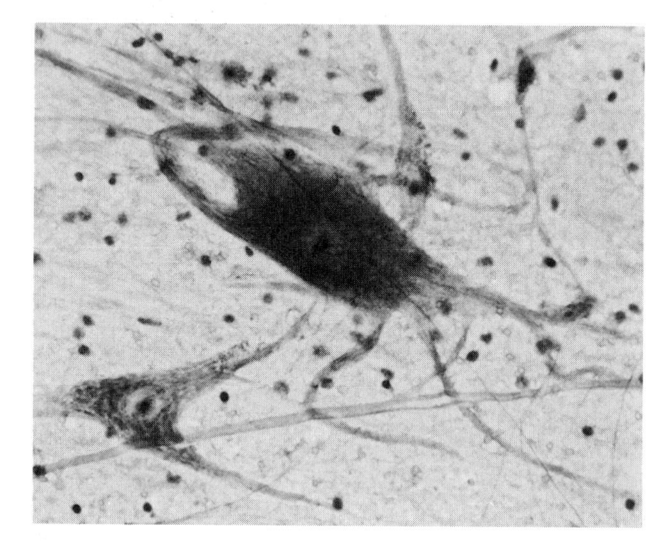

FIGURE 11.2
Light micrograph (LM) of giant multipolar neurons. (200×)

FIGURE 11.3
LM of Purkinje cells in the cerebellum. (400×)

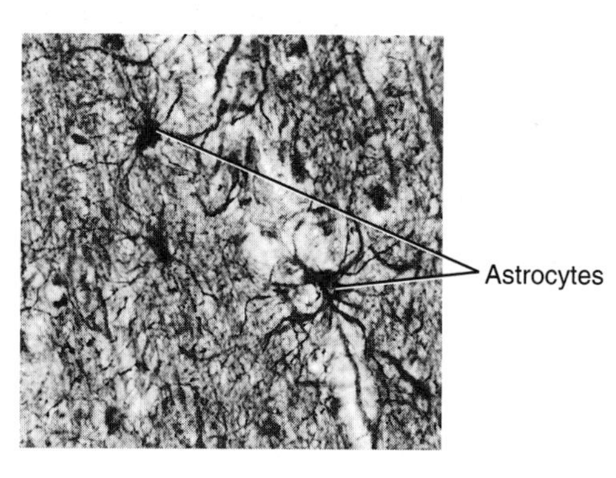

FIGURE 11.4
LM of fibrous astrocytes. (400×)

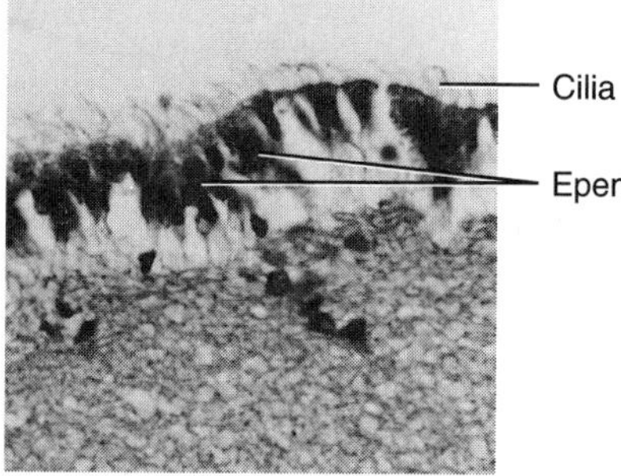

FIGURE 11.5
LM of ciliated ependymal cells lining the ventricle of the brain. (1000×)

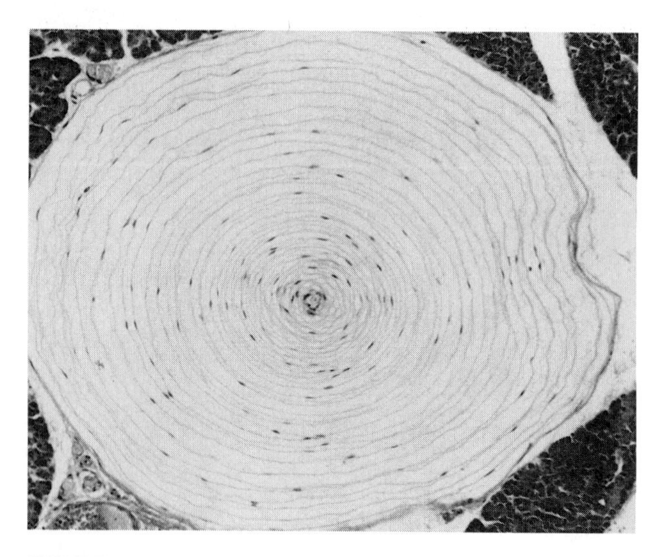

FIGURE 11.6
LM of a cross section of a Pacinian corpuscle displaying the spiral lamellated appearance. (400×)

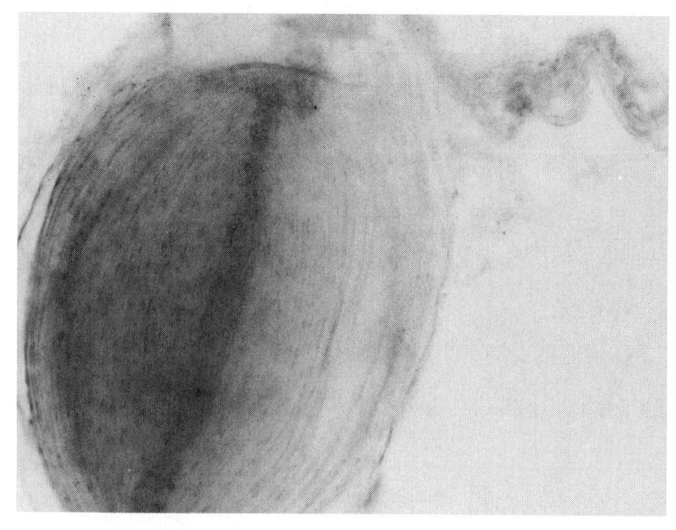

FIGURE 11.7
An isolated Pacinian corpuscle.

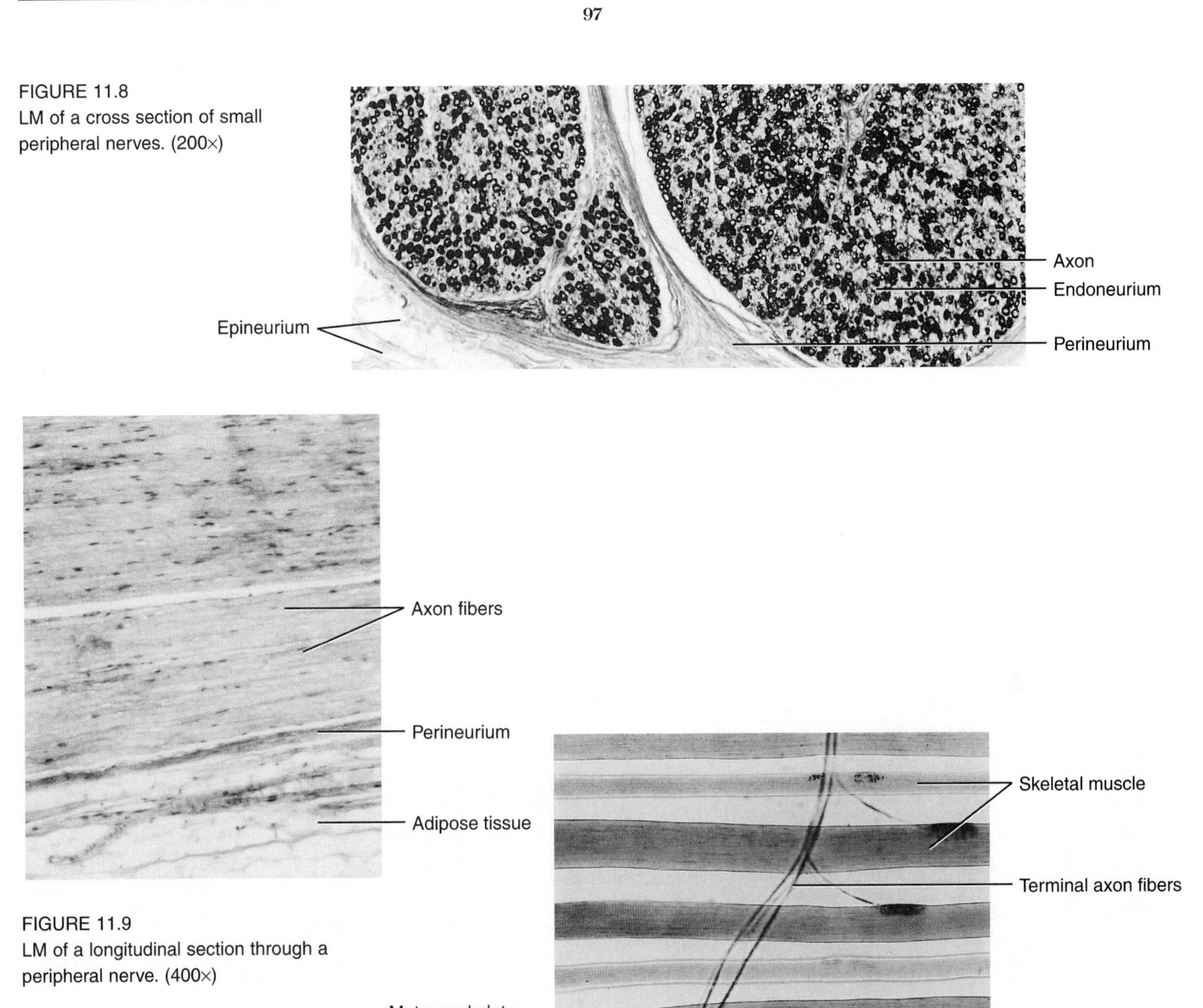

FIGURE 11.8
LM of a cross section of small peripheral nerves. (200×)

FIGURE 11.9
LM of a longitudinal section through a peripheral nerve. (400×)

FIGURE 11.10
LM of teased preparation of motor end plates on skeletal muscle fibers. (400×)

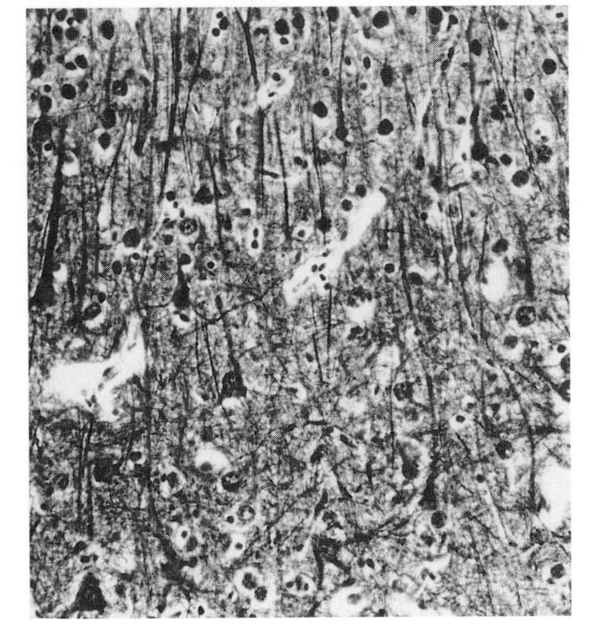

FIGURE 11.11
LM of Betz's cell as seen in a section of a cerebral cortex. (200×)

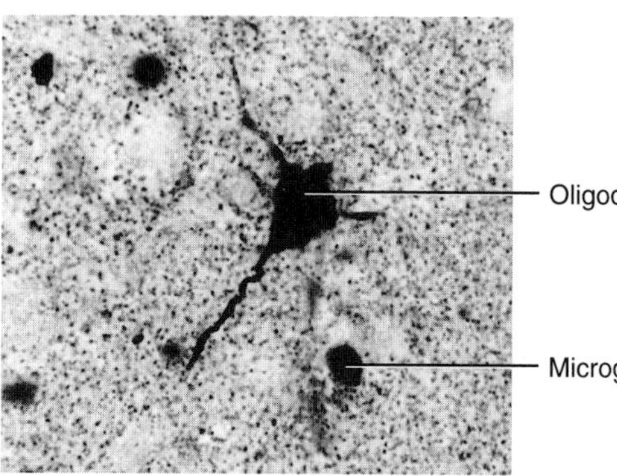

FIGURE 11.12
LM of an oligodendroglia and microglia cells. (1000×)

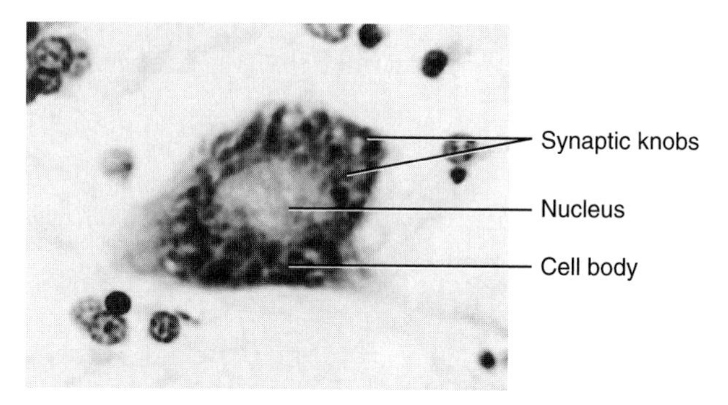

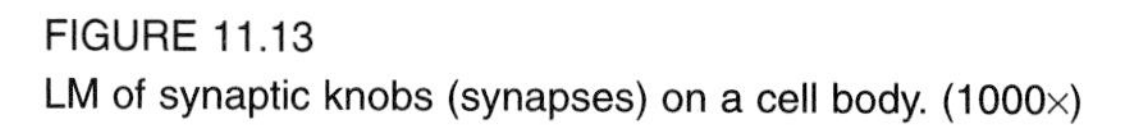

FIGURE 11.13
LM of synaptic knobs (synapses) on a cell body. (1000×)

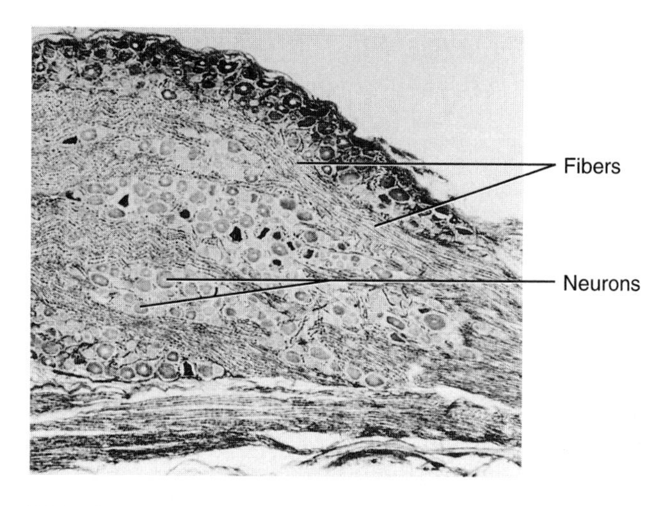

FIGURE 11.14
LM of a section through sympathetic ganglia. (100×)

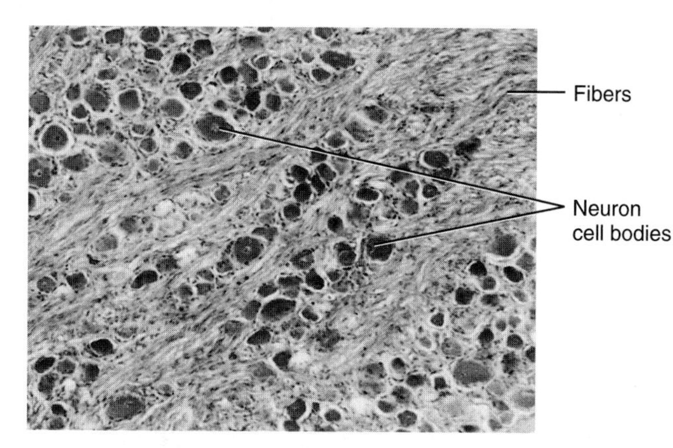

FIGURE 11.15
Sympathetic ganglia at a higher magnification displaying neurons, satellite cells, and fibers. (200×)

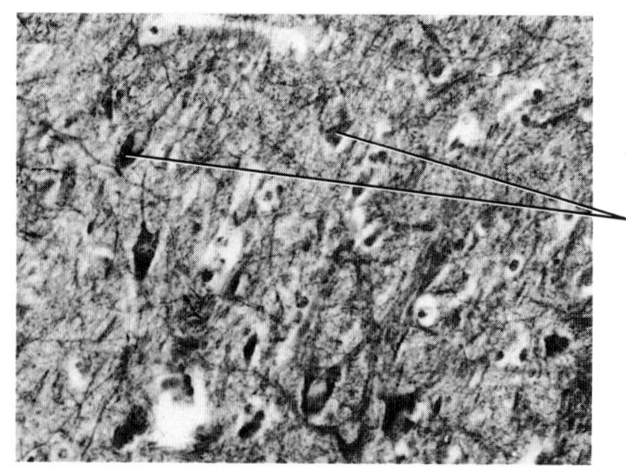

FIGURE 11.16
LM of neurons in the spinal cord. (100×)

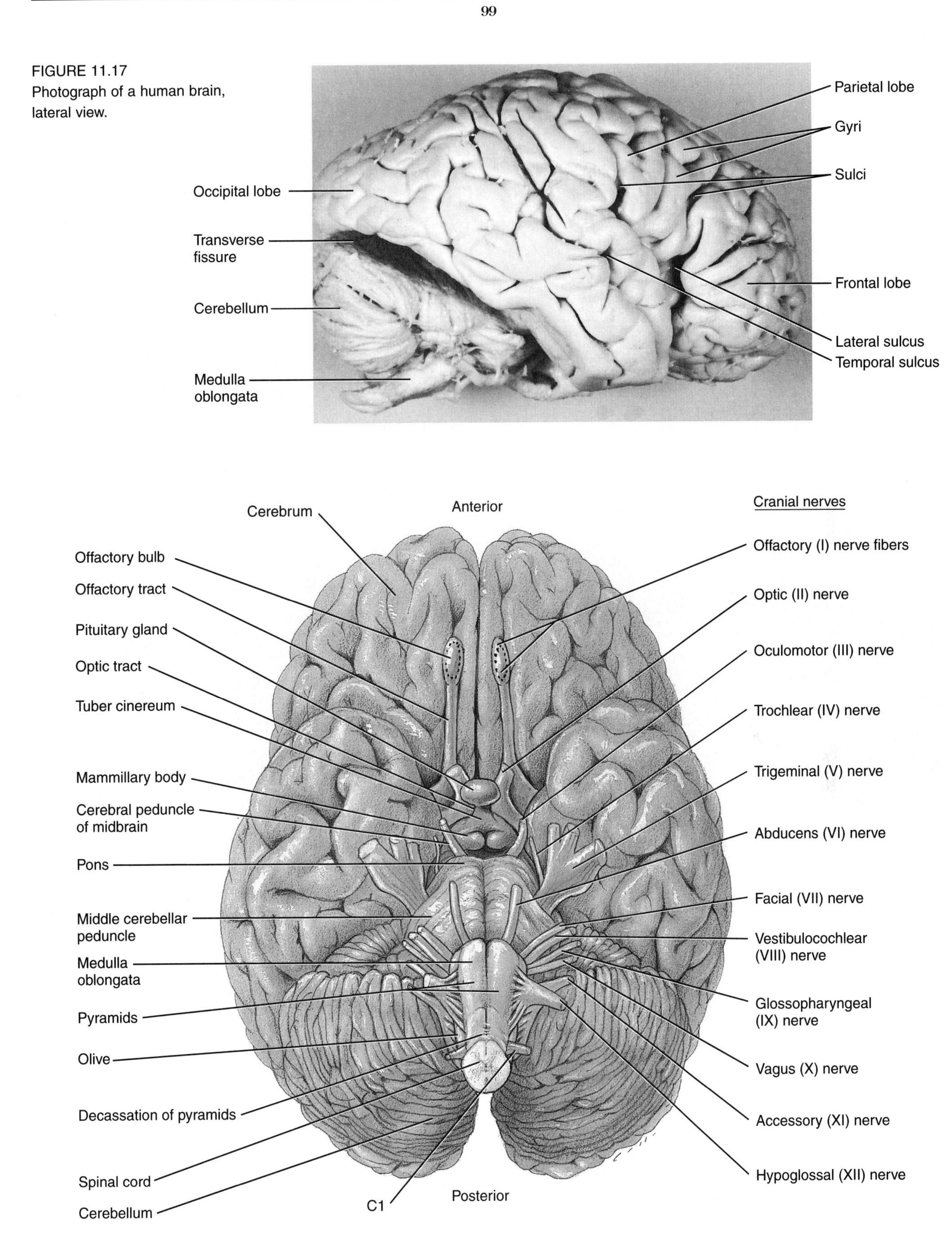

FIGURE 11.17
Photograph of a human brain, lateral view.

FIGURE 11.18
Diagrammatic presentation of the ventral surface of the brain and the anterior aspect of the brain stem and the cerebellum.

FIGURE 11.19
Photograph of the ventral aspect of the human brain and related structures.

Olfactory bulb
Frontal lobe
Optic chiasm
Infundibulum of pituitary
Medulla oblongata
Decussation of pyramids
Cerebellum
Temporal lobe

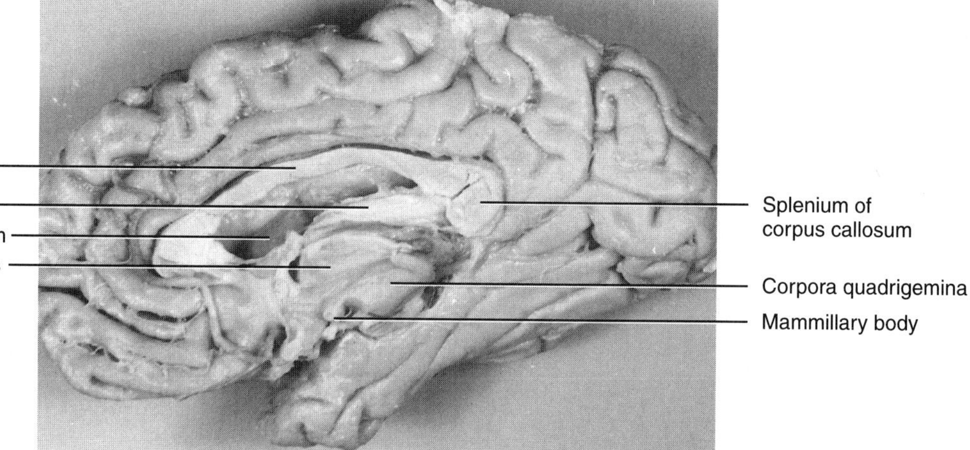

FIGURE 11.20
A midsagittal section of the human brain.

FIGURE 11.21
Human brain model in a sagittal section.

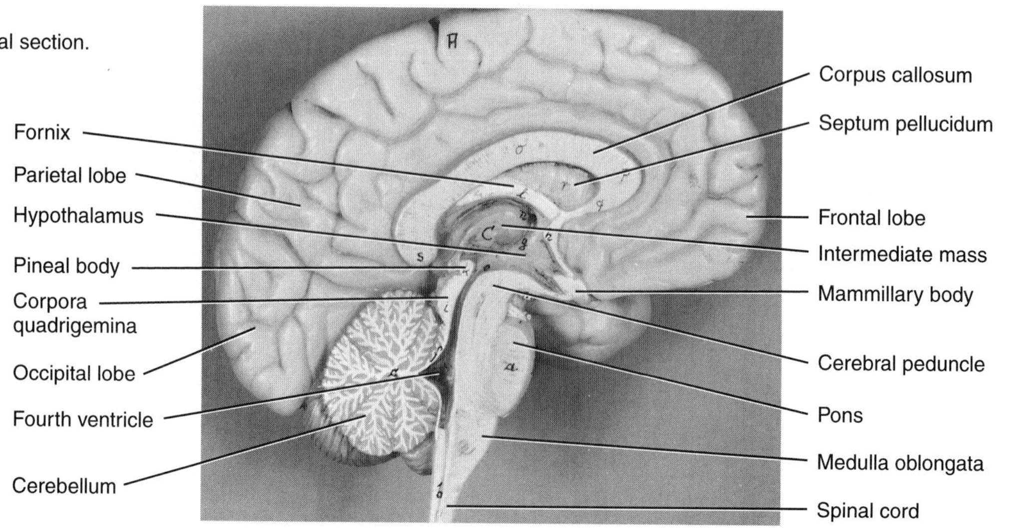

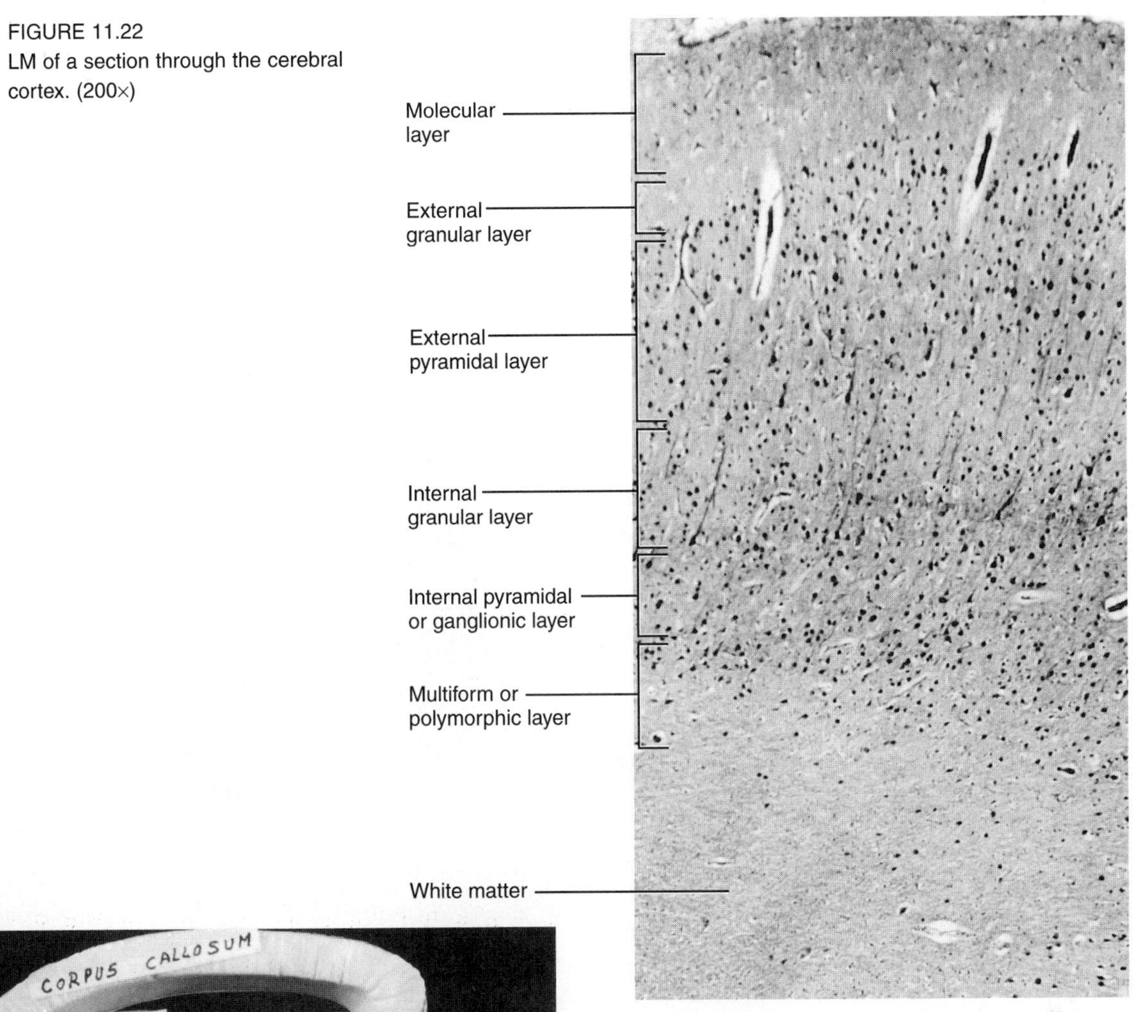

FIGURE 11.22
LM of a section through the cerebral cortex. (200×)

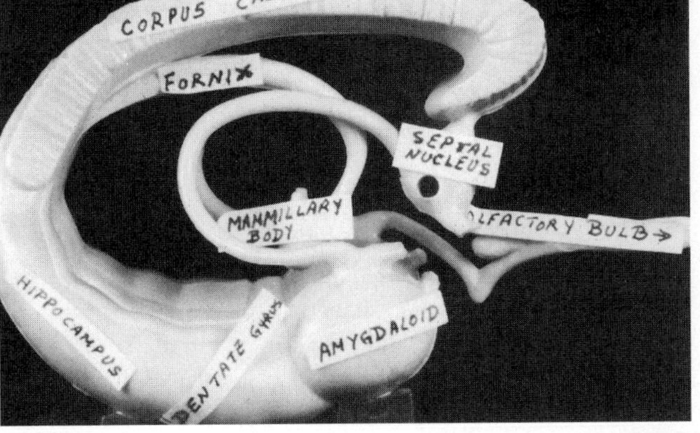

FIGURE 11.23
Model of the human limbic system.

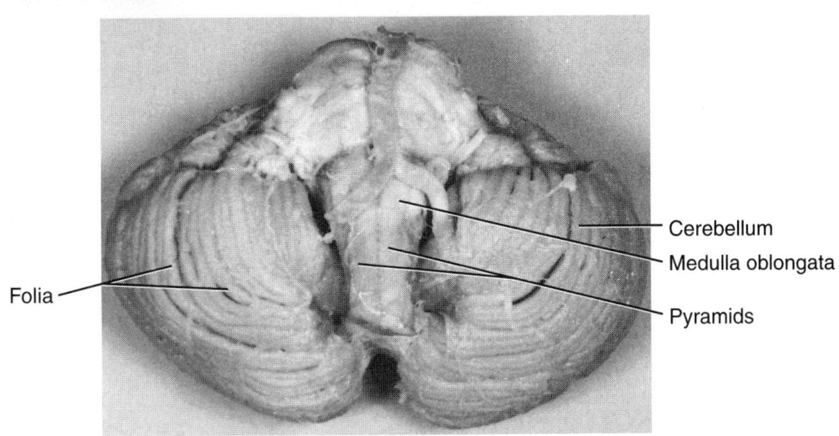

FIGURE 11.24
Human brain stem and cerebellum.

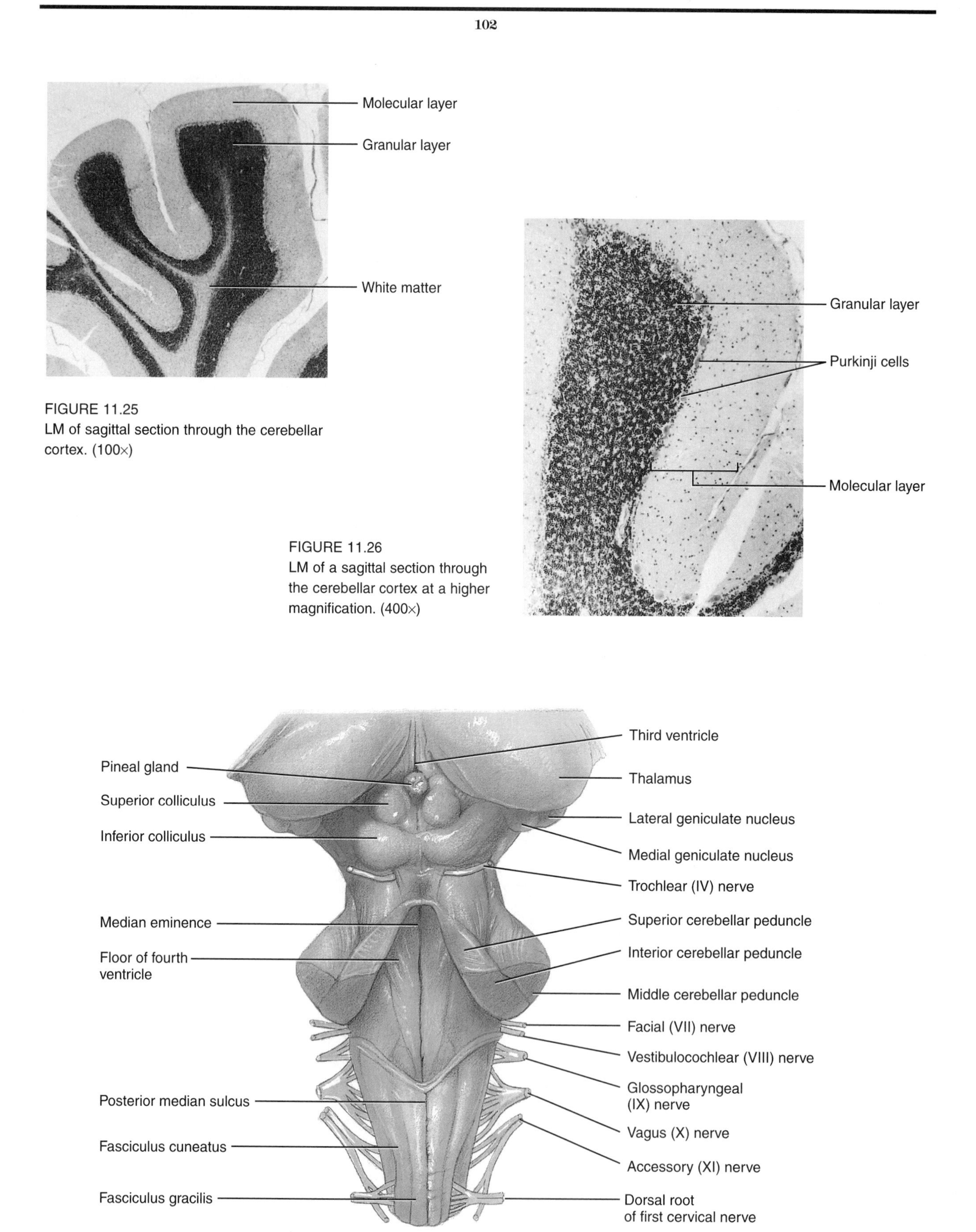

FIGURE 11.25
LM of sagittal section through the cerebellar cortex. (100×)

FIGURE 11.26
LM of a sagittal section through the cerebellar cortex at a higher magnification. (400×)

FIGURE 11.27
Diagrammatic presentation of the brain stem, posterior view.

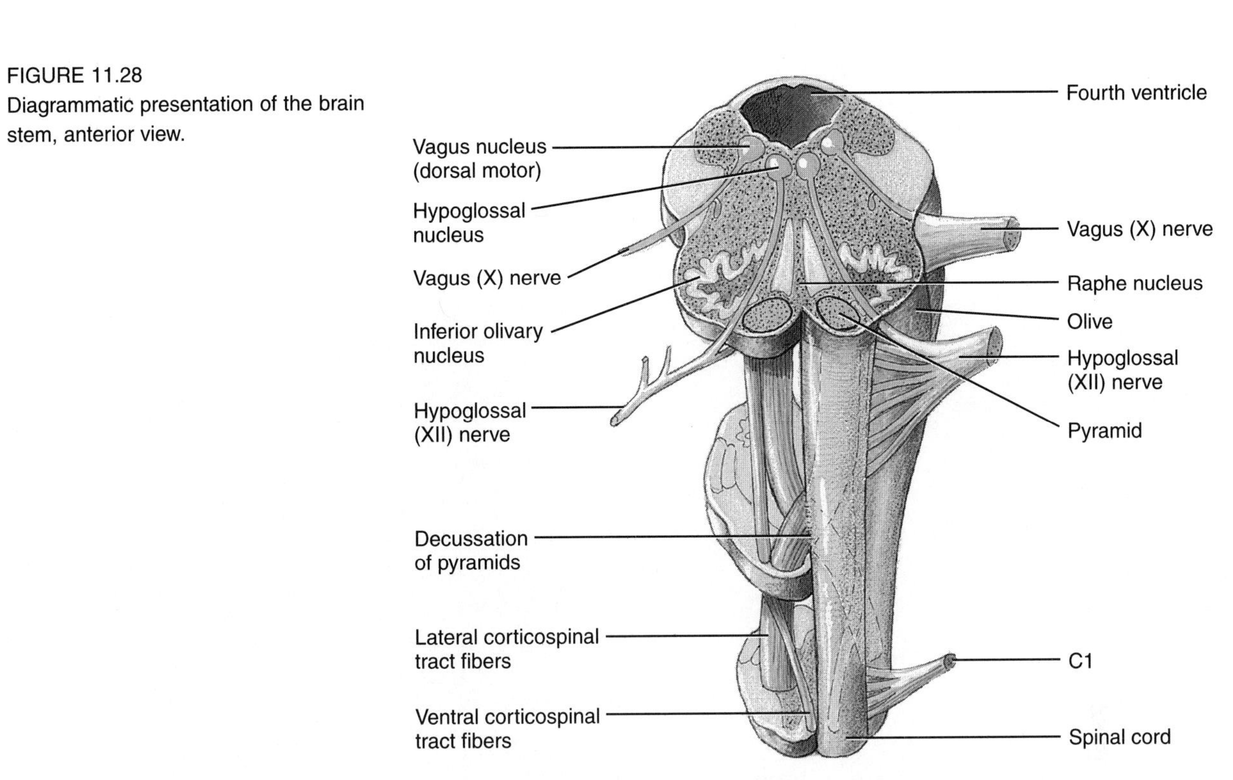

FIGURE 11.28
Diagrammatic presentation of the brain stem, anterior view.

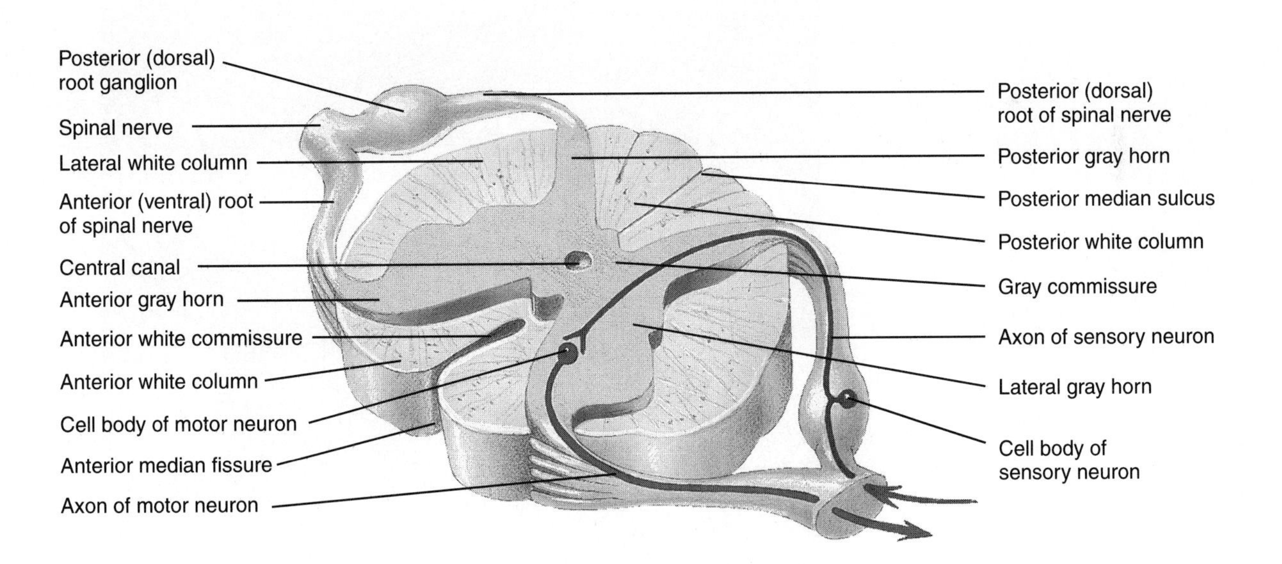

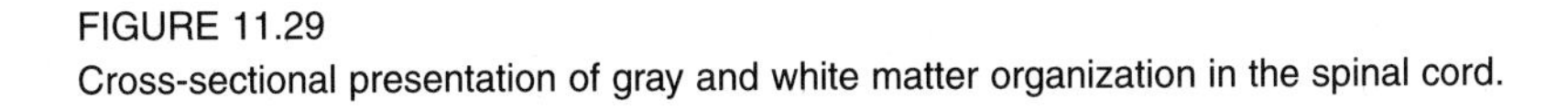
FIGURE 11.29
Cross-sectional presentation of gray and white matter organization in the spinal cord.

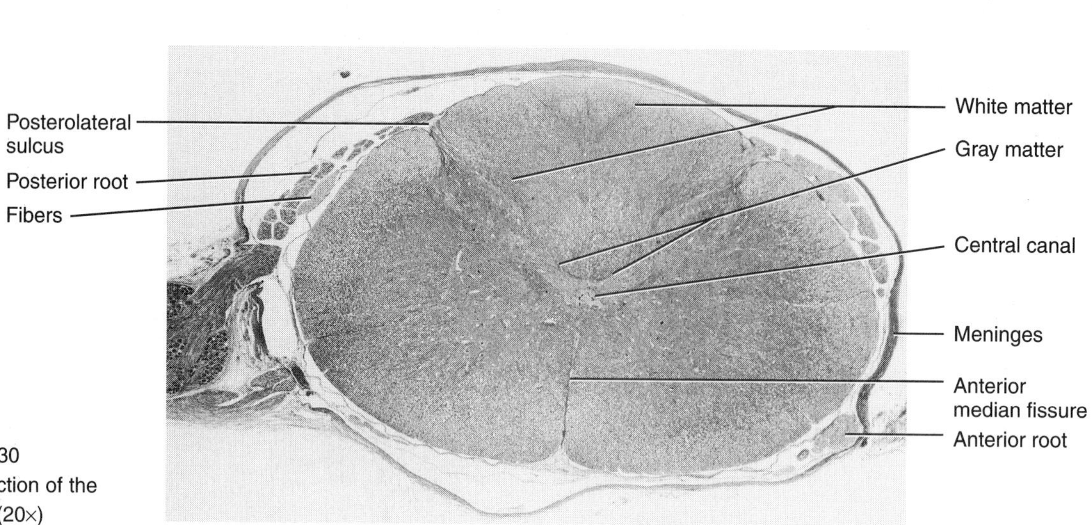

FIGURE 11.30
LM cross section of the spinal cord. (20×)

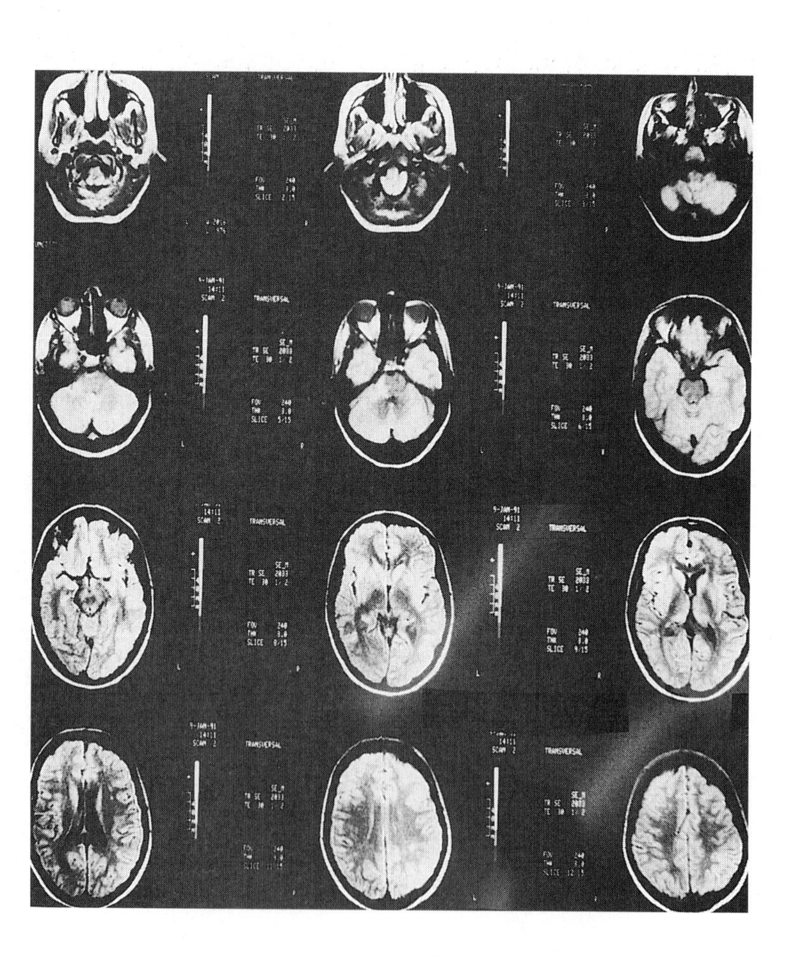

FIGURE 11.31
Series of planes of the brain shown by MRI (magnetic resonance imaging).

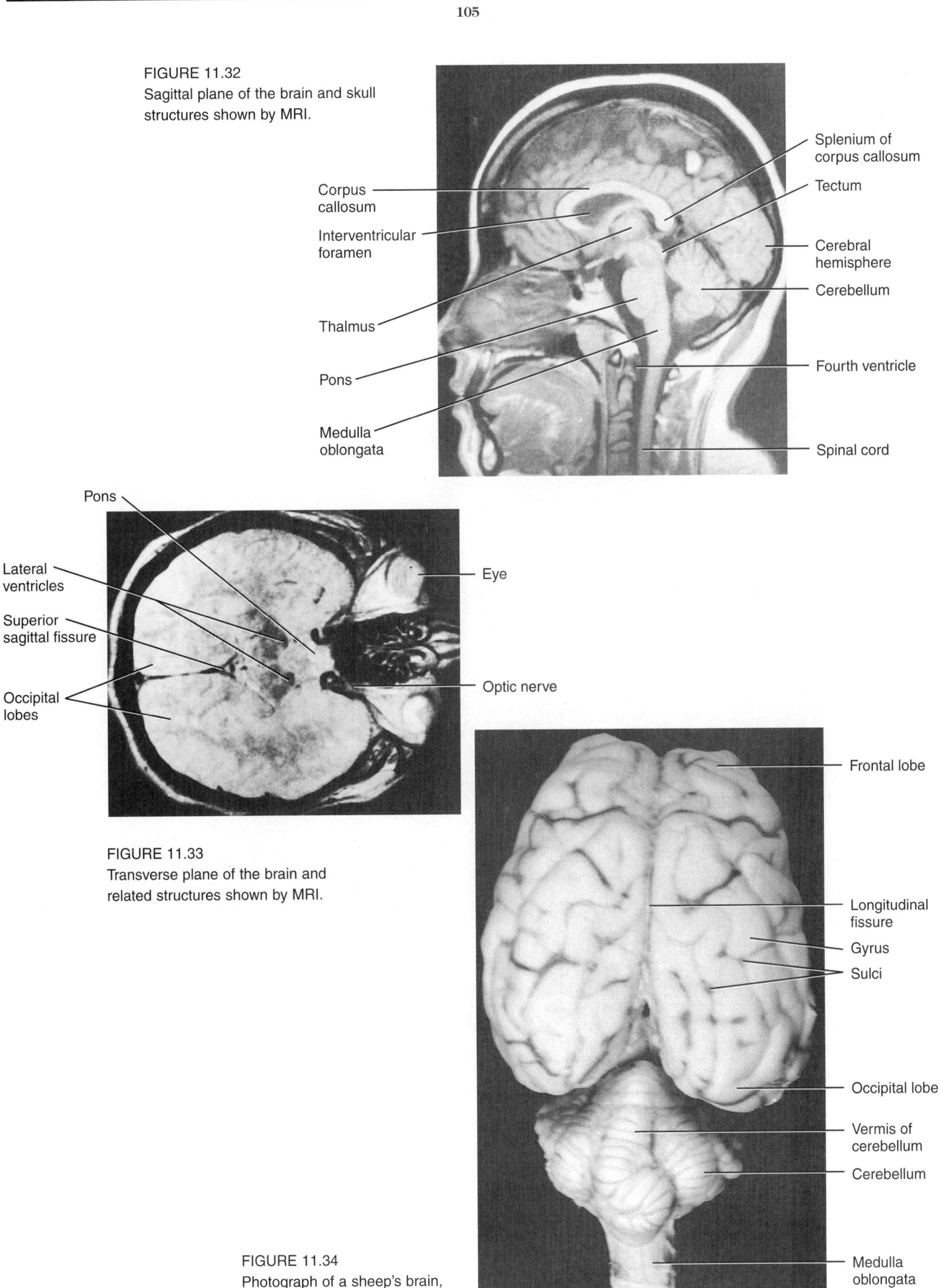

FIGURE 11.32
Sagittal plane of the brain and skull structures shown by MRI.

FIGURE 11.33
Transverse plane of the brain and related structures shown by MRI.

FIGURE 11.34
Photograph of a sheep's brain, dorsal surface.

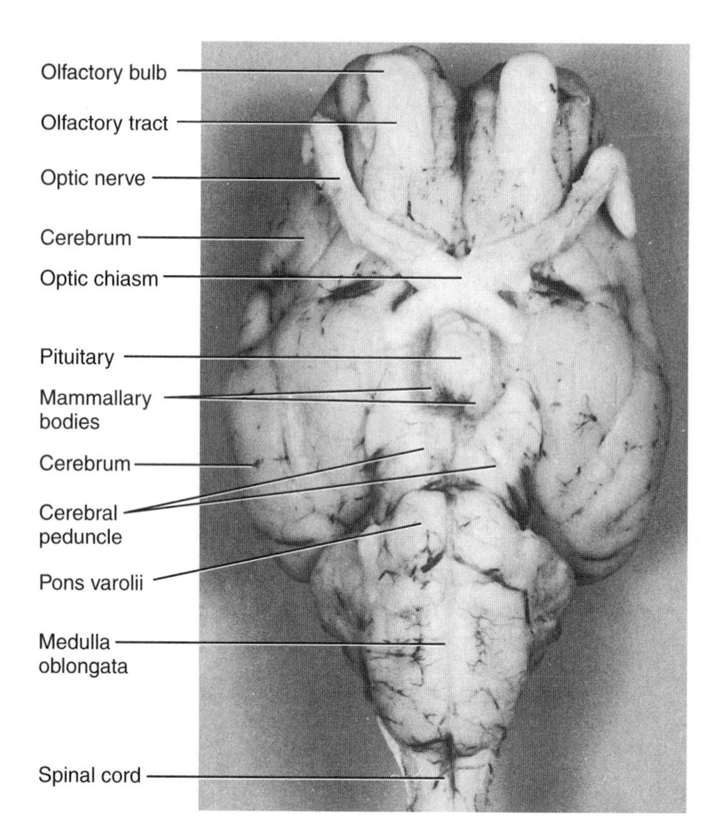

FIGURE 11.35
Photograph of a sheep's brain, ventral surface.

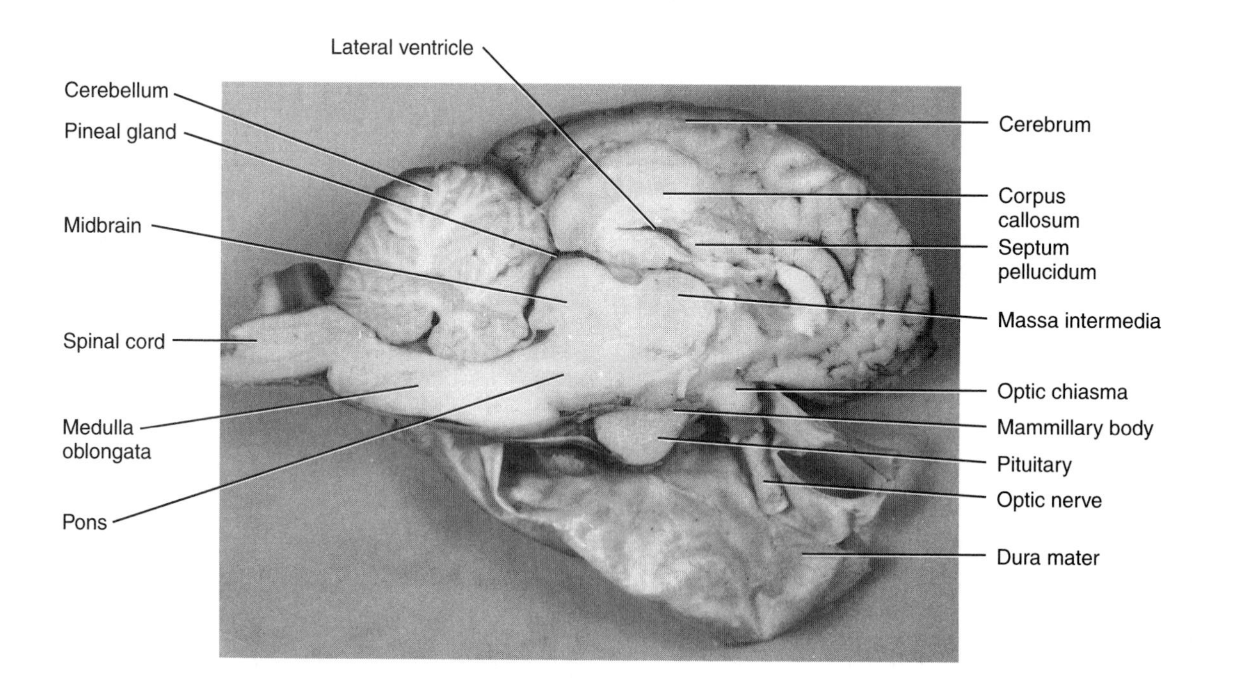

FIGURE 11.36
Photograph of a sheep's brain, midsagittal section.

Integumentary System

The integumentary system (skin and related structures) has the greatest mass of all the systems in the body. It functions as a barrier of protection from mechanical, chemical, and temperature-related injuries, in fluid homeostasis, temperature regulation, excretion of wastes, and neurosensory input to effector organs of the body. The integument consists of two components: the **skin,** which includes the **epidermis** and **dermis,** and the underlying **hypodermis.**

The epidermis is an epithelium several cell layers thick. It has an underlying noncellular basement membrane. The epidermis consists of squamous cells of various shapes. There are also several specialized cells within the epidermis.

Five different types of cells form different layers in the epidermis. These layers include the **stratum basale,** the **stratum spinosum,** the **stratum granulosum,** the **stratum lucidum** (only in thick skin), and the **stratum corneum.**

The **dermis** lies below the epidermis and is composed of **collagenous** and **elastic connective tissue:** fibroblasts, mast cells, macrophages, blood vessels, and nerve endings. The dermis consists of two microscopically different layers: the **reticular** and **papillary** layers. The **reticular** layer lies below the papillary layer and is composed of more densely arranged connective tissue. Cell types are less abundant in this region than in the papillary region. The **papillary** region of the dermis consists of less densely arranged connective tissue and a wider variety of cell types.

The **hypodermis,** or superficial fascia, lies below the dermis. It is composed of **loose areolar connective tissue** surrounding adipose tissue. It adheres to the dermis by means of collagenous fibers.

The skin appendages are modifications of the epidermis and consist of **nails, sweat glands** (**eccrine** and **apocrine**), **sebaceous glands,** and **hair.**

The **nails** are keratinized plaques and consist of several layers of **cornified epithelium.** The nail consists of the **nail bed, nail root, nail plate, nail fold, eponychium,** and **hyponychium.**

The **sweat** glands are of two types: eccrine and apocrine. The **eccrine glands** are simple tubular structures consisting of **cuboidal** to low **columnar epithelium.**

The **apocrine** glands are somewhat different. They are usually larger and the epithelium has only one cell type in its secretory parenchyma. The apocrine gland empties into a **hair follicle.** Thus the ducts of these glands are composed of **stratified squamous epithelium.**

The **sebaceous glands** are more complex than the simple tubular sweat glands. The **parenchymal epithelium** consists predominantly of polyhedral-shaped cells that are arranged in the form of an **alveoli.** The sebaceous alveoli are located in the dermis and are surrounded by thin connective tissue. The alveoli of the sebaceous glands usually drain their secretions into the superior portions of hair follicles.

Hair follicles are relatively complex skin appendages with three basic layers: a **connective tissue sheath** and **outer** and **inner epithelial root sheaths.** The hair follicle connective tissue sheath is composed of three layers: an **inner glassy membrane,** a **middle connective tissue layer,** and an outer layer of mixed collagenous and elastic fibers.

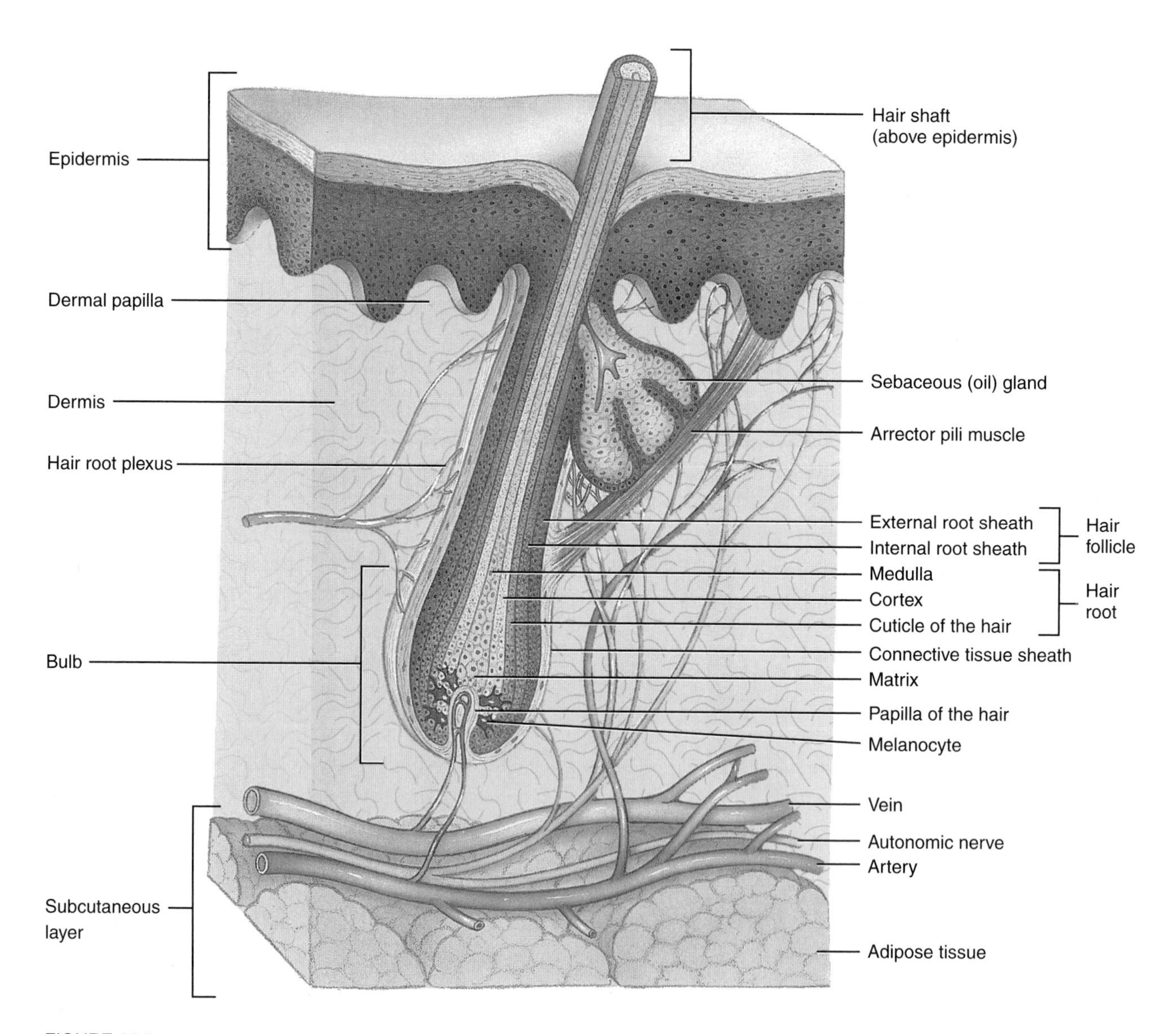

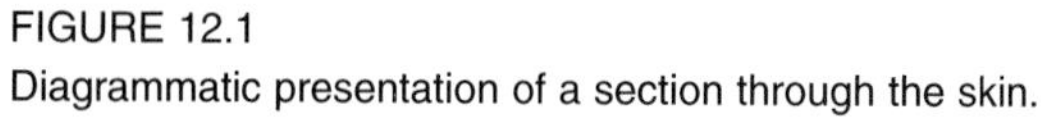
FIGURE 12.1
Diagrammatic presentation of a section through the skin.

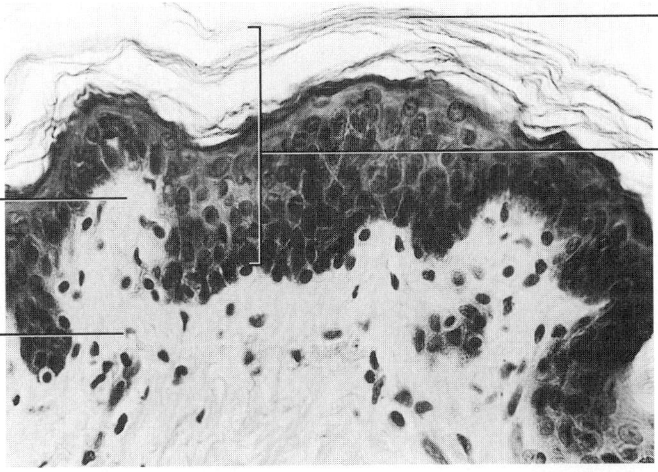

FIGURE 12.2
Light micrograph (LM) of thin skin from the abdominal region. The individual layers are difficult to discern. Stratum corneum is greatly reduced. (400×)

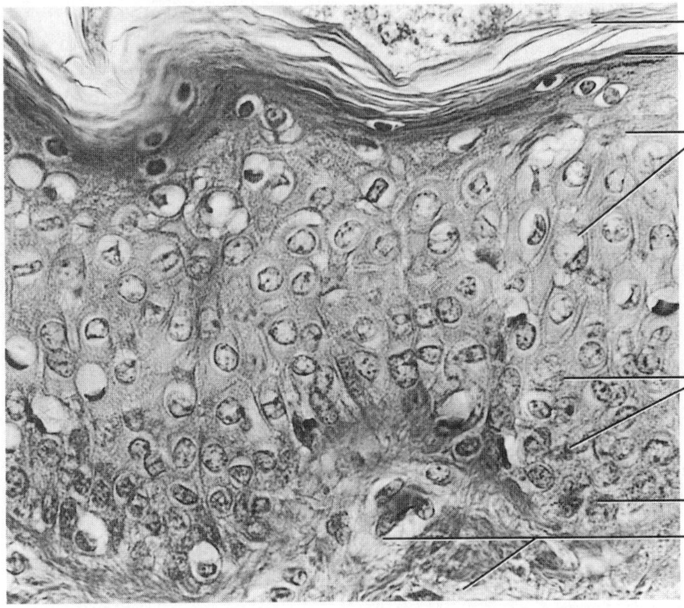

Stratum corneum (desquamating layer of epidermis)
Epidermis
Dermal papilla
Dermis

Stratum corneum
Stratum lucidum
Stratum granulosum
Stratum spinosum
Stratum basale
Dermis

FIGURE 12.3
LM of thick skin. Epidermis displays differentiating layers. (400×)

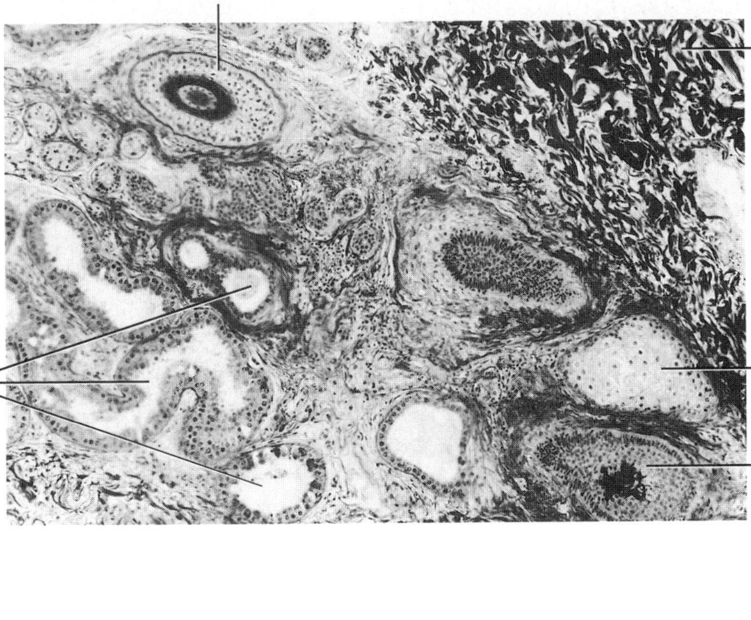

Hair follicle
Dense connective tissue
Apocrine sweat glands
Sebaceous gland
Hair follicle

FIGURE 12.4
LM of a section through the axillary skin. (400×)

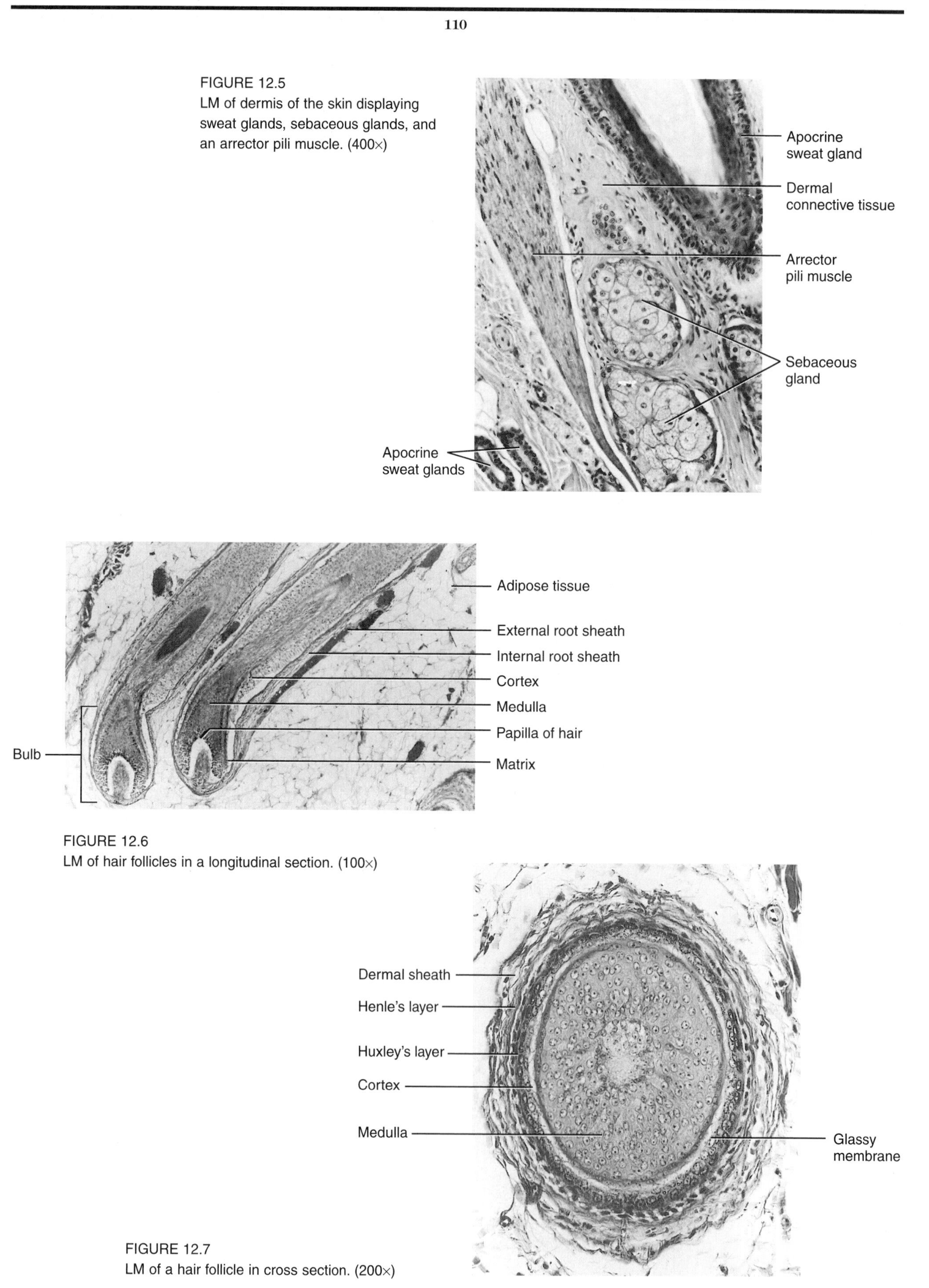

FIGURE 12.5
LM of dermis of the skin displaying sweat glands, sebaceous glands, and an arrector pili muscle. (400×)

FIGURE 12.6
LM of hair follicles in a longitudinal section. (100×)

FIGURE 12.7
LM of a hair follicle in cross section. (200×)

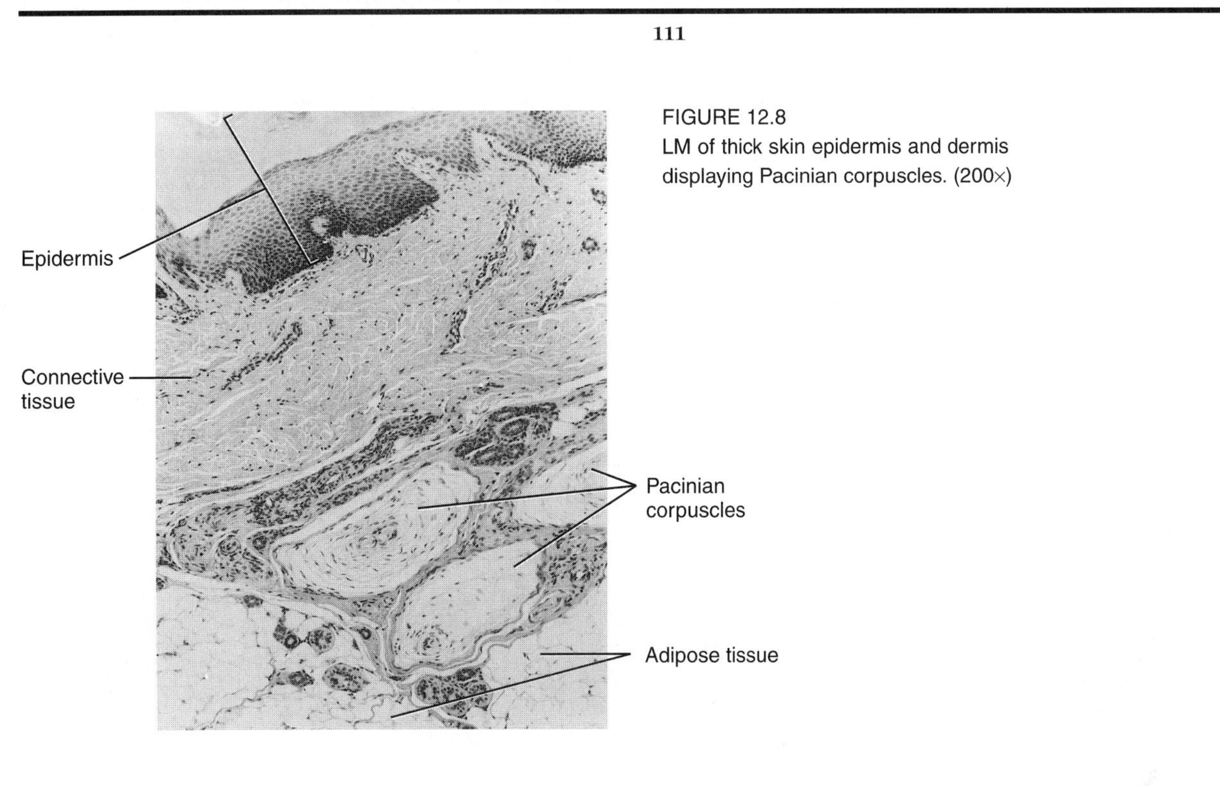

FIGURE 12.8
LM of thick skin epidermis and dermis displaying Pacinian corpuscles. (200×)

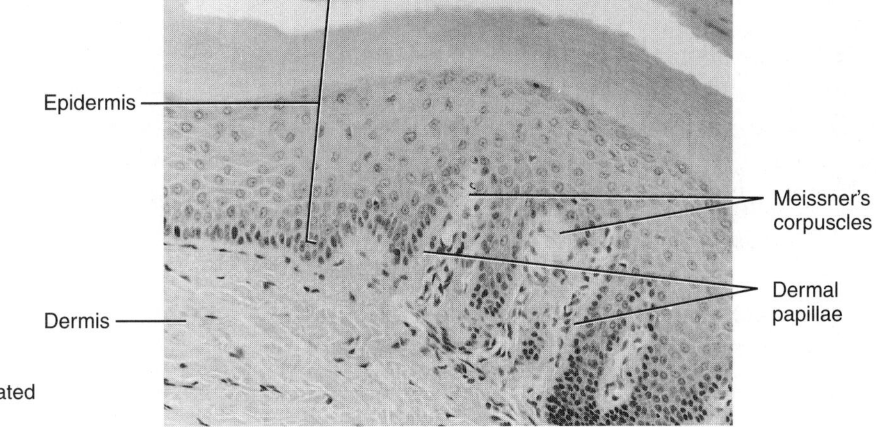

FIGURE 12.9
LM of Meissner's corpuscles located in the dermal papillae. (200×)

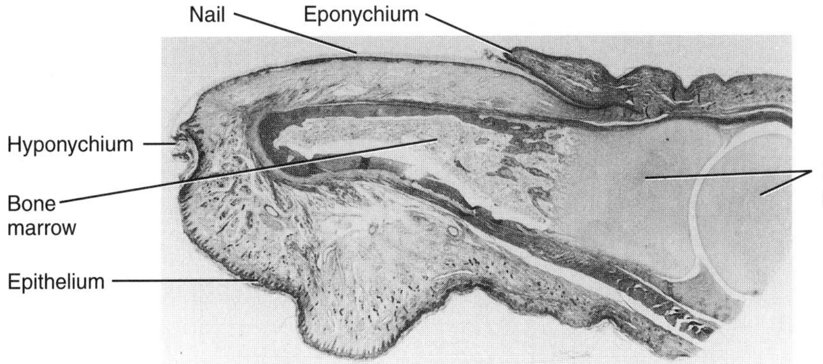

FIGURE 12.10
LM of fetal finger in a sagittal section displaying nail, nail fold, eponychium, and developing phalanges. (20×)

Cardiovascular System

The function of the **cardiovascular system** is to circulate the blood throughout the organ tissues of the body. Therefore it provides for nutrition and **hormone** delivery, gas exchange, and the removal of waste from the tissues.

The vasculature of the cardiovascular system is arranged in three fairly distinct layers or tunics: the outer layer or tunica adventitia, the middle layer or tunica media, and the inner layer or tunica intima. Each tunic has a preponderance of a particular type of tissue. The mass of the tissue within the tunica depends on whether the vasculature is **venous** or **arterial.**

Arterial Vasculature

Arteries are classified as **elastic arteries** (large), **muscular arteries** (medium), and **arterioles** (smallest arteries). Elastic arteries are found closer to the heart than either muscular arteries or arterioles. The anatomic progression, from proximal to distal, results in a morphological change from elastic arteries to arterioles.

Elastic Arteries **Elastic arteries** have three tunics: an **adventitia,** a **media,** and an **intima.** The tunica intima consists of two tissues: epithelium and connective. The intimal epithelium is a simple squamous epithelium called **endothelium.** The connective tissue component of the intima consists predominantly of elastic fibers, some forming incomplete laminae and others forming the internal **elastic lamina.** Along with the rest of the connective tissue matrix of elastic fibers and **amorphous** substance, there is a very small amount of collagen fibers.

The tunica media of elastic arteries is wider than the intima and is predominantly elastic tissue arranged concentrically in the form of **fenestrated laminae.**

The tunica adventitia of elastic arteries is composed of mixed irregular elastic and collagen tissues concentrically arranged superiorly to the external elastic lamina of the tunica media. Within the adventitia is the **vasa vasorum,** which consists of the blood vessels to the body wall of the artery.

Muscular Arteries The muscular artery tunics vary somewhat in terms of thickness and composition compared to the elastic arteries. The tunica adventitia is thicker, with respect to its own media in muscular arteries. The media of muscular arteries consists of **concentrically** arranged **smooth muscle fibers.** Elastic fibers are found intercellularly within the tunica media. Muscular arteries also have both internal and external elastic laminae, which serve as boundaries between intima and media and between media and adventitia, respectively. The internal elastic media may contain elastic fibers and a well-defined endothelium.

Arterioles The tunica intima of arterioles consists of an **endothelium** and an internal elastic membrane. The tunica media is relatively thick, with little connective tissue and few smooth muscle cells. The subendothelial layer is lacking in arterioles. The internal elastic membrane vanishes as the arterioles become smaller. The external elastic lamina is undifferentiated, and the adventitia is thin and relatively small in relation to the other tunics.

Capillaries are simpler blood vessels. They are predominantly made of a **tunica intima** and a fine **tunica media,** but a nearly nonexistent **tunica adventitia.** The intima of capillaries consists of a one-cell **endothelium** and the underlying **basal lamina.**

Venous Vasculature

The **venous** vasculature follows the same morphological arrangement as do the arteries. The walls of veins are thinner as a result of reduction in the mass of medial components. Venous vasculature lacks a distinct **external elastic lamina,** and the internal elastic lamina is less obvious in most veins than in arteries. In general, there is more variability in the venous vasculature. The veins may be classified as **venules, small** and **medium** veins, and **large** veins. On the basis of the size and thickness of the body wall, veins are small, medium, or large. The small and medium veins are grouped together.

Heart

The **heart** is a four-chambered valvular structure that is similar in micromorphology to the **vasculature.** The body wall of the heart has three distinct layers: an **endocardium,** a **myocardium,** and an **epicardium.**

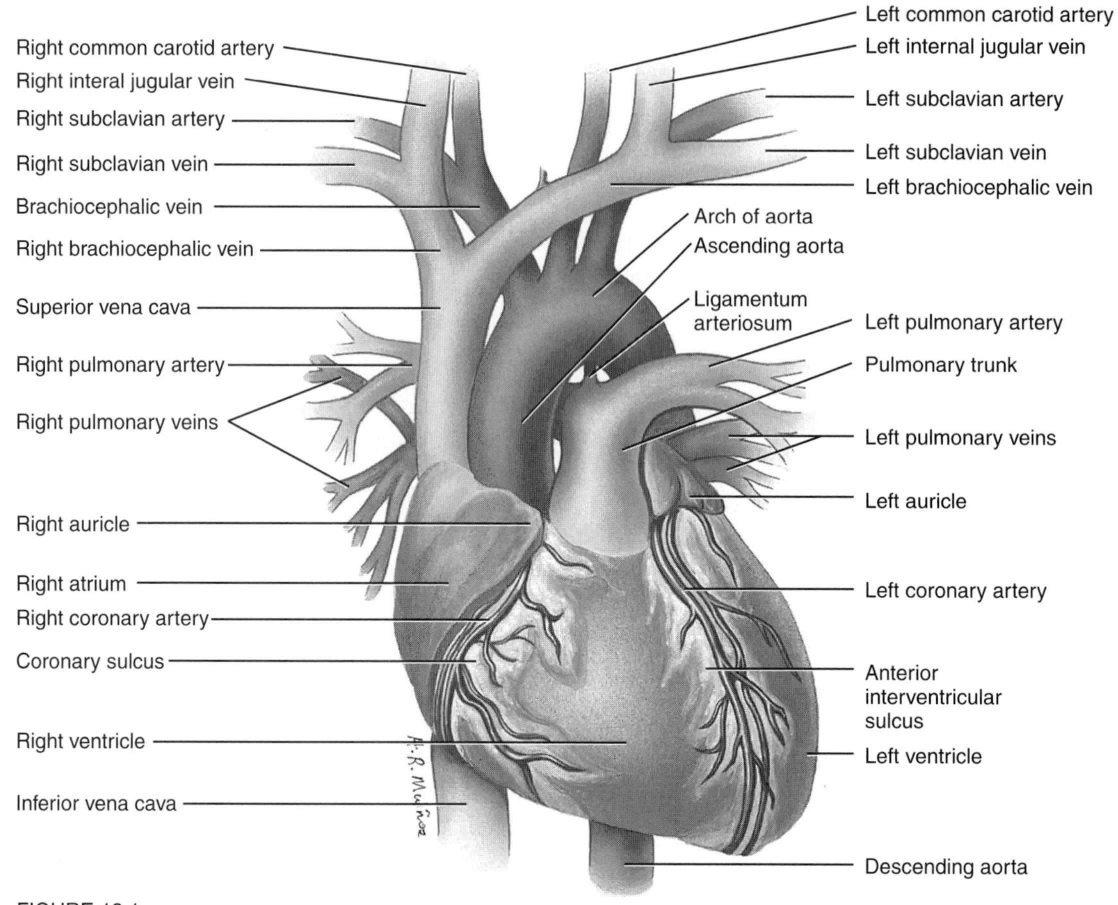

FIGURE 13.1
Diagrammatic presentation of a human heart and associated blood vessels.

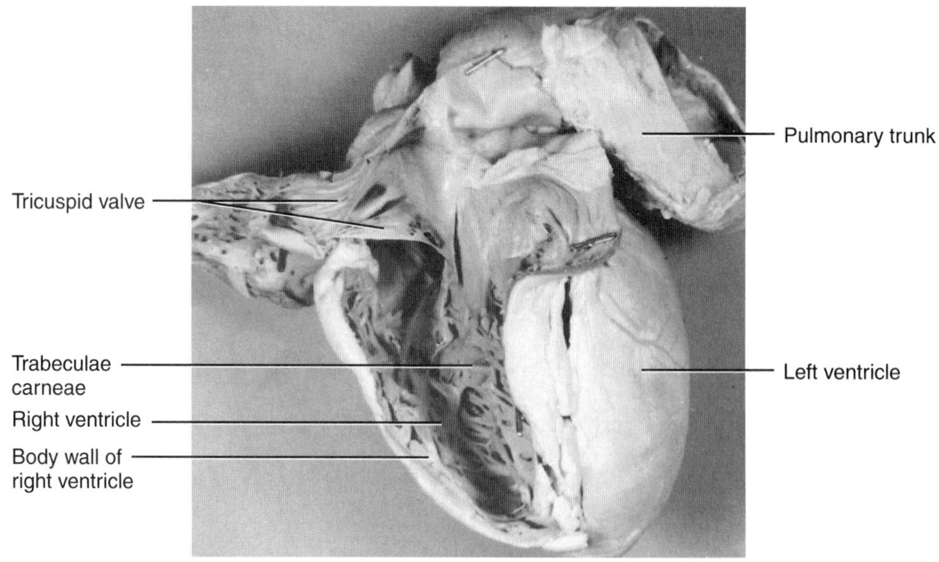

FIGURE 13.2
The human heart with part of the right ventricle body wall removed to show the internal structures.

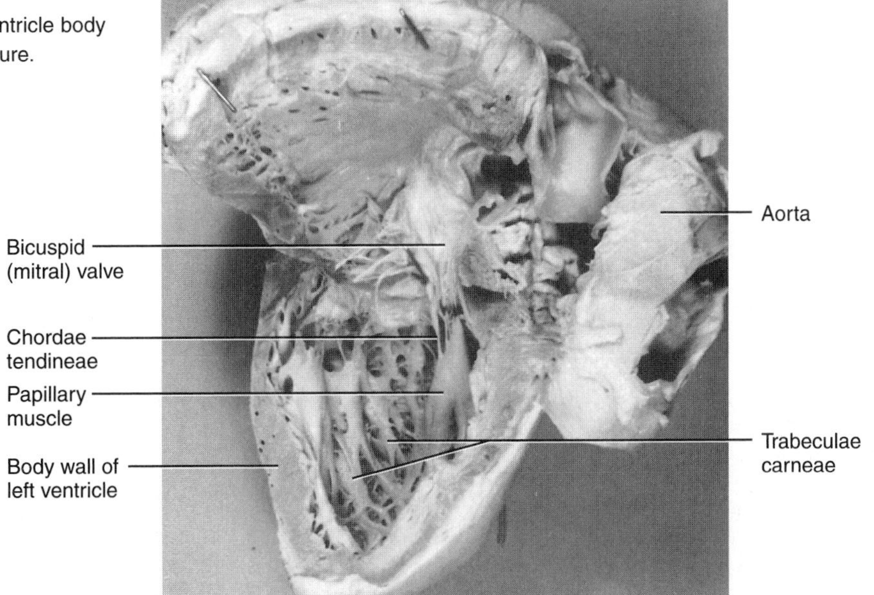

FIGURE 13.3
The human heart with part of the left ventricle body wall removed to show the internal structure.

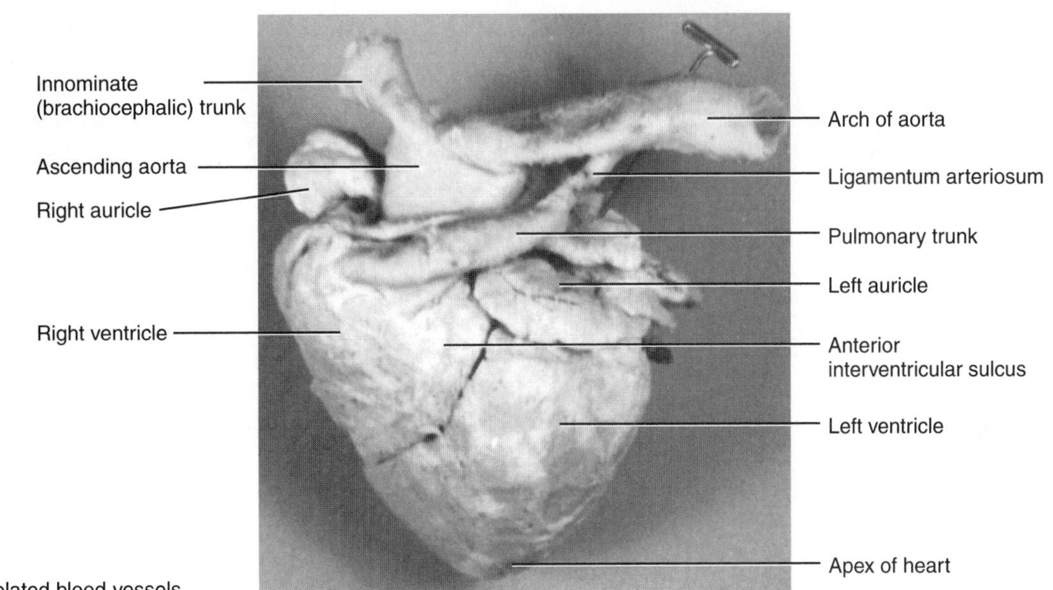

FIGURE 13.4
Sheep heart and related blood vessels.

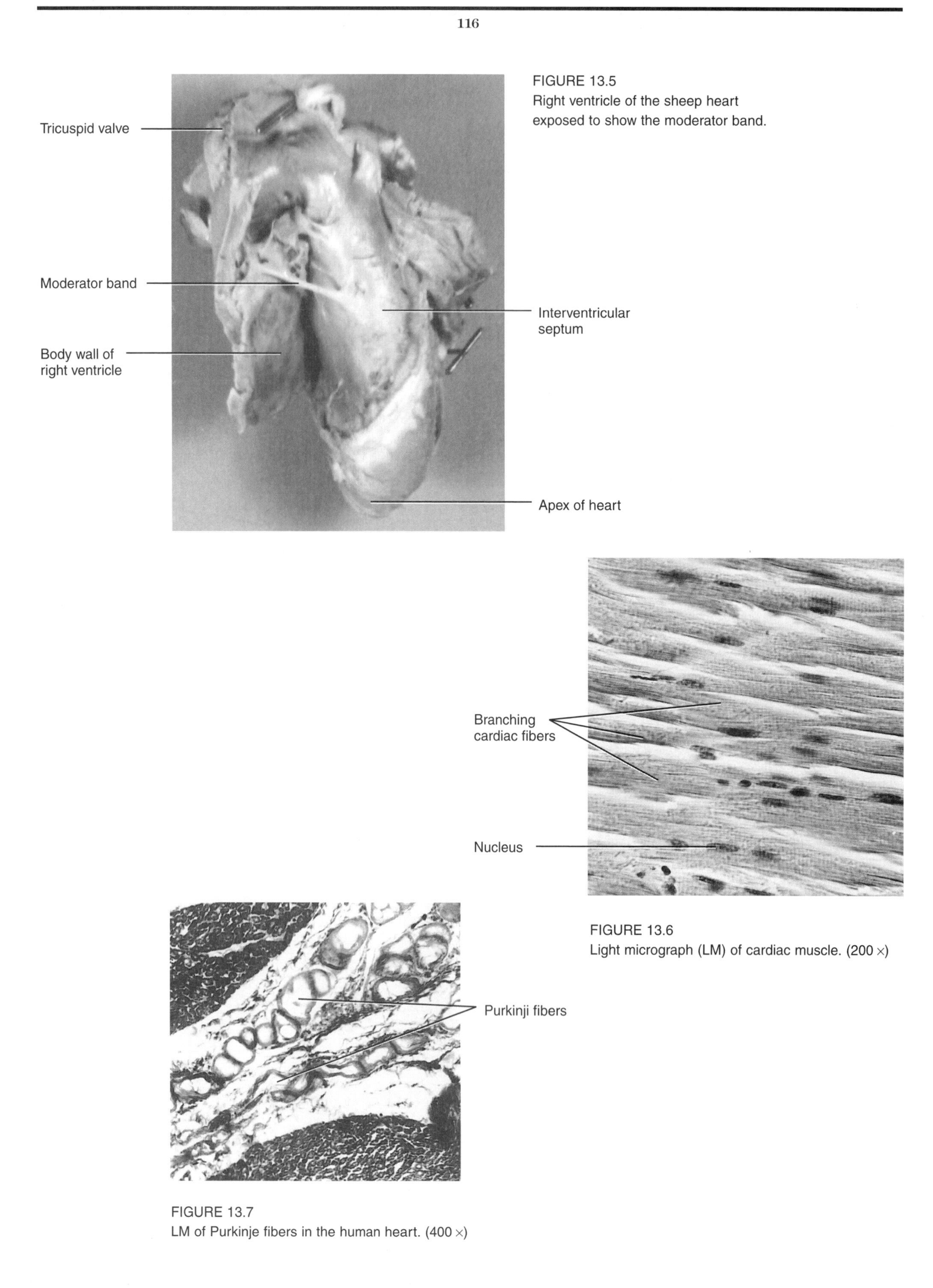

FIGURE 13.5
Right ventricle of the sheep heart exposed to show the moderator band.

FIGURE 13.6
Light micrograph (LM) of cardiac muscle. (200 ×)

FIGURE 13.7
LM of Purkinje fibers in the human heart. (400 ×)

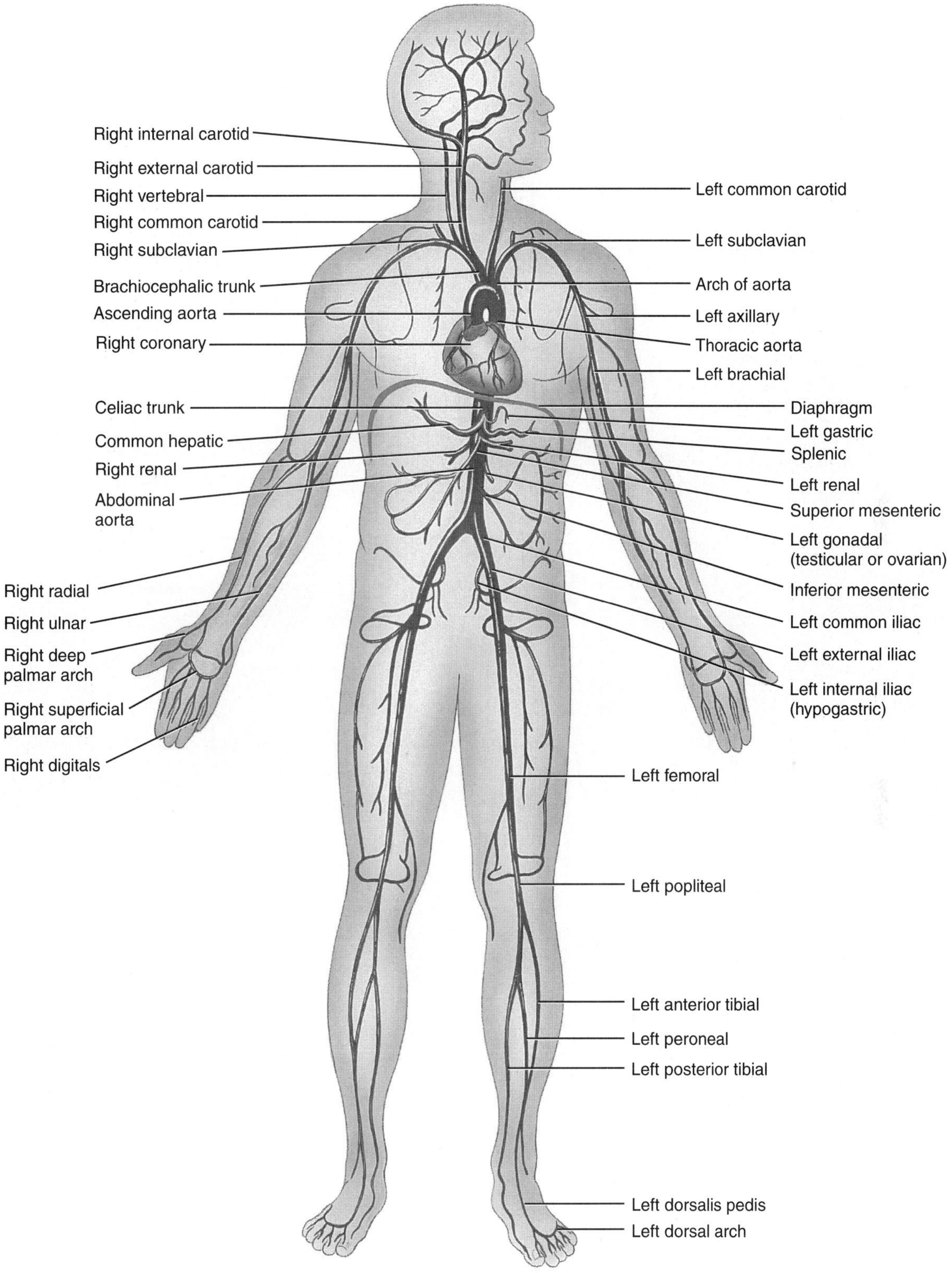

Right internal carotid
Right external carotid
Right vertebral
Right common carotid
Right subclavian
Brachiocephalic trunk
Ascending aorta
Right coronary
Celiac trunk
Common hepatic
Right renal
Abdominal aorta
Right radial
Right ulnar
Right deep palmar arch
Right superficial palmar arch
Right digitals

Left common carotid
Left subclavian
Arch of aorta
Left axillary
Thoracic aorta
Left brachial
Diaphragm
Left gastric
Splenic
Left renal
Superior mesenteric
Left gonadal (testicular or ovarian)
Inferior mesenteric
Left common iliac
Left external iliac
Left internal iliac (hypogastric)
Left femoral
Left popliteal
Left anterior tibial
Left peroneal
Left posterior tibial
Left dorsalis pedis
Left dorsal arch

FIGURE 13.8
Diagrammatic presentation of the arterial system.

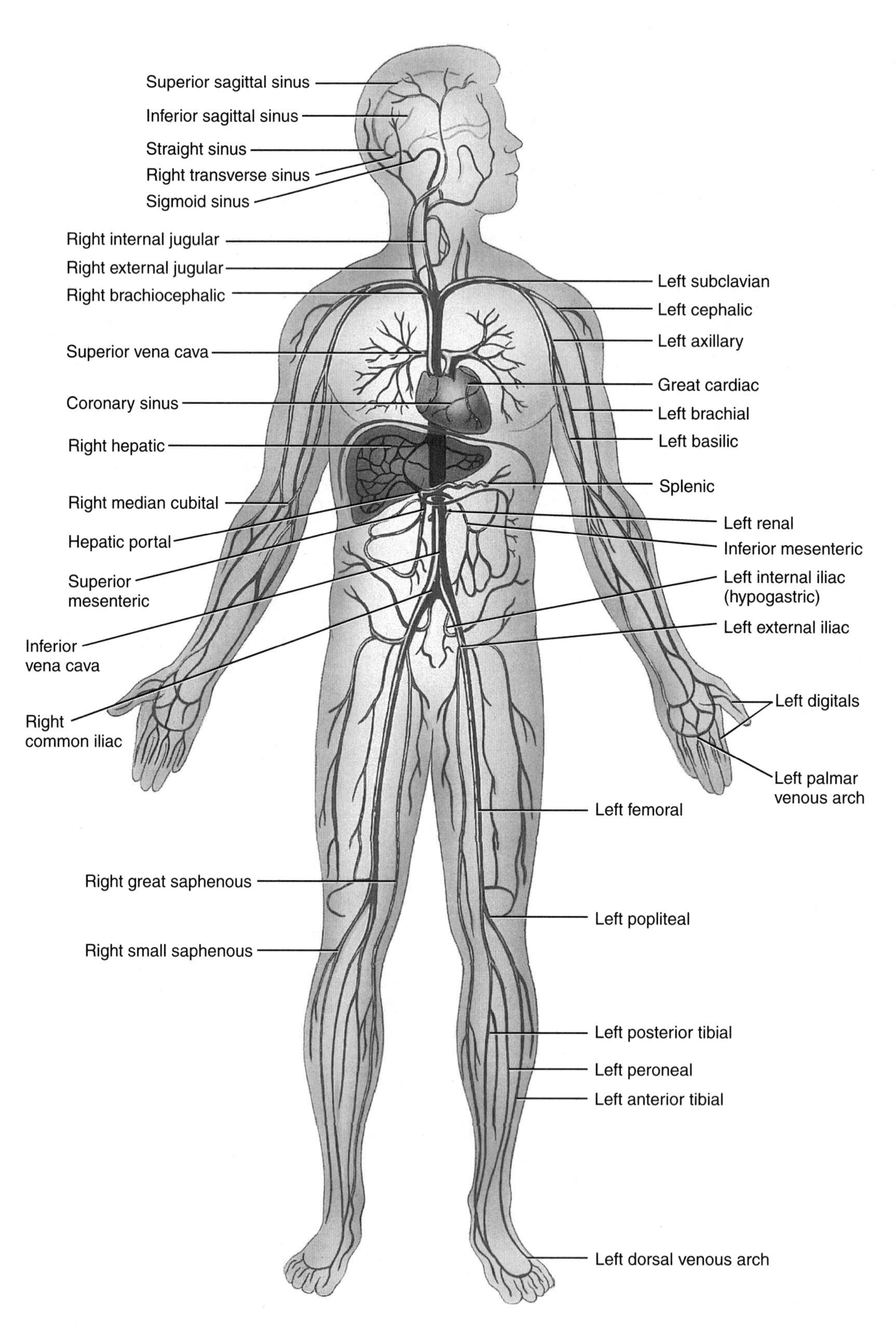

FIGURE 13.9
Diagrammatic presentation of the venous system.

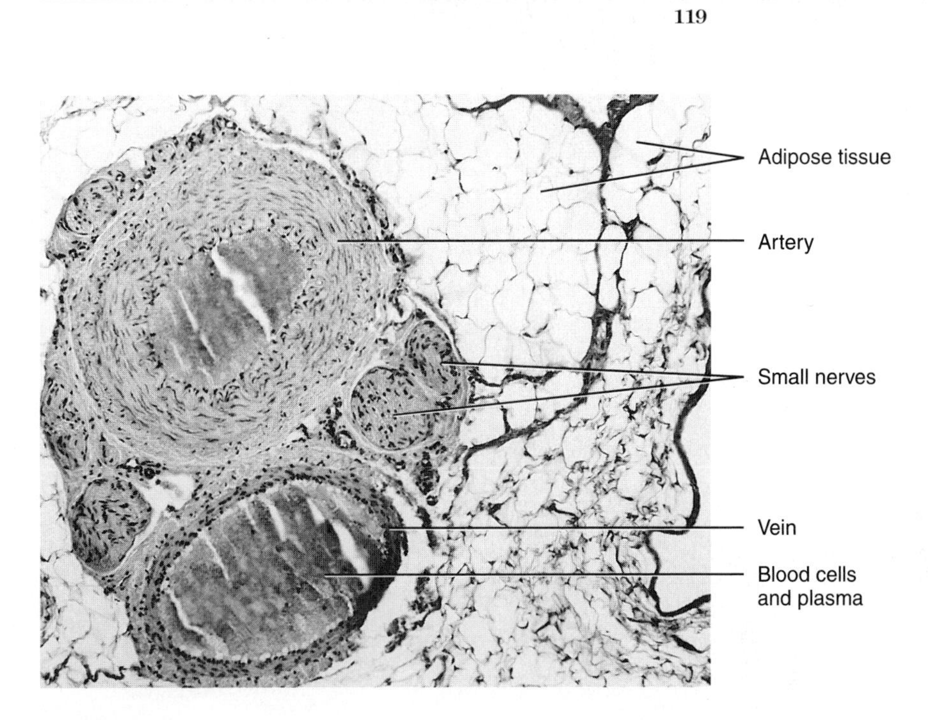

FIGURE 13.10
LM of a neurovascular bundle in cross section. (100 ×)

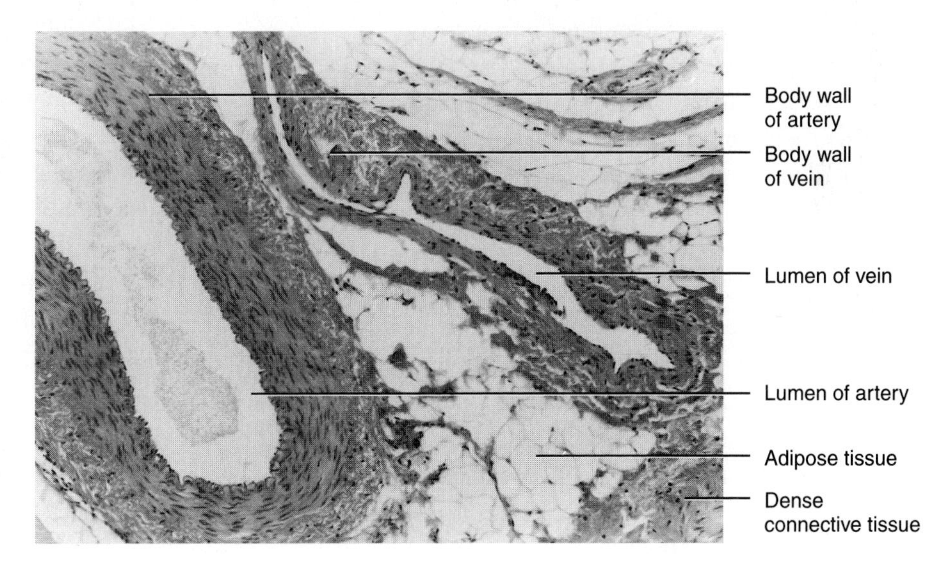

FIGURE 13.11
LM of an artery and a vein in cross section.

Lymphatic System

The **lymphatic system** functions as part of both the **immune** system and the **circulatory** system. As part of the immune system, it functions in the proliferation, development, and storage of immune cells and as a site for immune reactions to occur. As part of the circulatory system, it functions in the removal of excess interstitial fluid and contributes to fluid **homeostasis.**

The lymphatic system is organized on a gross anatomical level and on a microanatomical level. It consists of several cellular components organized to form solid, vascular, or mixed organs. On the gross level, the anatomical arrangement of the lymphatic system is diffused and consists of **lymph nodules, lymph nodes, tonsils, thymus, spleen, lymphatic vessels,** and **Peyer's patches.**

The **lymph nodules** are well-defined aggregates of lymphoid tissue within the mucosae of certain organs. Nodules are found in the **lamina propria** of the **gastrointestinal tract,** the **tongue,** the **tonsils,** the **lymph nodes,** and the tissue of the **respiratory** system. Lymph nodules are not encapsulated.

The **lymph nodes** are more complex than the nodules. A lymph node may have several lymph nodules in its **stroma** of connective tissue. Surrounding the node is a thin **connective tissue capsule.**

The **lymph node cortex** is the region beneath the capsule and is divided into an **outer cortex** and an inner **paracortex.** The outer cortex is composed of **lymphoid nodules.** Within the lymphoid nodules are **germinal centers** composed of proliferating **B lymphocytes.** The paracortex consists of **B** and **T lymphocytes.**

The **medulla** of the **lymph node** lies below the cortex, where the lymphoid tissue is less densely arranged. The cells of the medulla are organized to form **medullary cords.** The cords are composed of **lymphocytes, plasma cells,** and **macrophages.** Also present in the medulla are **reticular** fibers that interconnect with the trabeculae and the capsule.

Three types of **tonsils** can be identified. Based on their anatomic location, tonsils are **lingual** (at the base of the tongue), **palatine** (two, located between the **glossopalatine** and **pharyngopalatine arches**), and a single **pharyngeal** tonsil (located along the midline of the posterior wall of the **nasopharynx**).

The **thymus** is an encapsulated organ consisting of two **lobes.** The lobes are divided into **lobules** by **connective tissue septa.** Similar to lymph nodes, the thymic lobules have both **medullary** and **cortical** regions. The **cortex** of the thymus is relatively denser than the medulla. The cortical cells consist of **small, medium,** and **large lymphocytes.** The **medulla** of the thymus is less densely packed with lymphocytes.

The **spleen,** like the lymph nodes, tonsils, and thymus, is an encapsulated organ. The capsule is composed of dense connective tissue covered by **squamous mesothelium.** Also found in the capsule are elastin and smooth muscle. The capsule extends deep into the splenic substance to form **trabeculae.**

The splenic **parenchyma** consists of **red** and **white pulp.** The red pulp is associated with the venous components of the spleen. It consists of a network of **venous sinuses** and an array of **reticuloendothelial** cells and their components arranged to form **Billroth's** or **splenic cords.** The white pulp is associated with the arterial system within the spleen. It consists of lymphoid tissue adhering to the **adventitial** layers of the arteries. The lymphoid tissue is arranged into sheaths and gradually merges into lymphoid splenic nodules.

The **lymphatic vessels** consist of blind lymphatic **capillaries, collecting vessels,** and **lymphatic trunks.** The morphological structures are similar to the venous vasculature and blood capillaries.

Peyer's patches are aggregates of **lymphatic nodules** found in the **appendix** and the lower half of the **small intestine.** The patches consist of many **secondary nodules** forming large oval structures that lie below the epithelium. Peyer's patches are more prevalent in children and decrease in number in adults.

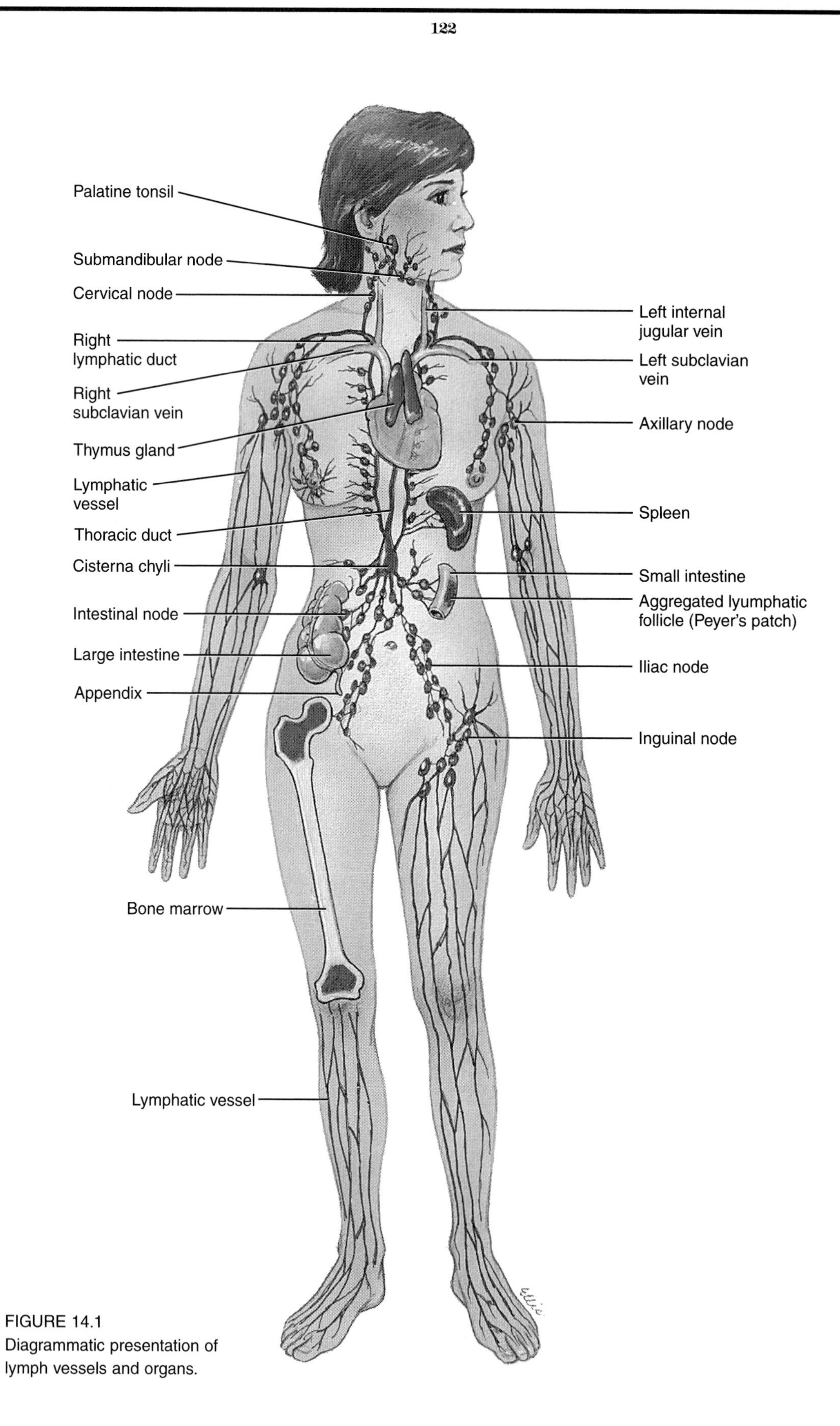

FIGURE 14.1
Diagrammatic presentation of lymph vessels and organs.

FIGURE 14.2
Diagrammatic presentation of a partially sectioned lymph node.

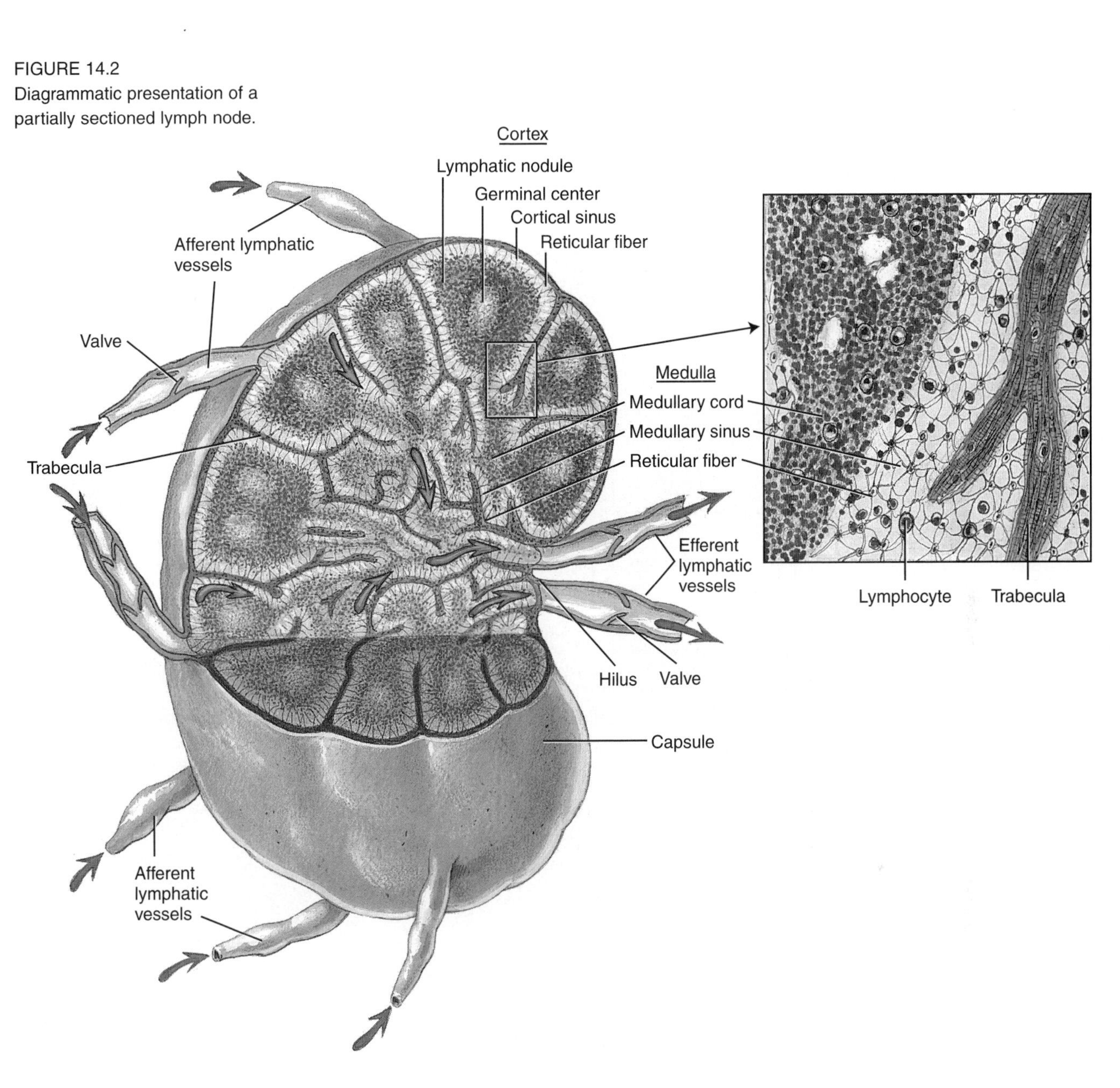

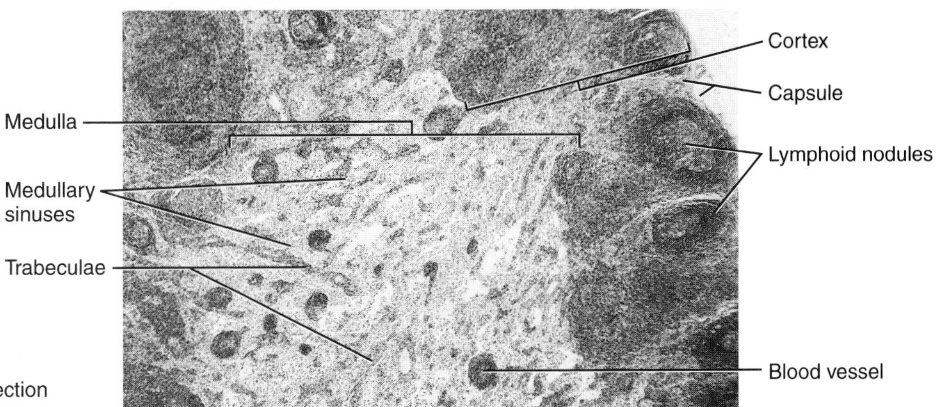

FIGURE 14.3
Light micrograph (LM) of a section through a lymph node. (20×)

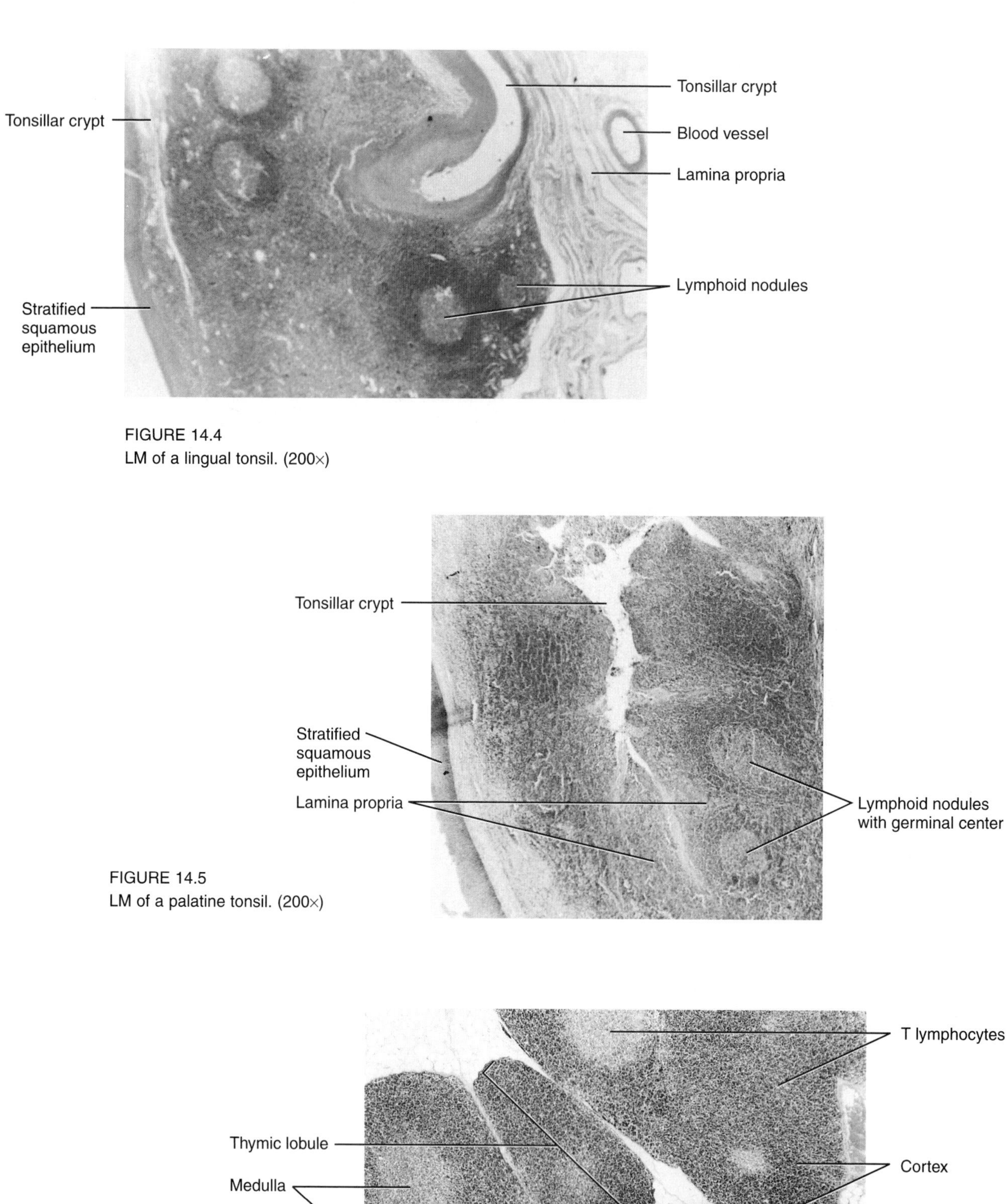

FIGURE 14.4
LM of a lingual tonsil. (200×)

FIGURE 14.5
LM of a palatine tonsil. (200×)

FIGURE 14.6
LM of thymus lobules. (200×)

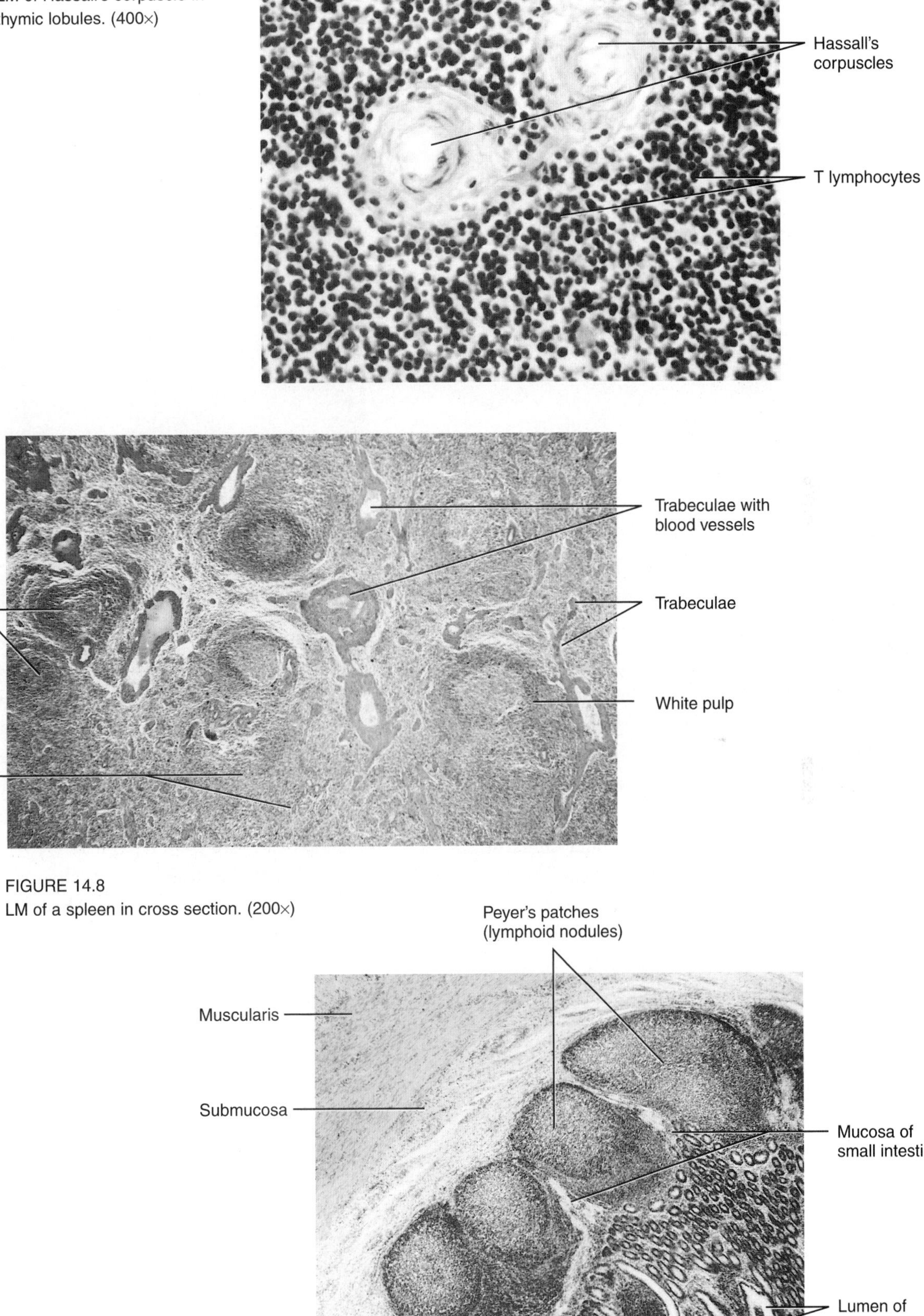

FIGURE 14.7
LM of Hassall's corpuscle in thymic lobules. (400×)

FIGURE 14.8
LM of a spleen in cross section. (200×)

FIGURE 14.9
LM of Peyer's patches in the mucosal layer of the ileum. (200×)

Digestive System

The large and fairly extensive **digestive system** begins with the **oral cavity,** extends through the **thoracic, abdominal,** and **pelvic cavities,** and terminates with the **anus.** The system has two basic components: the **digestive system proper** or **alimentary tract** and its associated **glands** and **organs.**

The **digestive system proper** is organized into **tunics,** and its gross morphology is tubular in nature. It consists of several histological components with tissue variance within the tunics.

Overall, the organs of the digestive system are the **oral cavity** (including the **salivary glands, teeth,** and **tongue**), **oral pharynx, esophagus, stomach, duodenum, jejunum, ileum, colon, rectum, anus, pancreas, liver, gallbladder,** and the **mucosal glands** of the **alimentary tract.**

Tongue The tongue is a relatively complex organ consisting of muscle fascicles arranged in three planes perpendicular to each other. The dorsal surface of the tongue is covered with a mucosa of keratinized stratified squamous epithelium. Below the epithelium of the mucosa lies the lamina propria, but the tongue, like the gingiva, lacks a submucosa. The mucosal epithelium is connected to the underlying muscle through intermeshing of the connective tissue of the lamina propria.

The dorsal surface of the tongue is covered with several types of epithelial projections or **papillae: fungiform, filiform, foliate** (**rare**), and **circumvallate** papillae. **Taste buds** are specialized **chemoreceptors** that are located primarily in the circumvallate and foliate papillae.

Glands Several types of **salivary glands** are associated with the oral cavity. These **tuboalveolar compound glands** are considered serous, mucous, or mixed and are classified as minor or major. The minor salivary glands are intrinsic to the oral mucosa and are found in the submucosal region of the tissue. The major salivary glands are the **parotid, submandibular** (**submaxillary**), and **sublingual glands.** Major salivary glands lie outside the oral cavity but are connected by glandular ducts.

Teeth The **teeth** contain both hard and soft tissue. The hard tissues of the teeth, which are arranged in layers, include the **cementum, dentin,** and **enamel.** The soft tissues are the **pulp,** the **periodontal membrane,** and the surrounding **gingiva.**

In the center of the tooth is a chamber called the **pulp chamber.** The pulp is essentially collagen connective tissue with a gelatinous matrix. Cell types within the pulp include **fibroblasts, mesenchymal cells, macrophages,** and **lymphocytes.** Odontoblasts are found at the junction of the pulp cavity and the dentin. Blood vessels, nerves, and lymphatics are also present in the pulp.

Esophagus The esophagus is a fairly straight, muscle-walled tubular structure. The mucosal lining is of nonkeratinizing stratified squamous epithelium. Also present in the body wall is the submucosa with esophageal gland, a muscularis with circularly and longitudinally arranged smooth muscle fibers, and an outer **adventitia.**

Stomach The body wall of the stomach is divided into four layers: the **serosa** (the outermost), **muscularis, submucosa,** and **mucosa.** The mucosa is lined with columnar cells. As the bolus enters the stomach through the **cardiac sphincter,** it is quickly reacted upon by digestive enzymes and **hydrochloric acid** (HCl) that is secreted by the gastric glands located in the gastric pits.

Intestinal Tract The intestine is a long, highly convoluted hollow tubular structure that extends from the **pyloric** end of the stomach to the **anus.** The intestine is also composed of four identifiable layers: the **adventitia** (serosa, if the intestine is covered by the peritoneal fold of the mesentery), **muscularis, submucosa,** and **mucosa.** The mucosal epithelium and the lamina propria below it form fingerlike projections (**villi**), folds, or a flat surface.

There are two major subdivisions of the intestine, the **small intestine** and **large intestine.** The 23-foot long small intestine is further subdivided into the **duodenum, jejunum,** and **ileum.** The large intestine or **colon** is about 5 feet long, and its diameter is almost twice that of the small intestine. It is subdivided into the **cecum, appendix, colon** (**ascending, transverse, descending,** and **sigmoid**), and **rectum.**

Rectum The rectum is similar to the colon but lacks the taeniae coli bands of smooth muscle. The epithelium changes from columnar to nonkeratinized stratified squamous epithelium at the rectoanal junction.

Liver The liver is the largest organ in the body and is located below the **diaphragm.** It is partially separated into four lobes: the **right, left, caudate,** and **quadrate lobes.** The right lobe is the largest. The lobes are divided into hexagonal lobules. Within the lobules are anastomosing **hepatic cords** composed of **hepatocytes** that radiate in all directions from the central vein. The hepatic sinusoids within the hepatic cords are lined with endothelial cells, reticulate fibers, and phagocytic **Kupffer cells.**

Gallbladder The gallbladder is a blind sac located under the large (right) lobe of the liver. The short **cystic duct** connects the hepatic duct to the gallbladder. The mucosa of the cystic duct folds toward the lumen, forming **spiral valves** with smooth muscle in the body walls.

Bile produced in the liver is emptied into the hepatic ducts and forced backward into the gallbladder, where it is concentrated and stored.

Pancreas The pancreas is retroperitoneal and is the second largest gland in the body. It functions as both an **endocrine gland** and an **exocrine gland.** The pancreas is divided into small lobules separated by delicate strands of connective and adipose tissue. The lobules contain pancreatic islets called the **islets of Langerhans,** which are small endocrine cell concentrations. These islets produce the hormones **glucagon, insulin, somatostatin, vasoactive intestinal peptide** (**VIP**), **substance P, motilin** (motilin is also secreted by the gut), and **pancreatic gastrin.**

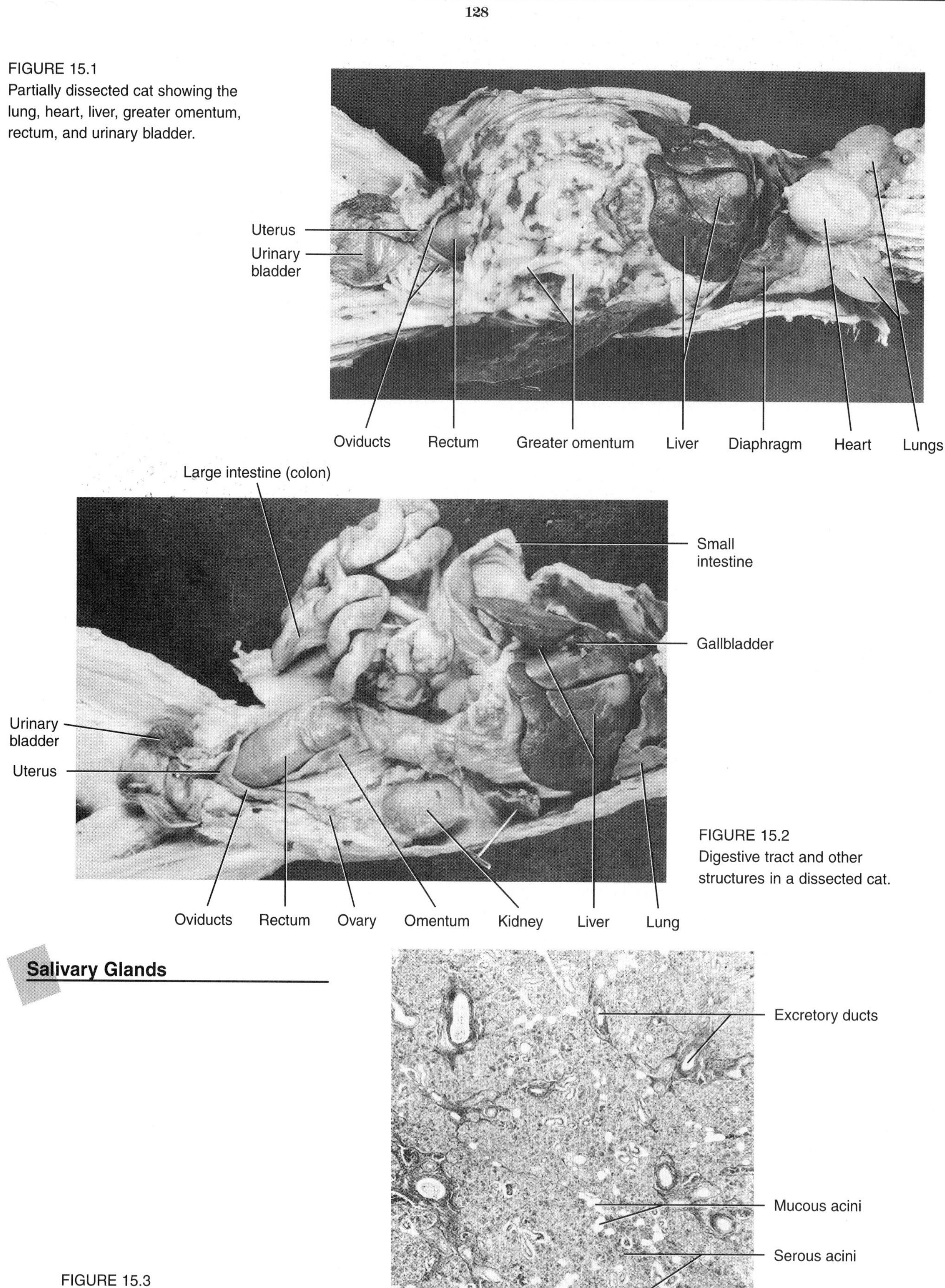

FIGURE 15.1
Partially dissected cat showing the lung, heart, liver, greater omentum, rectum, and urinary bladder.

FIGURE 15.2
Digestive tract and other structures in a dissected cat.

Salivary Glands

FIGURE 15.3
Light micrograph (LM) of a cross section of submaxillary (submandibular) gland. (100×)

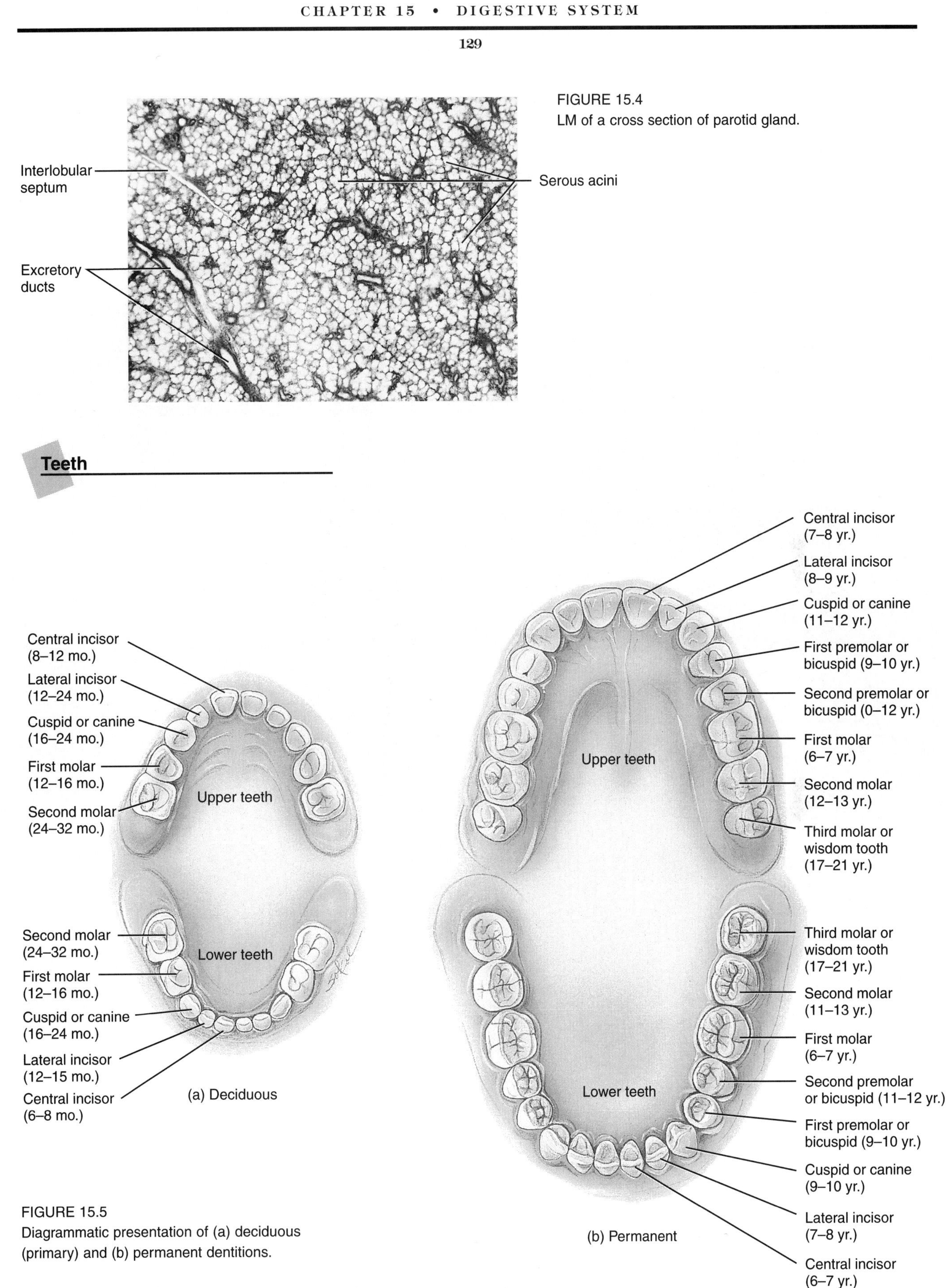

FIGURE 15.4
LM of a cross section of parotid gland.

Teeth

FIGURE 15.5
Diagrammatic presentation of (a) deciduous (primary) and (b) permanent dentitions.

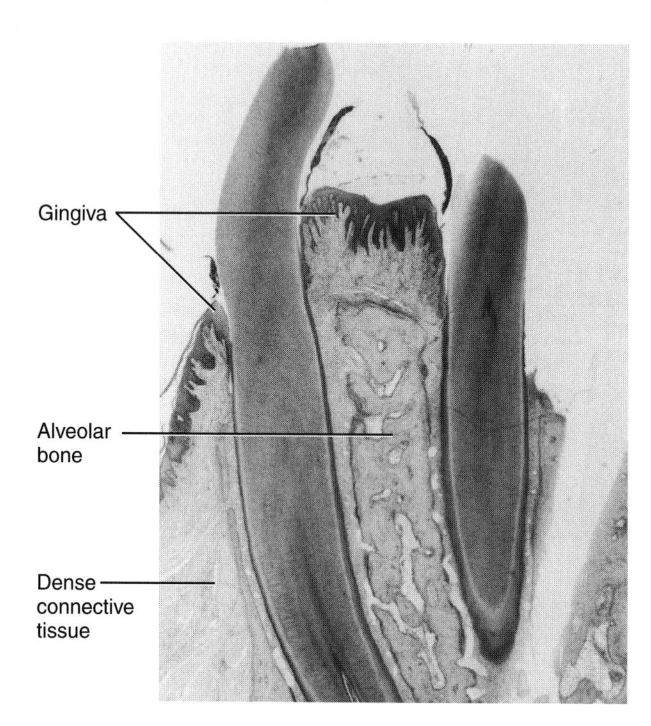

FIGURE 15.6
LM of two erupting teeth and supporting structures in a sagittal section. (200×)

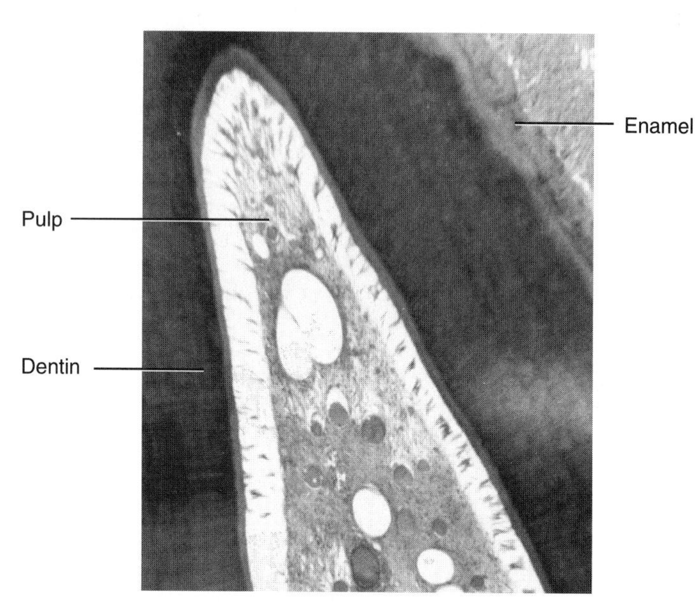

FIGURE 15.7
LM of a tooth in sagittal section. (200×)

Tongue

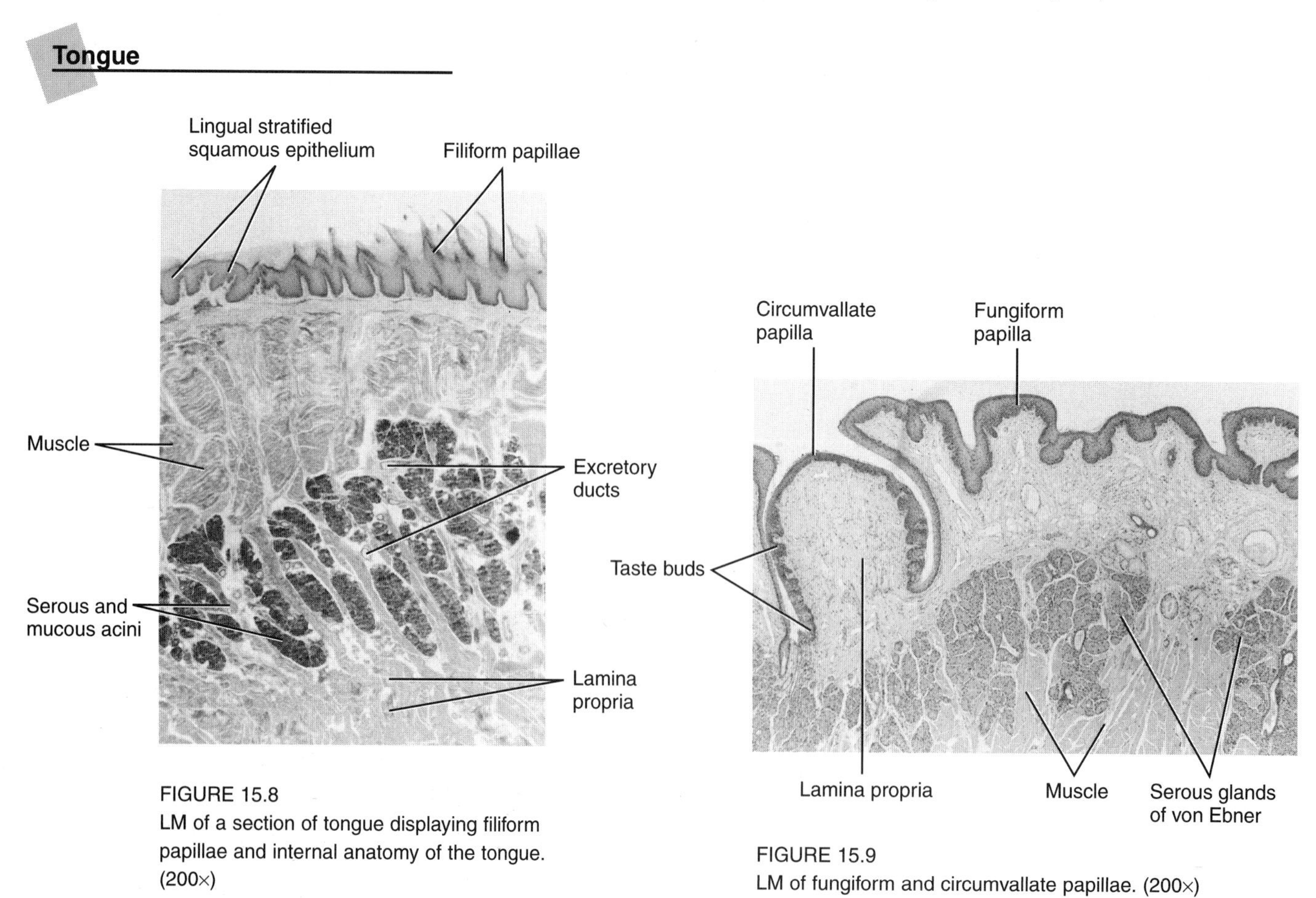

FIGURE 15.8
LM of a section of tongue displaying filiform papillae and internal anatomy of the tongue. (200×)

FIGURE 15.9
LM of fungiform and circumvallate papillae. (200×)

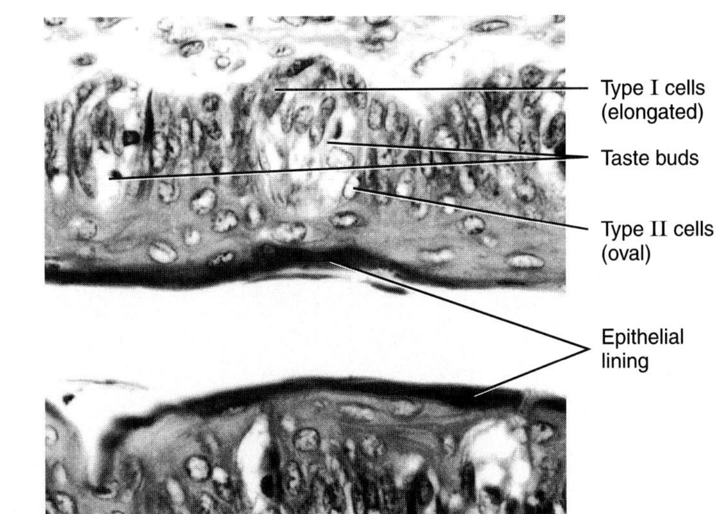

FIGURE 15.10
LM of taste buds in section.

Esophagus

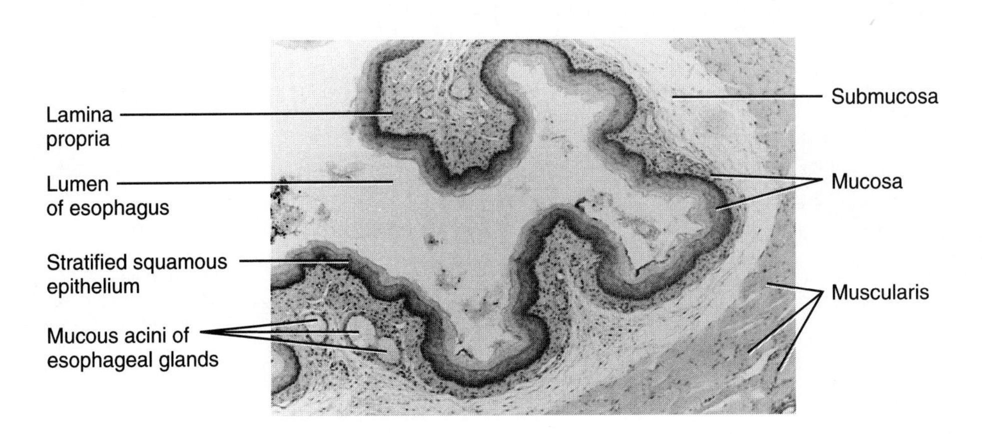

FIGURE 15.11
LM of cross section of the esophagus. (20×)

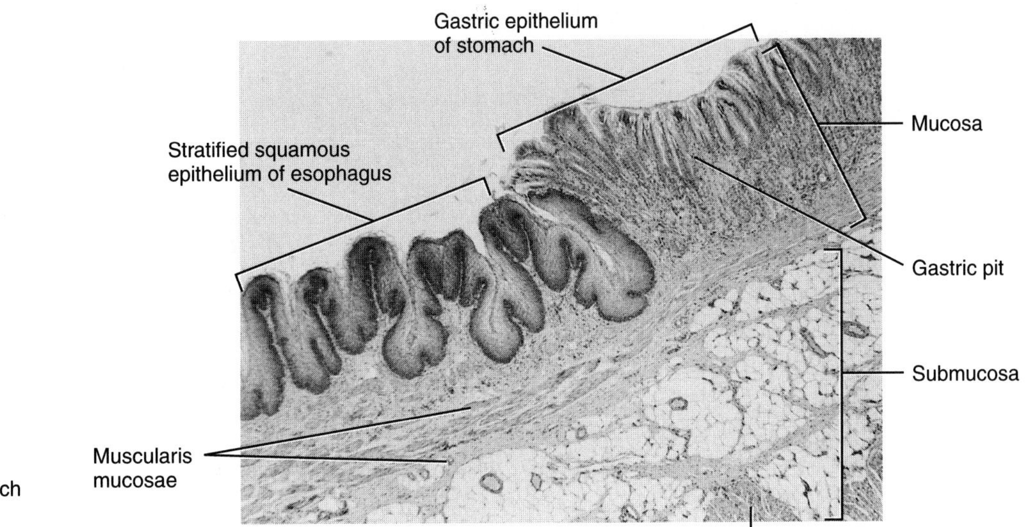

FIGURE 15.12
LM of the esophagus–stomach junction. (100×)

Intestine

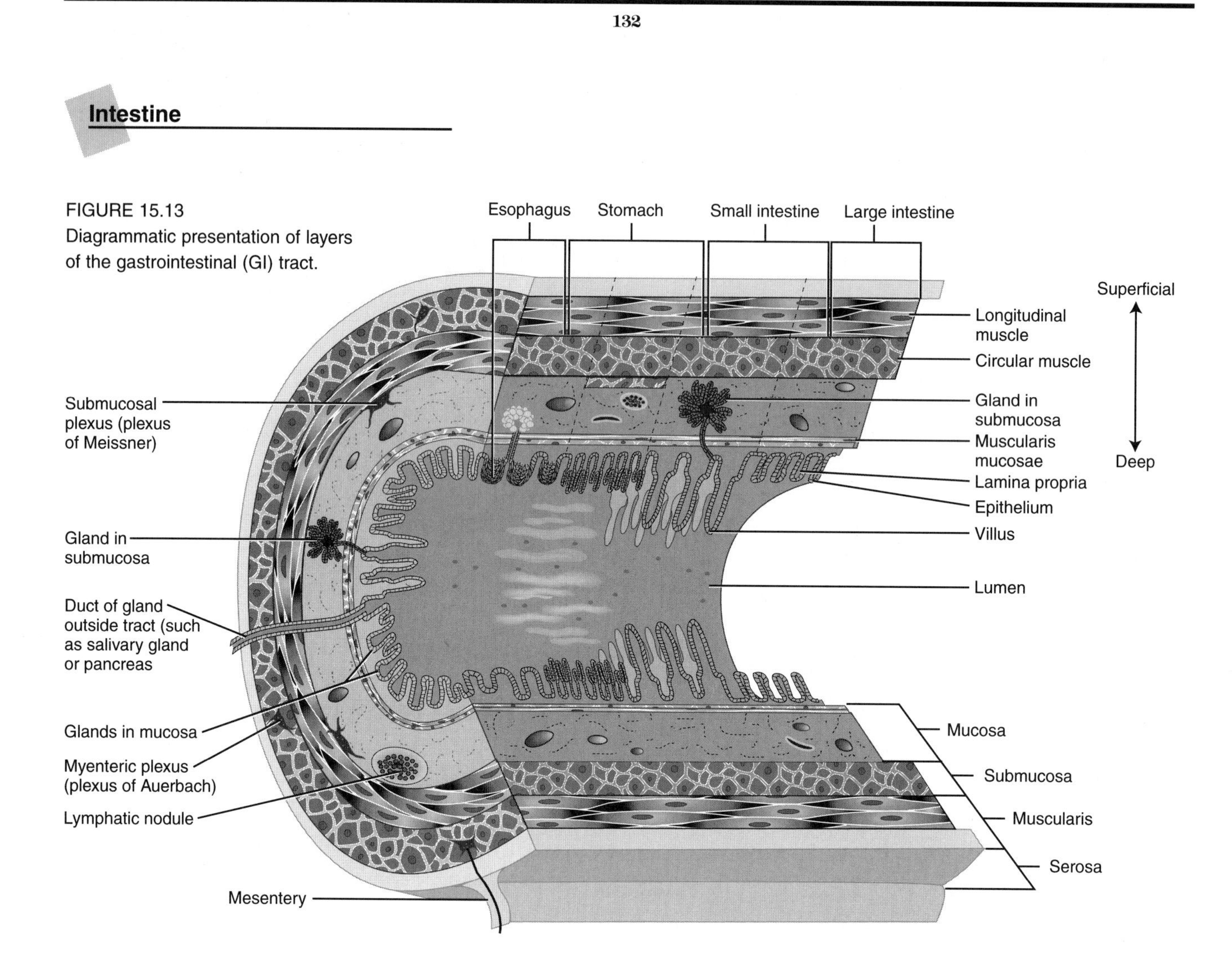

FIGURE 15.13
Diagrammatic presentation of layers of the gastrointestinal (GI) tract.

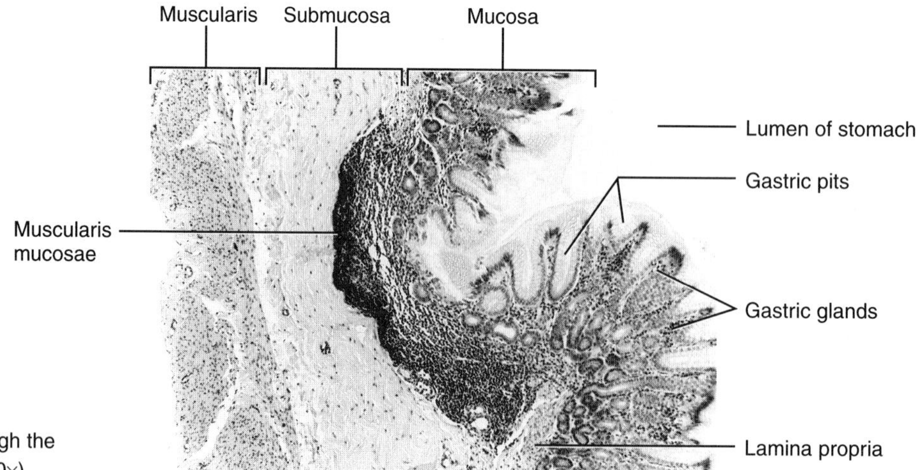

FIGURE 15.14
LM of a section through the cardiac stomach. (100×)

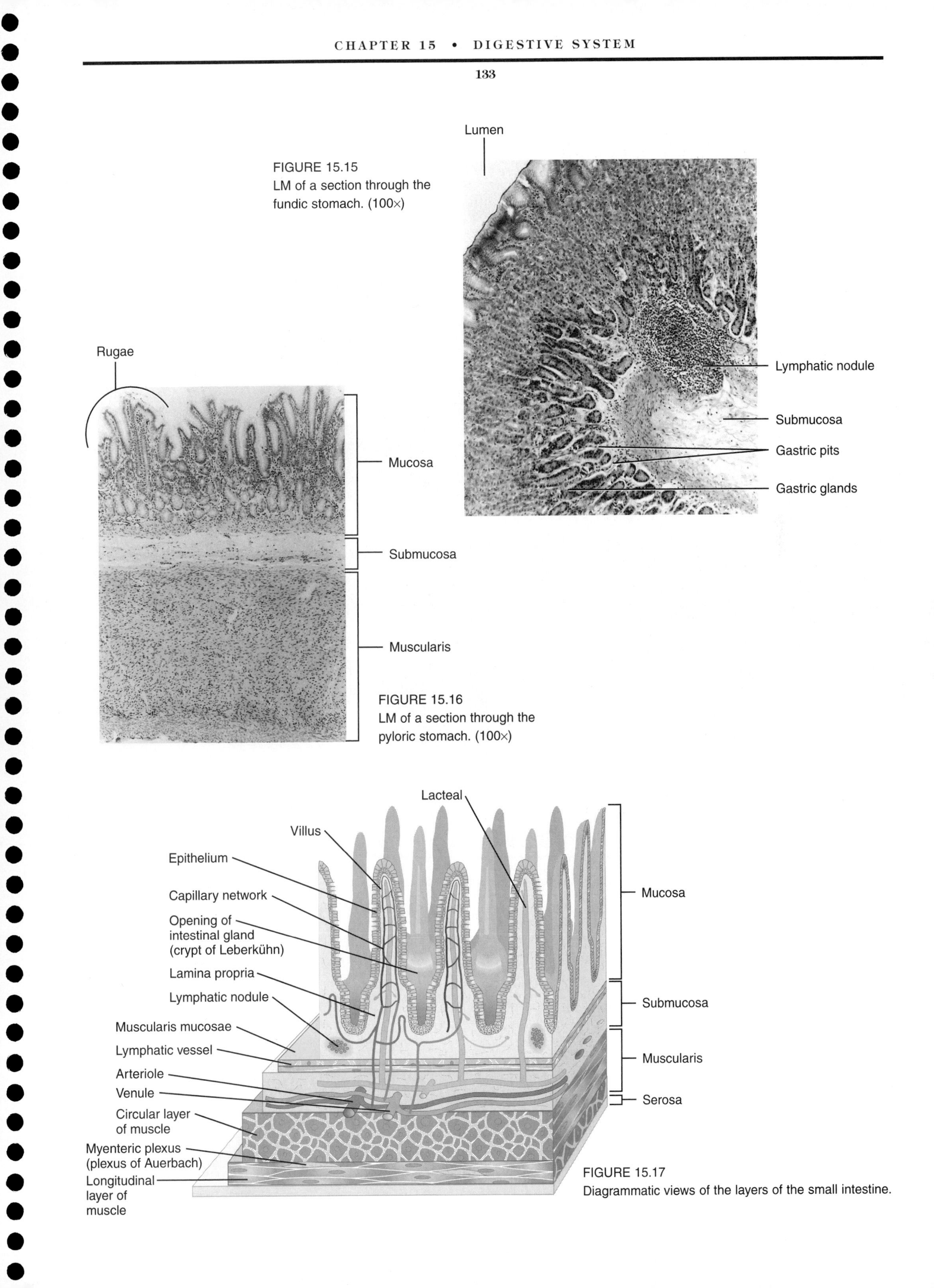

FIGURE 15.15
LM of a section through the fundic stomach. (100×)

FIGURE 15.16
LM of a section through the pyloric stomach. (100×)

FIGURE 15.17
Diagrammatic views of the layers of the small intestine.

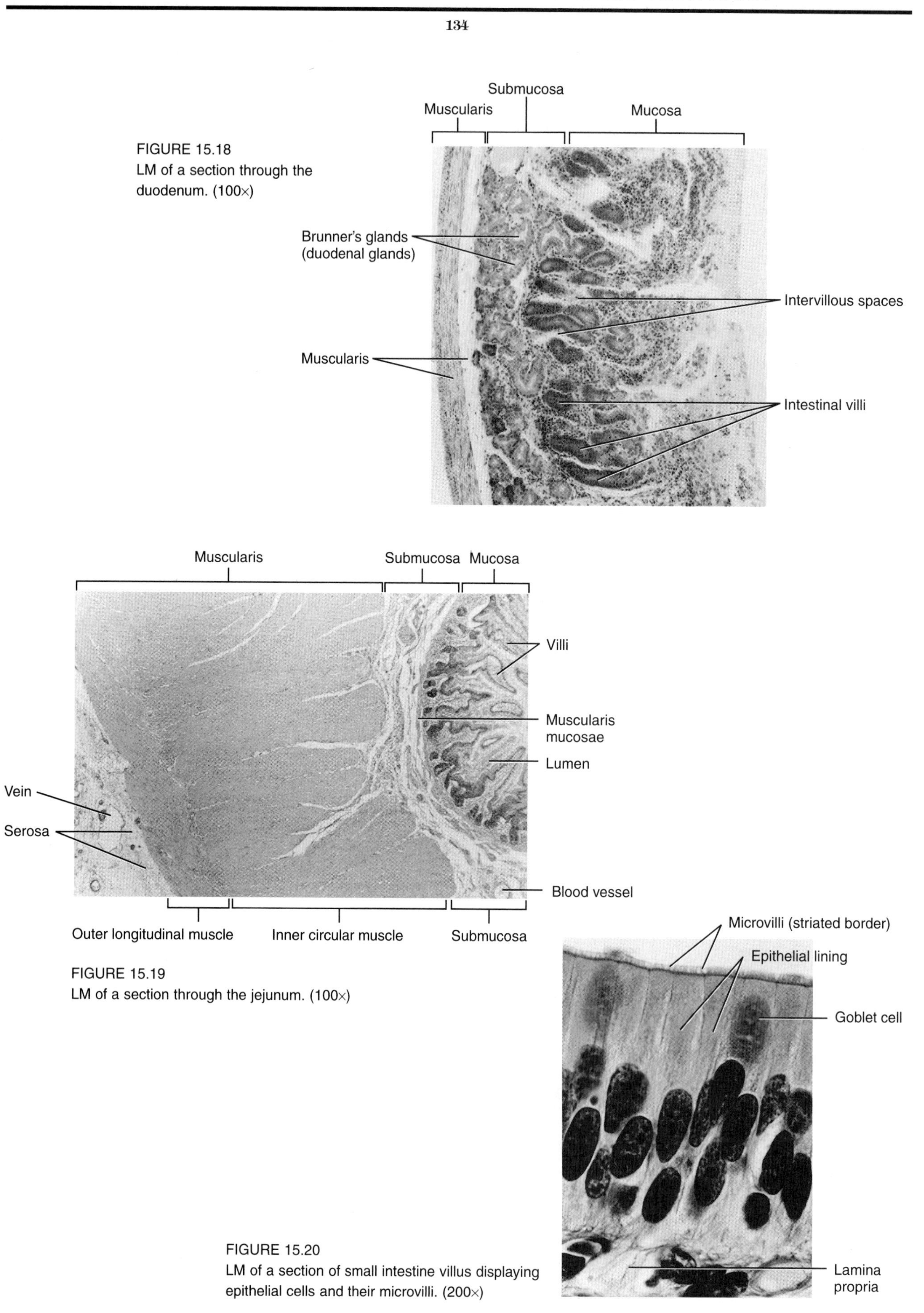

FIGURE 15.18
LM of a section through the duodenum. (100×)

FIGURE 15.19
LM of a section through the jejunum. (100×)

FIGURE 15.20
LM of a section of small intestine villus displaying epithelial cells and their microvilli. (200×)

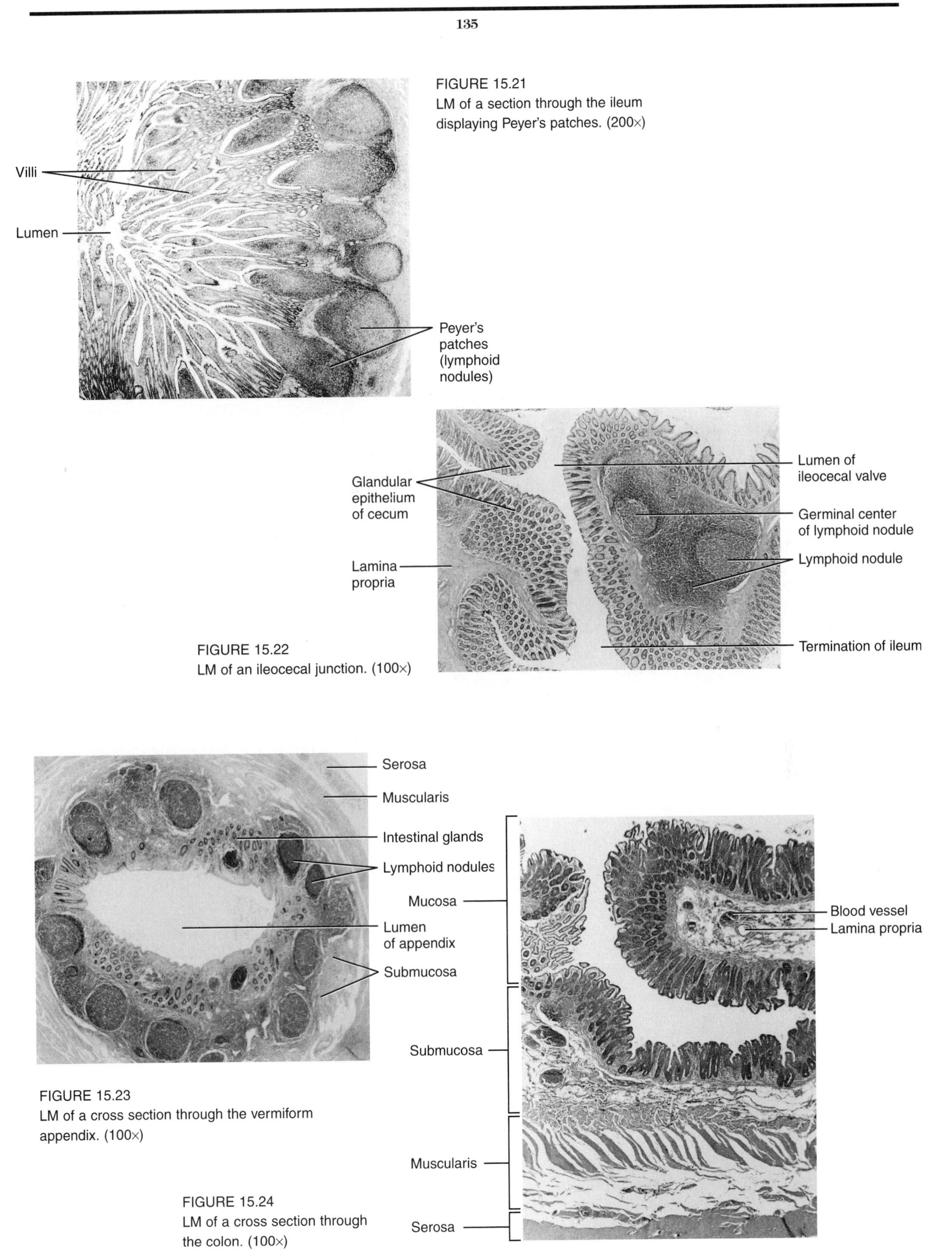

FIGURE 15.21
LM of a section through the ileum displaying Peyer's patches. (200×)

FIGURE 15.22
LM of an ileocecal junction. (100×)

FIGURE 15.23
LM of a cross section through the vermiform appendix. (100×)

FIGURE 15.24
LM of a cross section through the colon. (100×)

Goblet cells
Lamina propria
Blood vessels

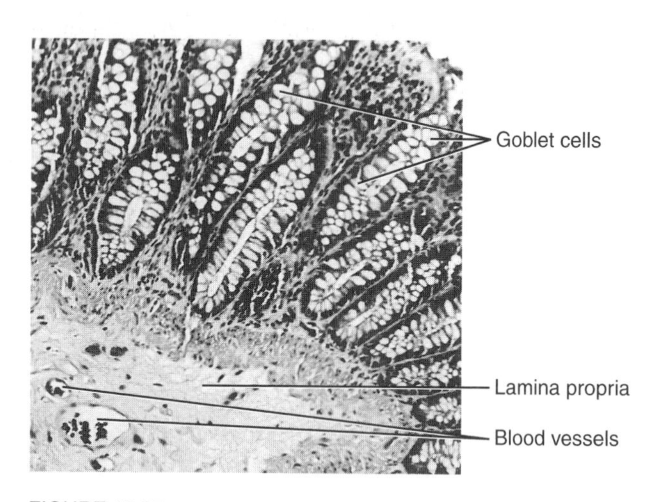

FIGURE 15.25
Cross section of the body wall of the colon at a higher magnification. (200×)

Mucosa
Submucosa
Muscularis
Serosa

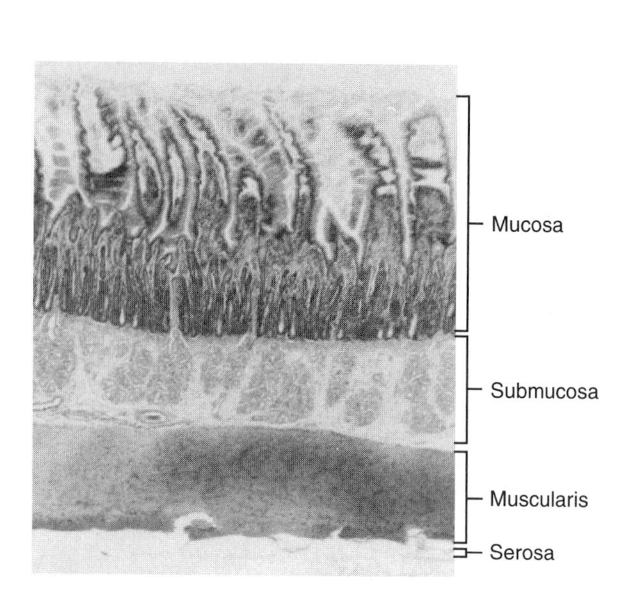

FIGURE 15.26
Cross section of the body wall of the rectum.

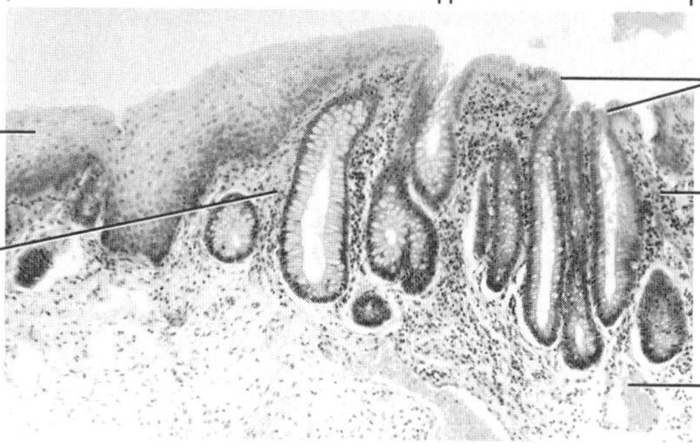

Anal canal
Rectum
Anal stratified squamous epithelium
Lamina propria of anal mucosa
Rectal epithelium
Lamina propria of rectal mucosa
Submucosa

FIGURE 15.27
Cross section of the rectoanal junction. (200×)

Liver

Lumen of central vein
Central vein
Hepatocytes
Hepatic sinusoids

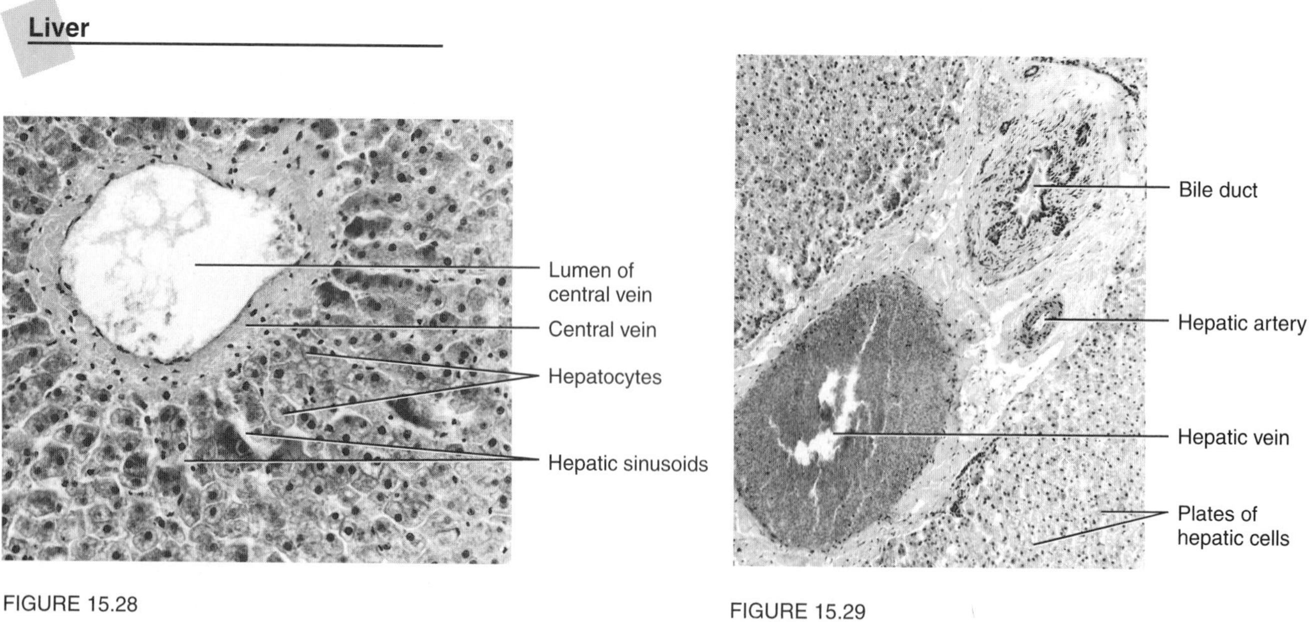

FIGURE 15.28
Liver lobule in cross section. (200×)

Bile duct
Hepatic artery
Hepatic vein
Plates of hepatic cells

FIGURE 15.29
LM of a portal tract between liver lobules. (400×)

Gallbladder

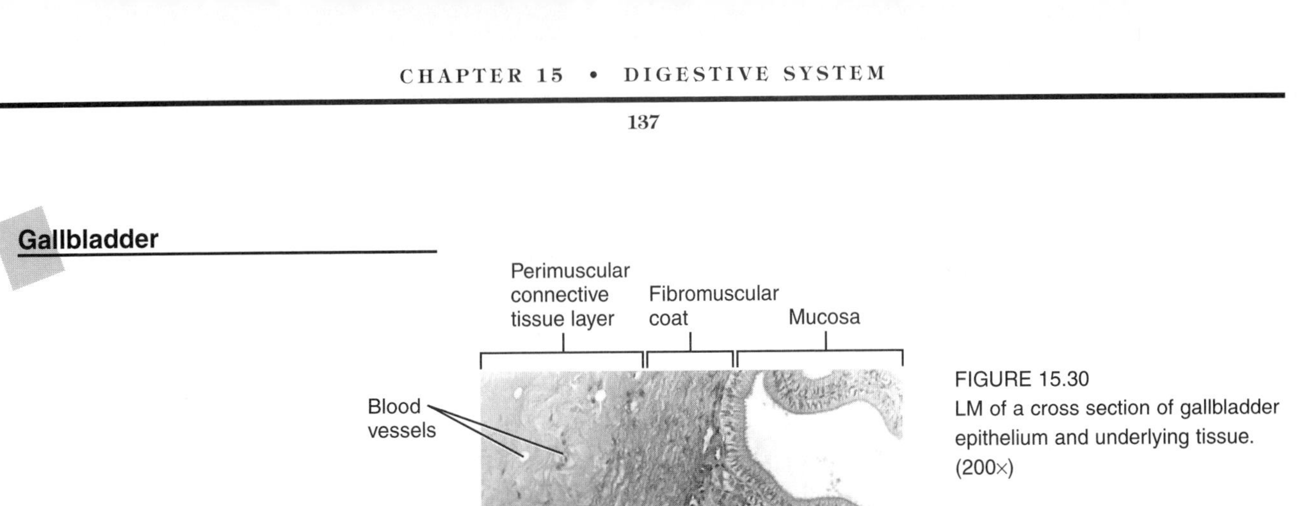

FIGURE 15.30
LM of a cross section of gallbladder epithelium and underlying tissue. (200×)

Pancreas

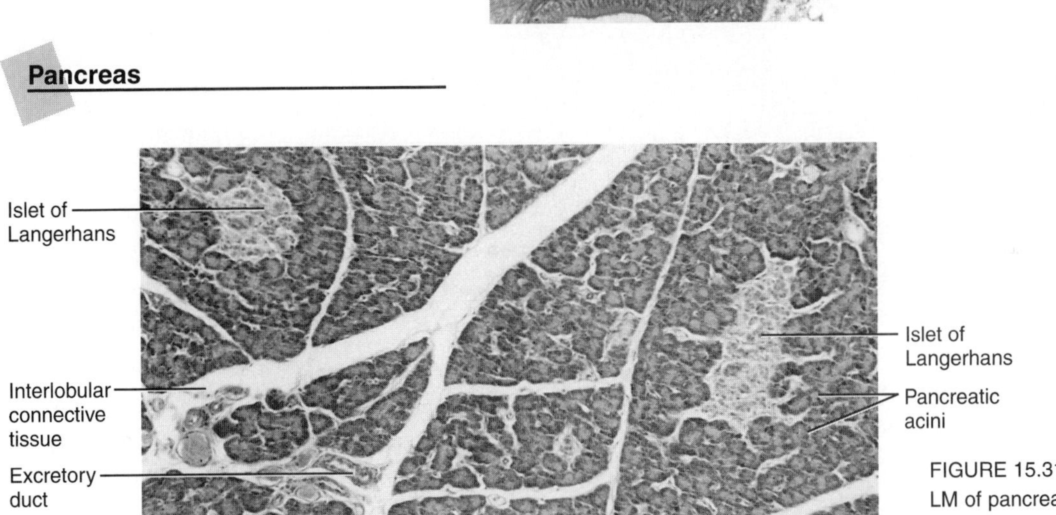

FIGURE 15.31
LM of pancreatic lobules and islets of Langerhans. (100×)

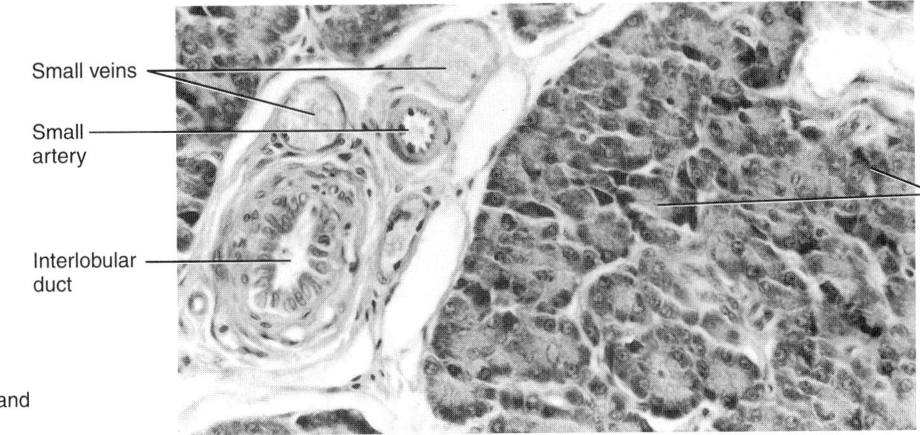

FIGURE 15.32
LM of pancreatic acini and excretory ducts. (200×)

Respiratory System

The function of the **respiratory system** is to take in oxygen and eliminate carbon dioxide from the blood. Oxygen is utilized in cellular respiration, and carbon dioxide is a by-product of cellular metabolism. This gaseous exchange occurs on two levels: **external respiration** occurs in the lungs, and **internal respiration** occurs at the cellular level between cells and the blood. The exchange of gases is a diffusion process. Oxygen diffuses or moves from a higher concentration to a lower concentration, as in alveolar air. **Carbon dioxide** diffuses from the blood to the **alveolar** air, which has a lower concentration of carbon dioxide. Morphologically, the respiratory system can be divided into two parts: (1) the upper respiratory structures, including the **interconnected cavities**—the **nasal cavity, paranasal sinuses, nasopharynx,** and **pharynx;** and (2) the lower respiratory tract, which includes the **larynx, trachea, lungs,** and associated structures of the **bronchi, bronchioles** and **alveolar ducts, alveoli,** and surrounding **blood vessels.**

The Upper Respiratory Tract: Nose and Nasal Cavity

A septum internally divides the nose into right and left **nasal cavities.** Within each cavity are three **turbinate bones** or **conchae.** The nasal cavities, including the conchae, are covered by respiratory epithelium supported by an underlying connective tissue, the lamina propria. The lamina propria is highly vascularized with thin-walled blood vessels and is called **cavernous** or **erectile** tissue. The epithelial lining of the nasal cavities is predominantly ciliated columnar cells. The **paranasal air sinuses** associated with the nasal cavities—**maxillary, frontal, ethmoidal,** and **sphenoidal**—act as resonance chambers for sound. The sinuses and the nose contain chambers for sound.

Nasopharynx The nasopharynx is located above the soft **palate** and behind the **posterior nares.** The nasopharynx opens into the large **pharyngeal cavity.** The posterior and lateral walls of the pharynx are muscular to allow for the flexibility to dilate and constrict.

The Lower Respiratory Structures

Leading into the pharynx are the two **nasopharynges, one oropharynx, one laryngopharynx,** two openings for the **auditory (Eustachian)** tubes, and one opening for the **esophagus.** Also present on the posterior wall of the pharynx is the **pharyngeal tonsil (adenoid).**

Larynx The larynx is a cartilaginous structure that connects the cavities of the **pharynx** and **trachea.** The cartilages that comprise the body wall of the larynx are the **thyroid, cricoid, corniculates, cuneiforms, arytenoids,** and **epiglottis.** The cartilages of the larynx are connected to the **hyoid bone** by three thin, flat membranes: the **thyrohyoid, quadrate,** and **cricoid.**

Trachea and Bronchi The trachea is a tube 10 to 12 cm long with about 20 spaced, horseshoe-shaped cartilaginous rings that keep the trachea from collapsing. At the lower end, the trachea bifurcates into the right and left main or **primary bronchi.** The main bronchus enters the lung and divides into smaller bronchi, the **secondary** and **tertiary bronchi.** The tertiary bronchi supply air to ten segments in each lung.

Bronchioles There is no abrupt transition in the body wall structure as tertiary bronchi differentiate into **bronchioles.** The bronchioles lack cartilage and glands in their body walls and are surrounded by only a thin **fibrous adventitia.**

Alveolar Ducts and Sacs The alveolar ducts and sacs (**alveoli**) are at the terminal ends of the respiratory tree. The alveolar ducts are narrow, thin-walled tubes lined by simple squamous epithelium. The alveolar ducts open into a single alveolus or into clusters of alveolar sacs (clusters of alveoli). The alveoli are supported by elastic and reticular fibers, **capillary plexus,** simple squamous epithelial cells, **macrophages (dust cells), septal cells,** and a basement membrane.

Blood Supply and Blood Vessels of the Lungs The lungs have a double blood supply. In pulmonary circulation, the deoxygenated blood from the **right ventricle** of the heart is pumped into the lungs by means of the **pulmonary artery.** It is oxygenated and returned to the **left auricle** of the heart by the pulmonary vein.

The second blood flow system through the lungs is maintained by the **bronchial arterial** system. Small branches of blood vessels originating from the **aorta** enter the lungs and supply oxygenated blood to the tissues of the lungs and the **pleura.**

FIGURE 16.1
Diagrammatic presentation of a sagittal section through the respiratory structures in the skull.

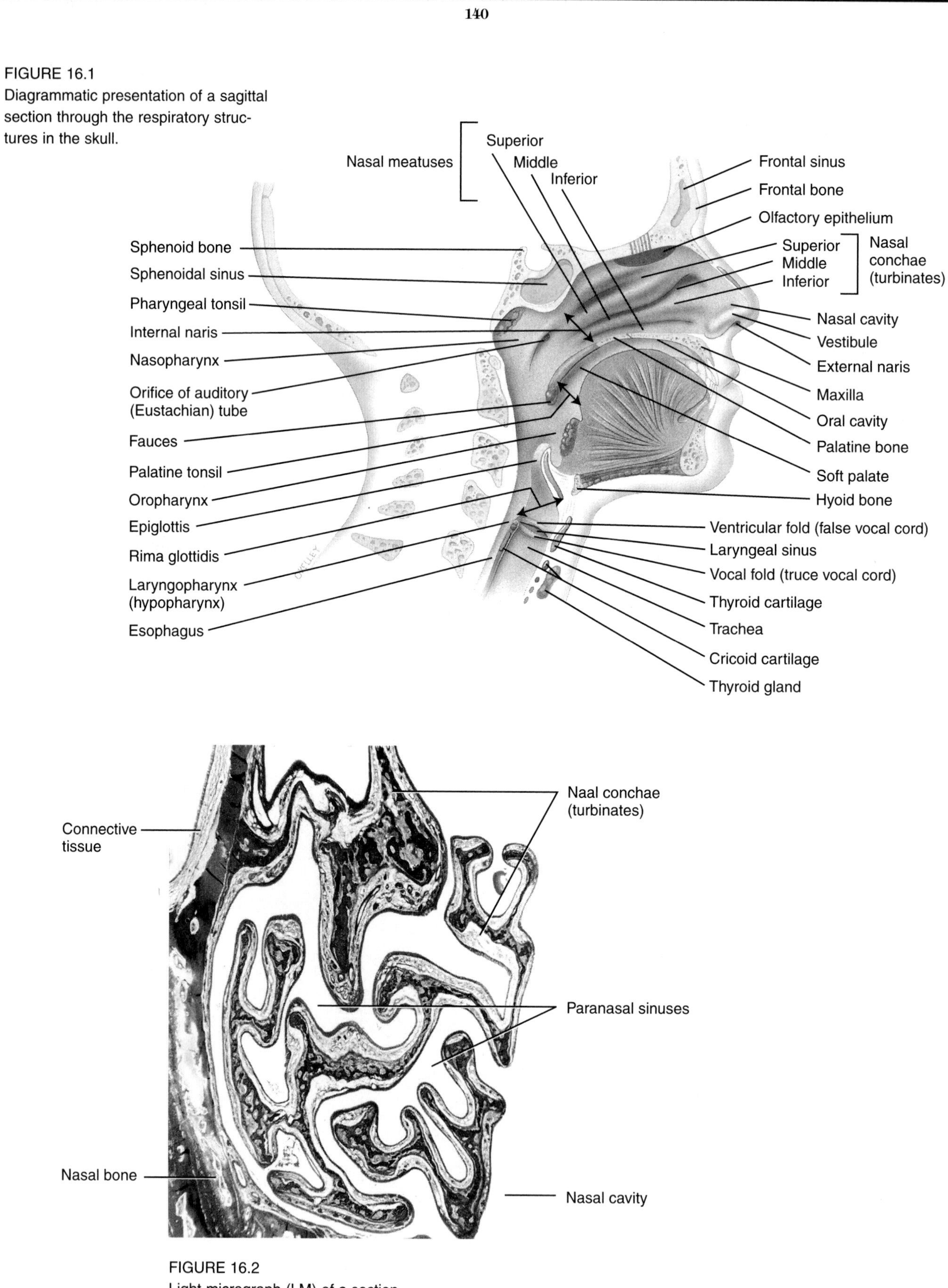

FIGURE 16.2
Light micrograph (LM) of a section through the nasal cavity. (20×).

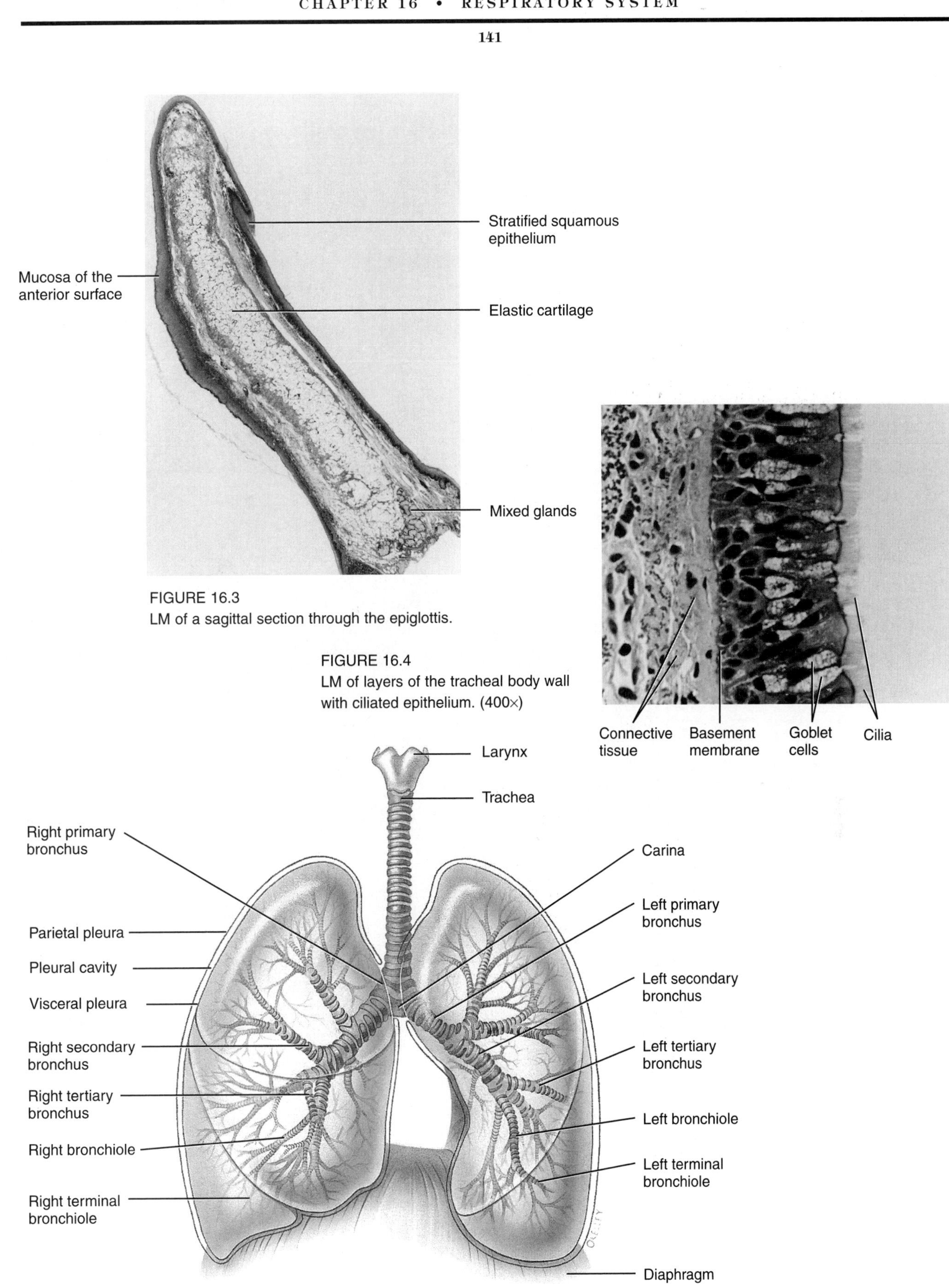

FIGURE 16.3
LM of a sagittal section through the epiglottis.

FIGURE 16.4
LM of layers of the tracheal body wall with ciliated epithelium. (400×)

FIGURE 16.5
Diagrammatic presentation of the trachea, bronchi, and lungs.

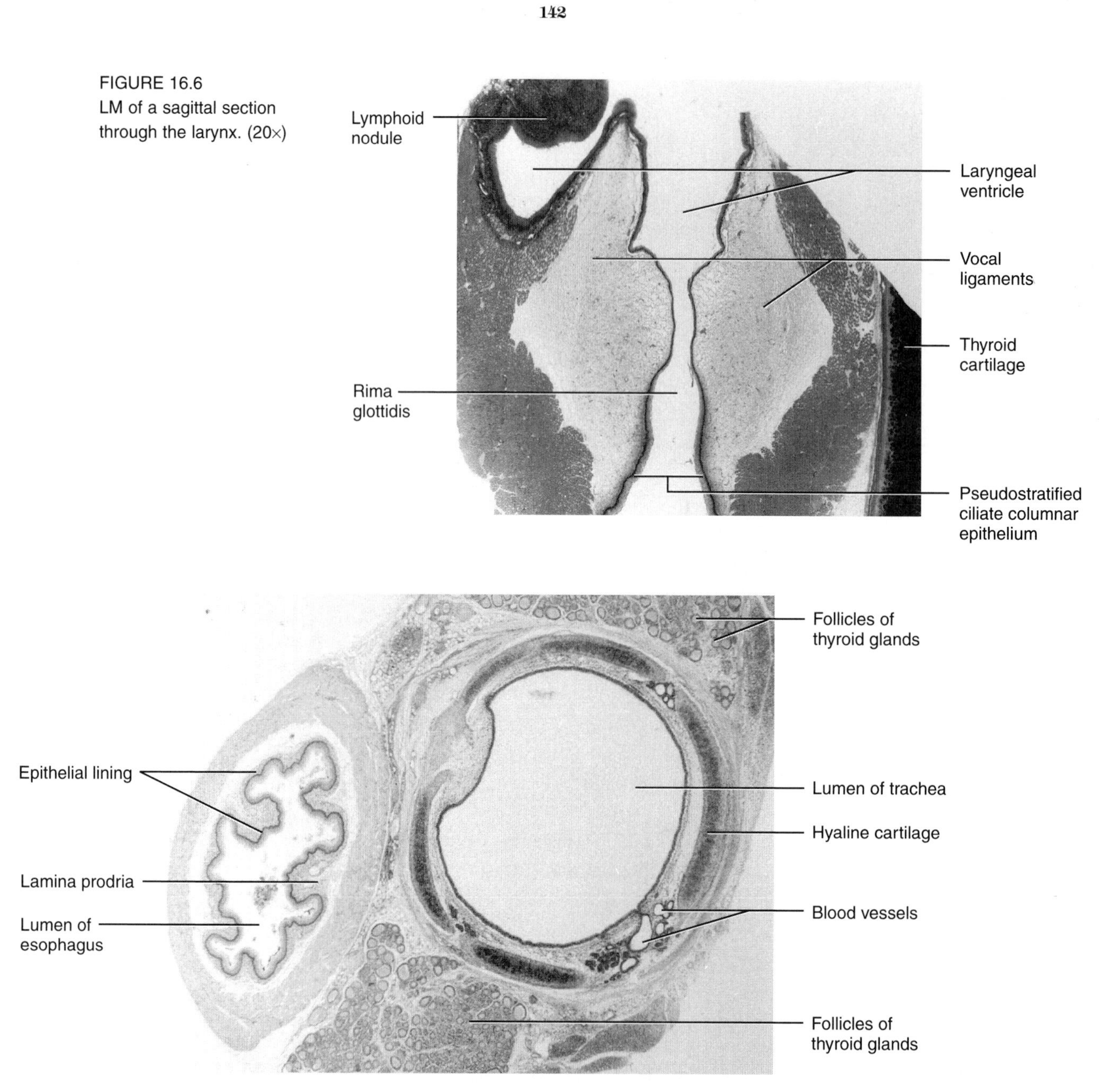

FIGURE 16.6
LM of a sagittal section through the larynx. (20×)

FIGURE 16.7
LM of the esophagus and trachea in cross section. (20×)

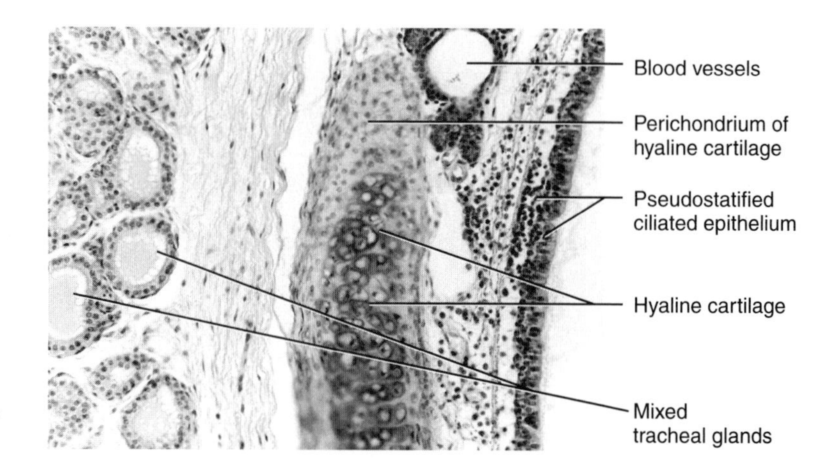

FIGURE 16.8
LM of the primary bronchus body wall in cross section. (200×)

FIGURE 16.9
LM of lung tissue. (100×)

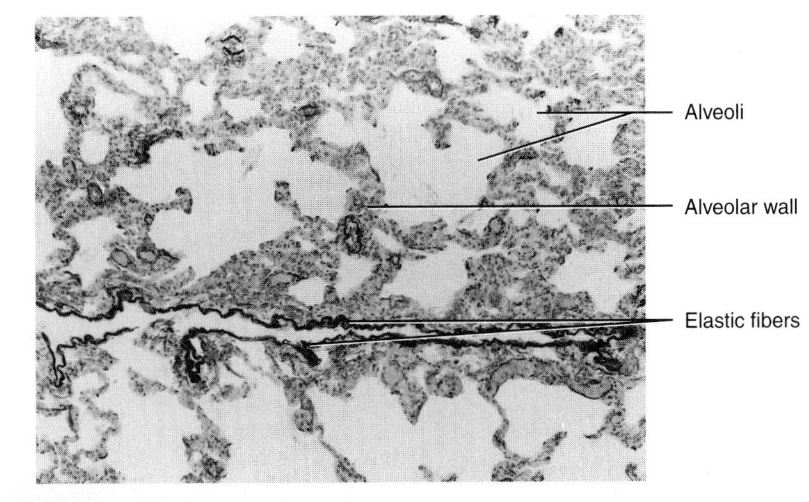

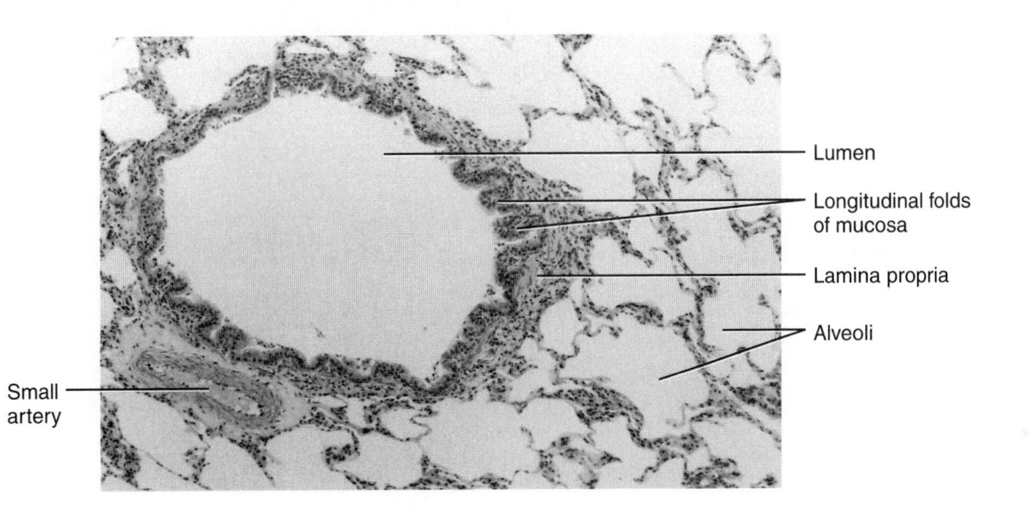

FIGURE 16.10
LM of a bronchiole in cross section. (200×)

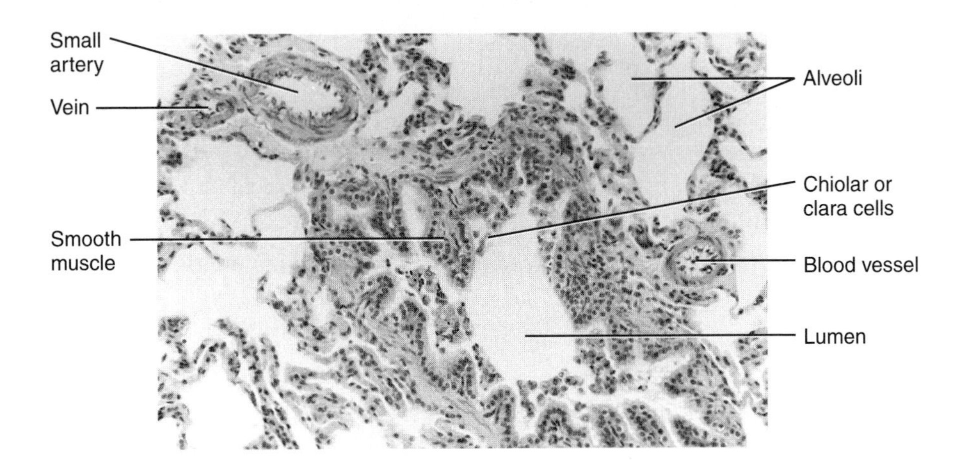

FIGURE 16.11
LM of a respiratory bronchiole in cross section. (200×)

Urinary System

Cellular metabolic waste products, especially **nitrogenous compounds,** excess **electrolytes, water,** and other **toxic** substances, are eliminated by the **urinary system.** The process of excretion and urine formation involves the filtration of blood plasma by the kidneys. The filtration within the kidneys is brought about by the filtration units called **nephrons.** The kidneys are able to function because they have an extremely efficient blood supply and a steady blood pressure. The blood pressure is maintained by **renin,** a hormone that is produced by the **juxtaglomerular cells** (**JG cells**) of the kidney. Kidneys are also essential to the secretion of **erythropoietin,** a hormone that stimulates bone marrow erythrocyte production.

The human kidneys are about 10 to 12 cm in length and 3.5 to 5 cm in width. They are enclosed in a fibrous capsule and are located in the upper posterior part of the abdomen, one on each side of the upper lumbar vertebrae. The **hilus** (**hilum**), a shallow depression on the medial aspect of the kidney, forms the entry point for the blood vessels. Exiting from the hilus is an excretory duct, the **ureter,** which transports waste products and excess water from the kidneys to the urinary bladder.

Blood Supply to the Kidneys

The **renal artery,** a branch of the abdominal aorta, supplies blood to the kidneys. The renal artery enters the kidney at the hilus and divides into **segmental arteries** that later branch into smaller **interlobar arteries.** The interlobar arteries divide into **arcuate arteries** in the **corticomedullary border** area. The arcuate arteries ascend toward the cortex and branch off into **interlobular arteries.** Close to the **Bowman's capsule** of the nephron, the interlobular arteries give rise to **afferent arterioles** of the **glomeruli.** The **efferent arterioles** receive blood from the glomeruli, which in turn pass the blood to the **vasa recta** and **peritubular** capillaries. Blood exits the kidney through the **renal vein** after passing through **interlobular, arcuate, interlobar,** and **segmental veins.**

Nephrons

There are over one million nephrons in each kidney. A nephron is a long, highly convoluted tubular structure that starts blindly as a **Bowman's capsule** and terminates by joining an **excretory duct.** The filtration of blood takes place in the Bowman's capsule. The filtrate passes into the tubule, where selective absorption begins and continues throughout the tubule. Finally, the filtrate is altered to form **urine,** which is excreted by the collecting tubules into the **renal pelvis.**

The urine is conveyed from the renal pelvis to the **ureters.** By the muscular contraction of the ureters, the urine is propelled into the **urinary bladder,** where it is stored and later excreted by the process of **micturition.**

Ureters and the Urinary Bladder

The ureters are tubular structures, approximately 25 to 30 cm in length, that connect the renal pelvis to the urinary bladder. The ureters have a well-established **mucosa** with **transitional epithelium** supported by a **basement membrane** and **lamina propria.**

The body wall of the urinary bladder is similar to the ureter in cross section. However, the **transitional epithelium** of the mucosa is thicker (six to eight layers when relaxed, and two to three layers when distended) in the bladder.

Urethra

The terminal tubular structure of the urinary system is the urethra, with marked differences between the sexes. In the male, the urethra is approximately 15 to 20 cm long. In the female, the urethra is much shorter, approximately 4 cm in length. Stratified squamous epithelium with patches of pseudostratified or stratified columnar epithelium lines the urethra.

FIGURE 17.1
Diagrammatic presentation of a nephron and associated blood vessels.

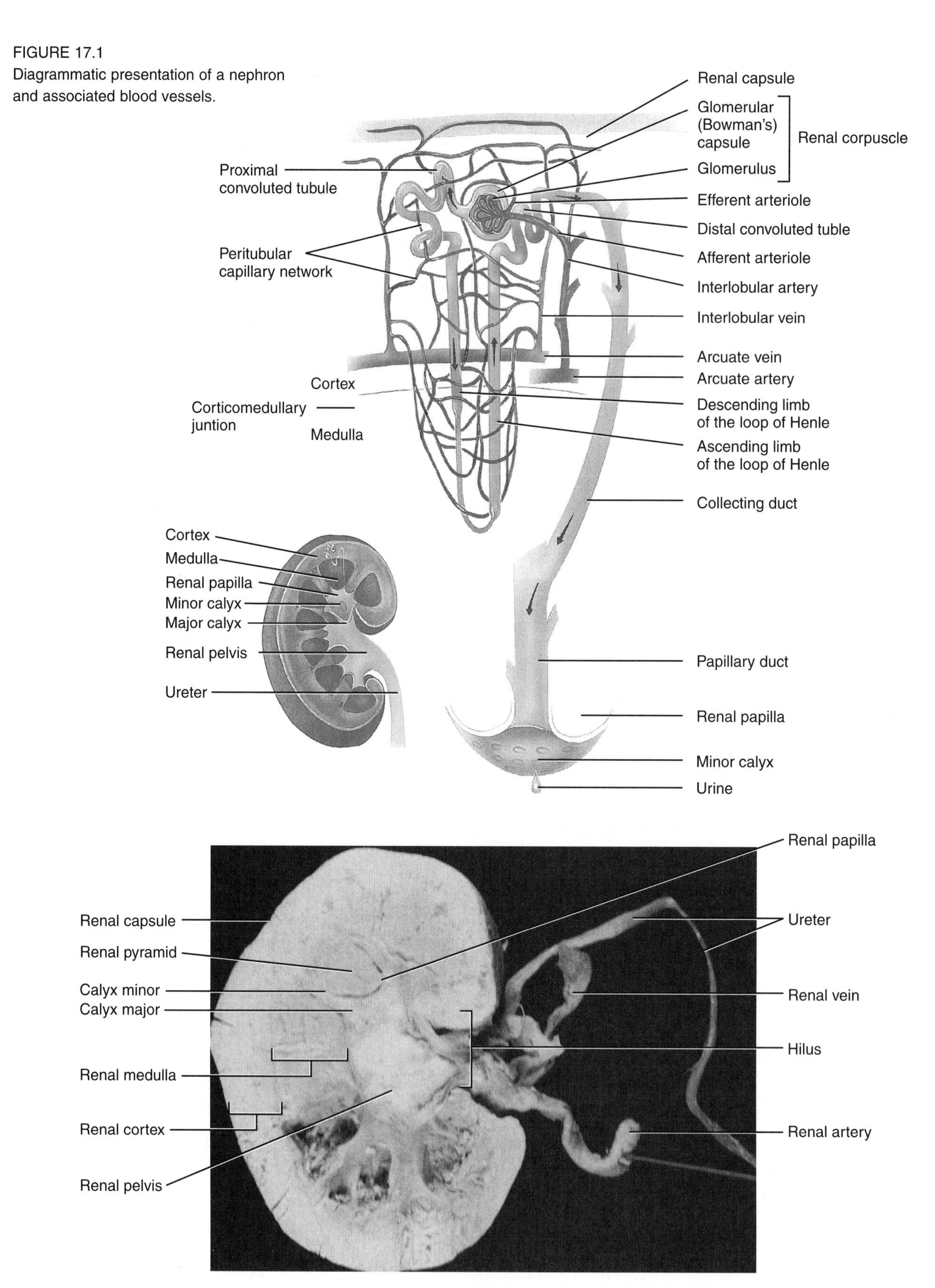

FIGURE 17.2
Photograph of a sheep kidney, sagittal section.

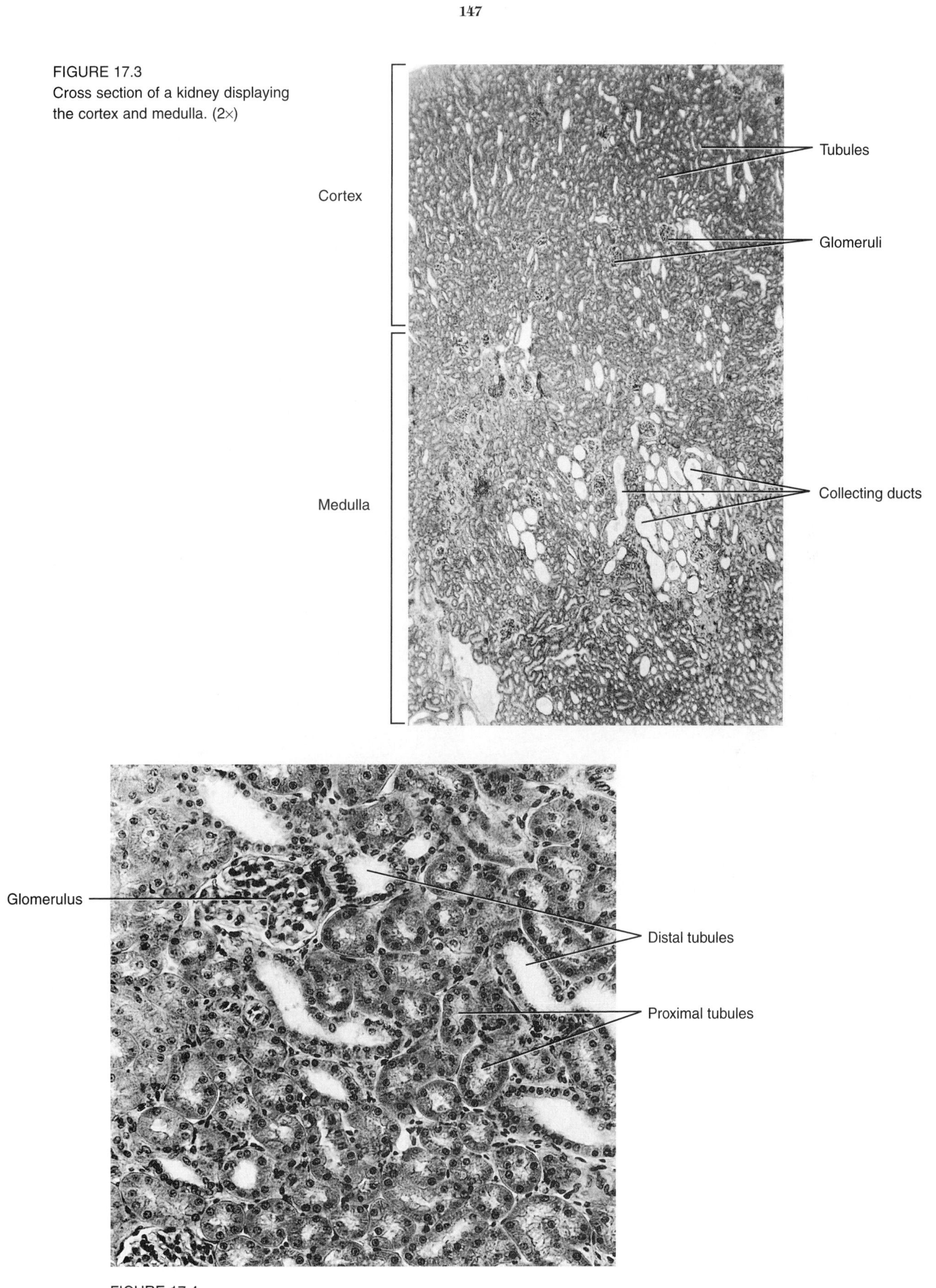

FIGURE 17.3
Cross section of a kidney displaying the cortex and medulla. (2×)

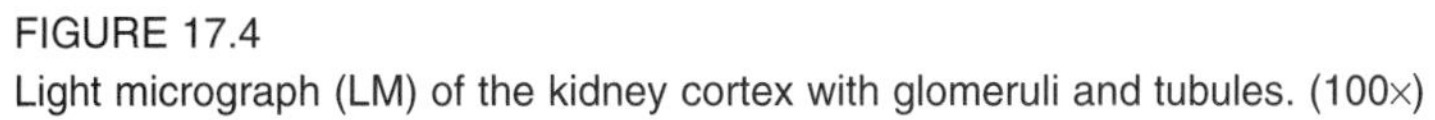

FIGURE 17.4
Light micrograph (LM) of the kidney cortex with glomeruli and tubules. (100×)

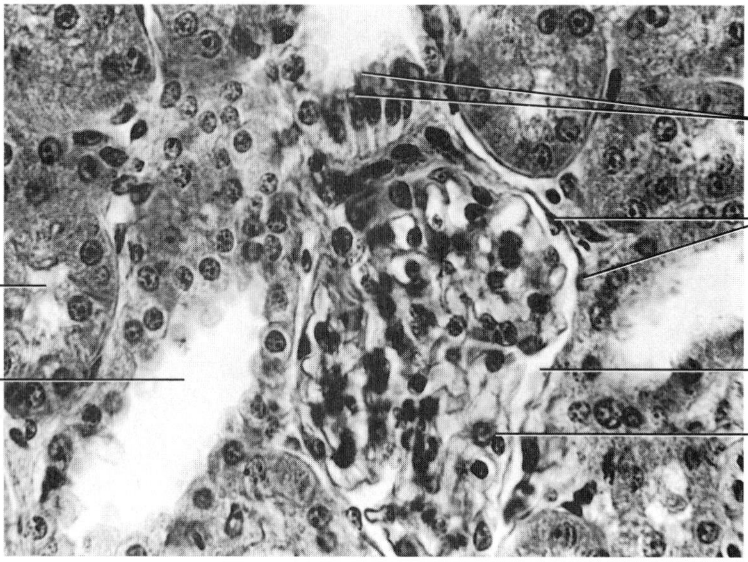

FIGURE 17.5
LM of a renal corpuscle and surrounding tubules. (200×)

Macula densa cells
Simple squamous cells
Proximal tubules
Distal tubules
Capsular space
Podocyte

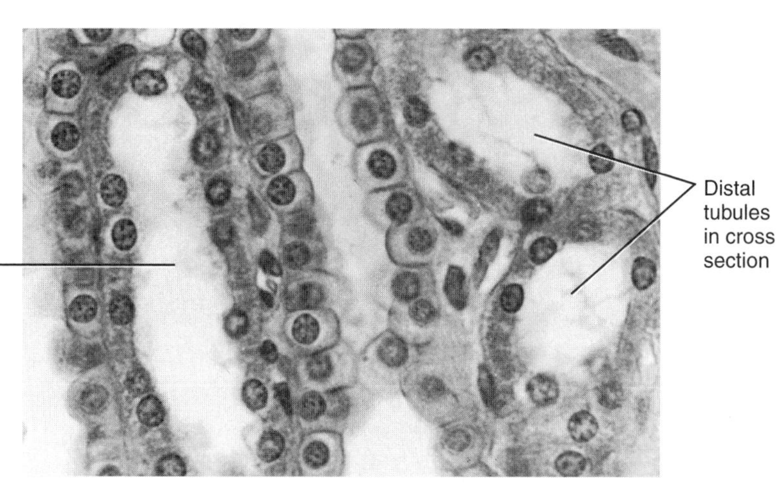

FIGURE 17.6
LM of distal tubules as seen in a longitudinal section. (200×)

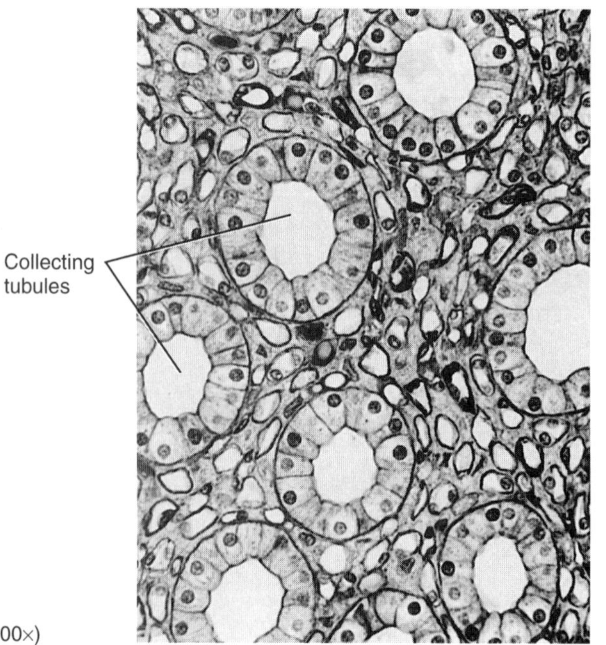

FIGURE 17.7
LM of collecting tubules in cross section. (200×)

FIGURE 17.8
LM of the collecting ducts in the cortex of the kidney.

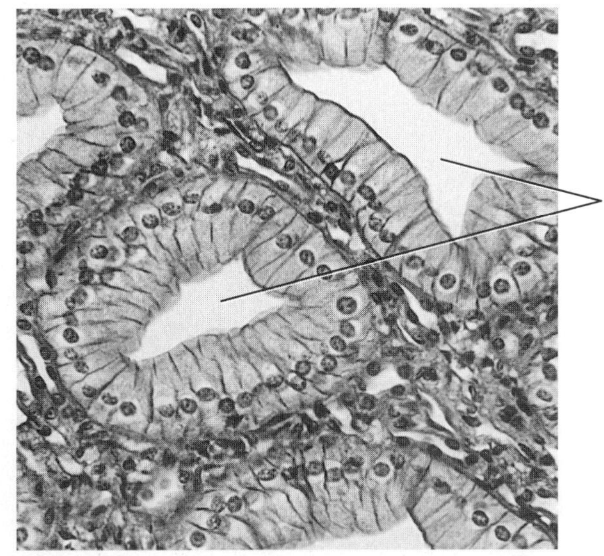

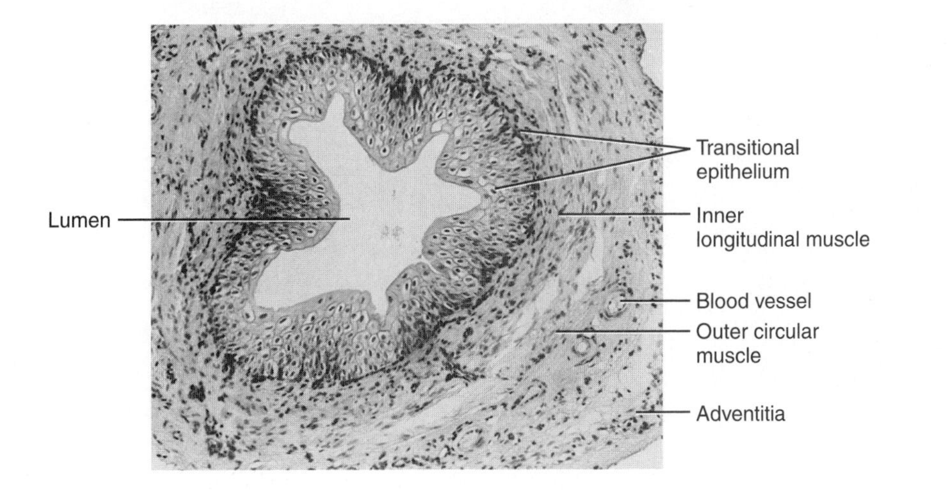

FIGURE 17.9
LM of a cross section through the ureter. (100×)

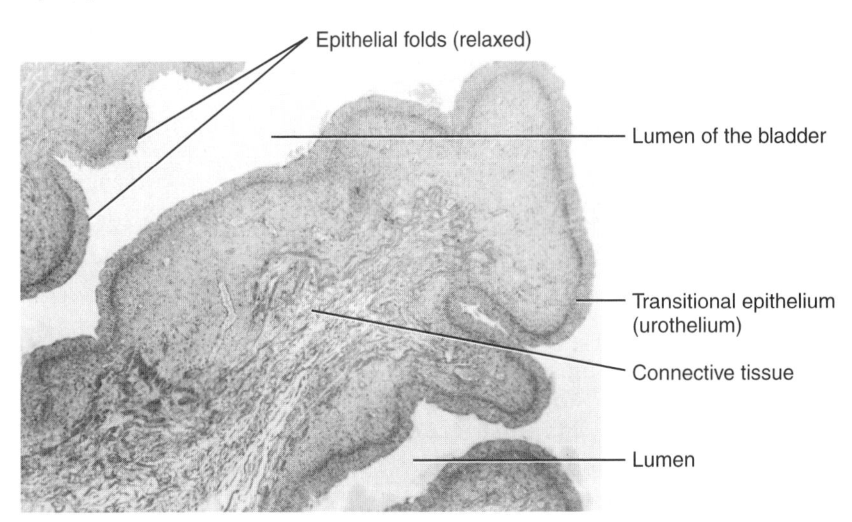

FIGURE 17.10
LM of the general structure of the urinary bladder wall. (100×)

Endocrine System

The **endocrine system** includes the ductless glands that secrete their products (hormones) directly into the blood or lymph system. The endocrine glands in general are separate entities: the **pineal, pituitary, thyroid, parathyroid,** and **adrenal.** However, scattered masses of endocrine cells may be present in the tissue of the exocrine glands: for example, the **islets of Langerhans** in the **pancreas,** the **Leydig cells** in the testis interstitium, and the **corpora lutea** in the **ovaries.** These combinations of organs and glands are classified as **mixed glands.**

The endocrine glands are simple glands with glandular cells surrounded by connective tissue and an elaborate system of **fenestrated** blood capillaries that course through the endocrine tissue. Embryologically, the endocrine glands are derivatives of all three germinal layers:

1. The **pituitary** or **hypophysis, adrenal medulla,** and **chromaffin bodies** are of **ectodermal** origin.

2. The testes, **ovaries,** and **adrenal cortex** are derivatives of the **mesoderm.**

3. The **parathyroid, thyroid,** and **islets of Langerhans** are derived from the **endoderm.**

Chemically, the hormones may be **cholesterol** derivatives (**steroids**), **amino acids, proteins, glycoproteins,** or **peptides.** Because some of the hormones are lipid derivatives, they can diffuse through the cell membrane into the cell and bind with their receptor molecules in the cytoplasm or the nucleus. Hormones that are not lipid soluble bind to their respective receptor molecules on the membrane of target cells and bring about their hormonal action. A good example of such an interaction is the binding of a protein or peptide hormone (**first messenger**) to its receptor on the membrane, activating adenylate cyclase, an enzyme that converts **adenosine triphosphate** (**ATP**) into cyclic adenosine monophosphate (**cAMP**). The cAMP acts as a second messenger in the cytoplasm and mediates hormonal action in the cell.

FIGURE 18.1
Diagram showing the location of endocrine glands in the body.

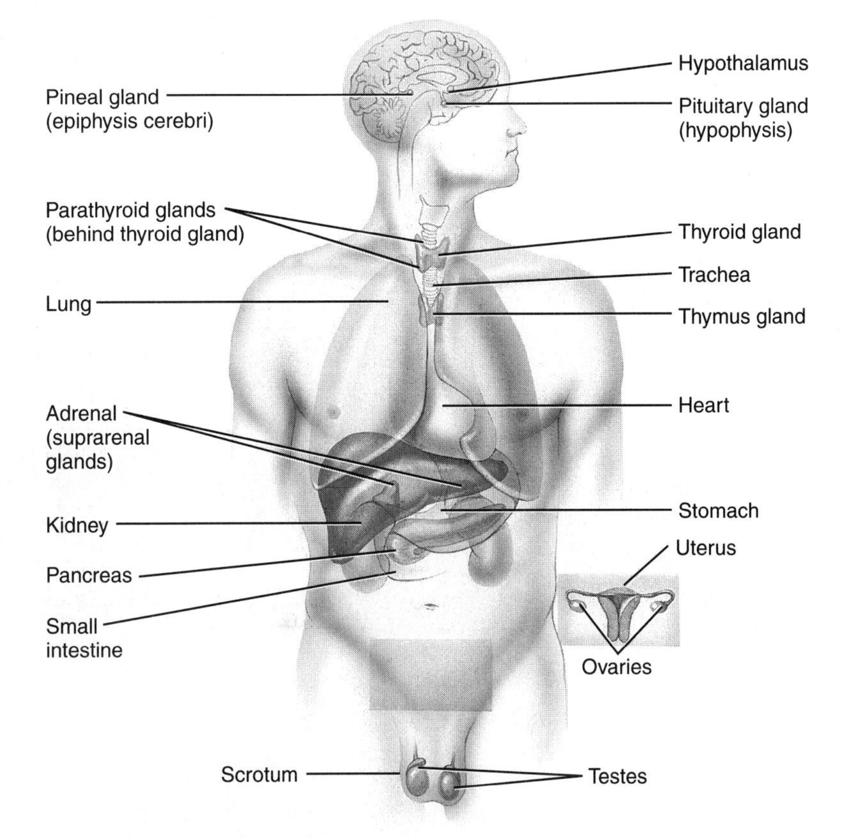

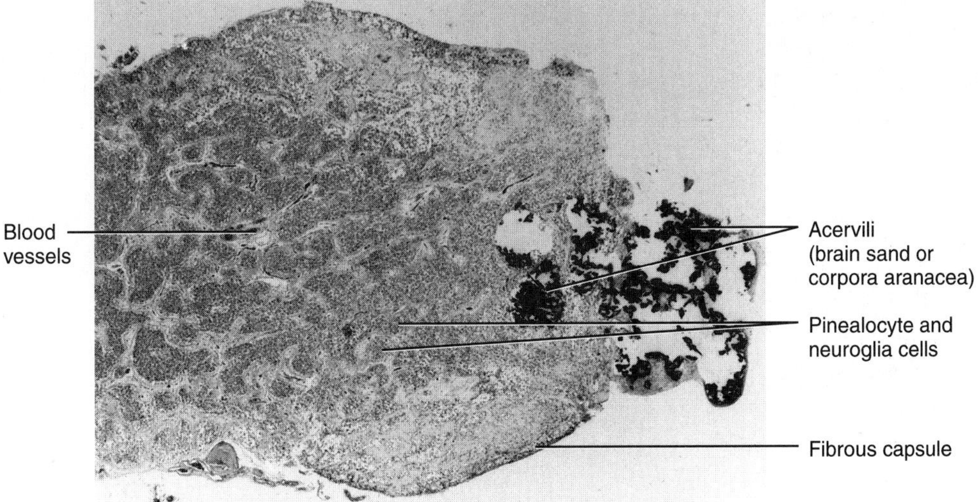

FIGURE 18.2
Light micrograph (LM) of a pineal gland displaying calcium and magnesium deposits, corpora aranacea (brain sand). (100×)

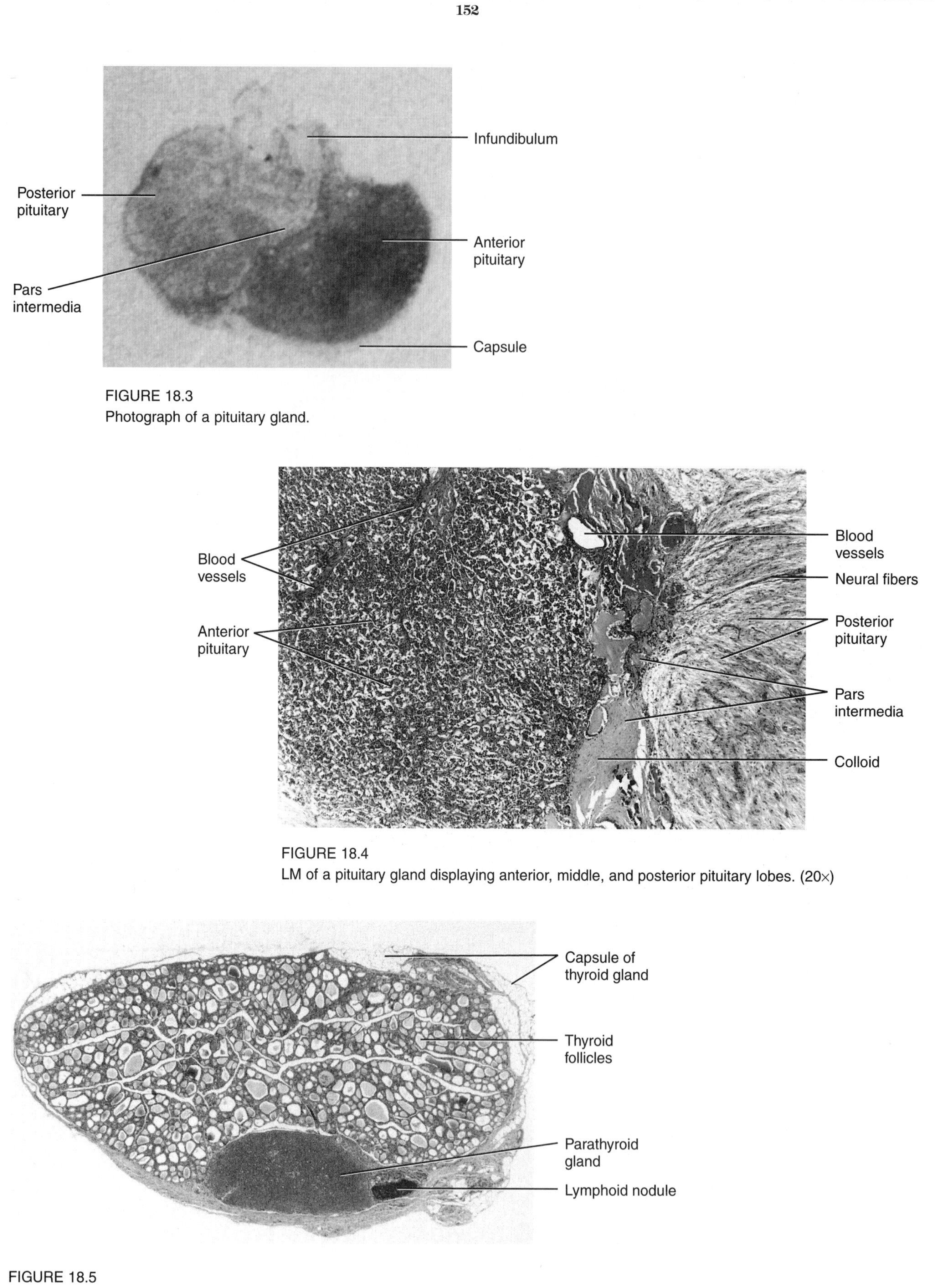

FIGURE 18.3
Photograph of a pituitary gland.

FIGURE 18.4
LM of a pituitary gland displaying anterior, middle, and posterior pituitary lobes. (20×)

FIGURE 18.5
LM of thyroid and parathyroid glands. (20×)

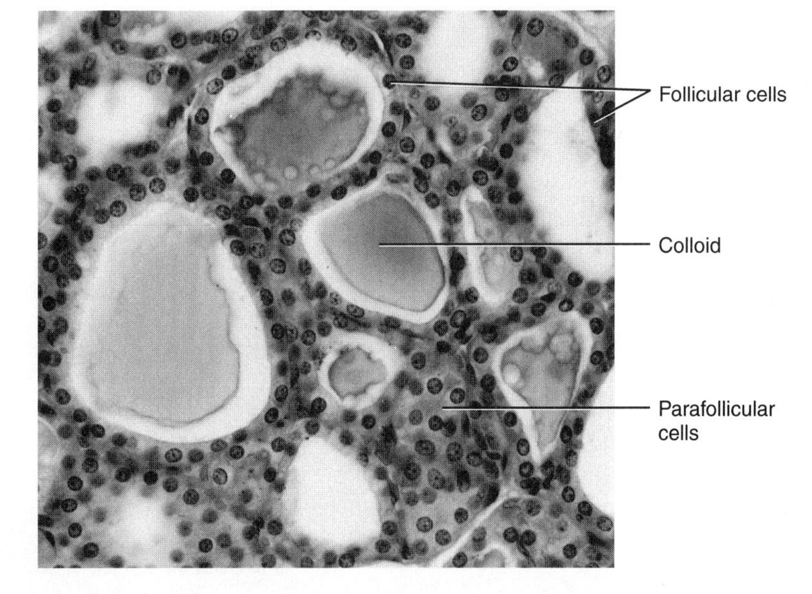

FIGURE 18.6
LM of a thyroid gland. (200×)

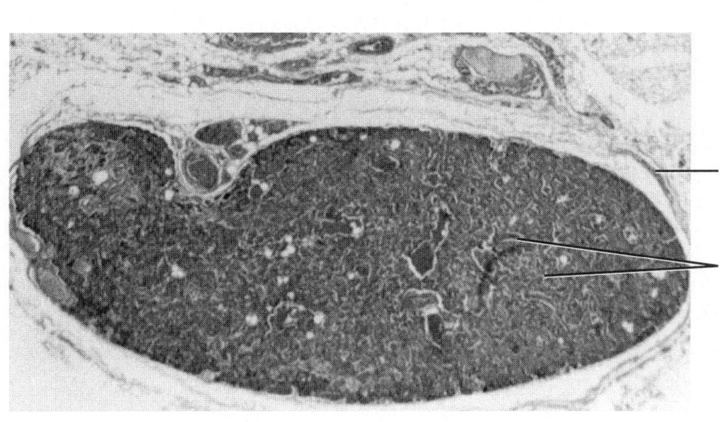

FIGURE 18.7
LM of a parathyroid gland. (20×)

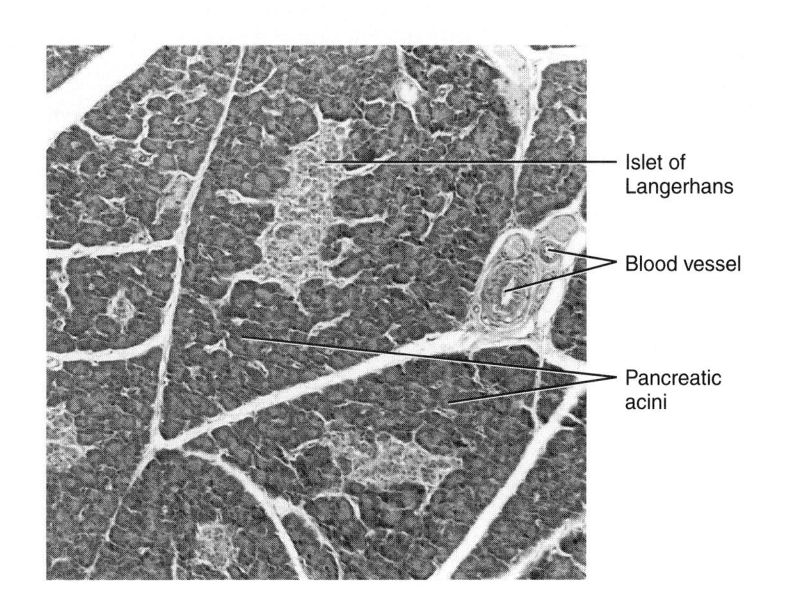

FIGURE 18.8
LM of a pancreas displaying pancreatic acini and islets of Langerhans. (100×)

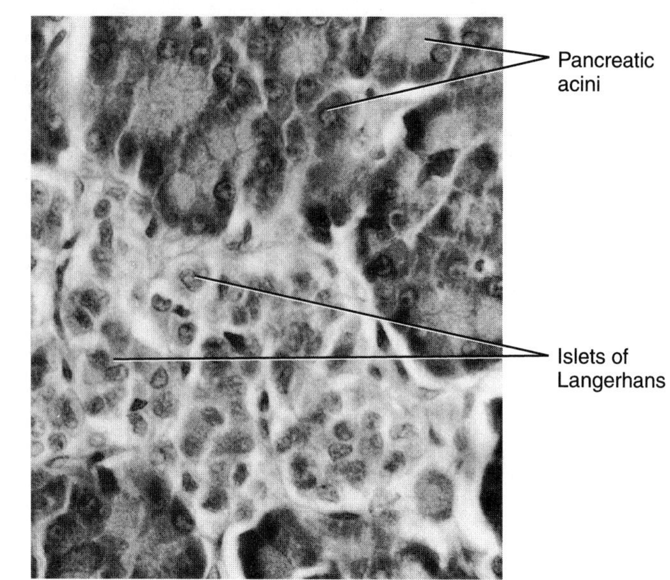

FIGURE 18.9
LM of islets of Langerhans. (400×)

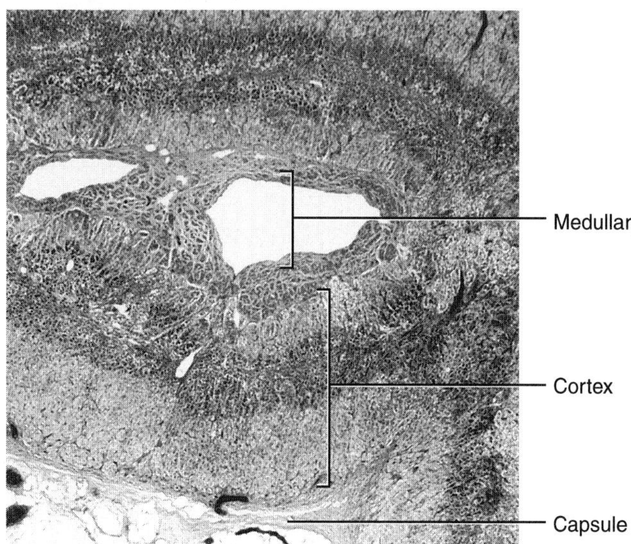

FIGURE 18.10
LM of the adrenal gland. (20×)

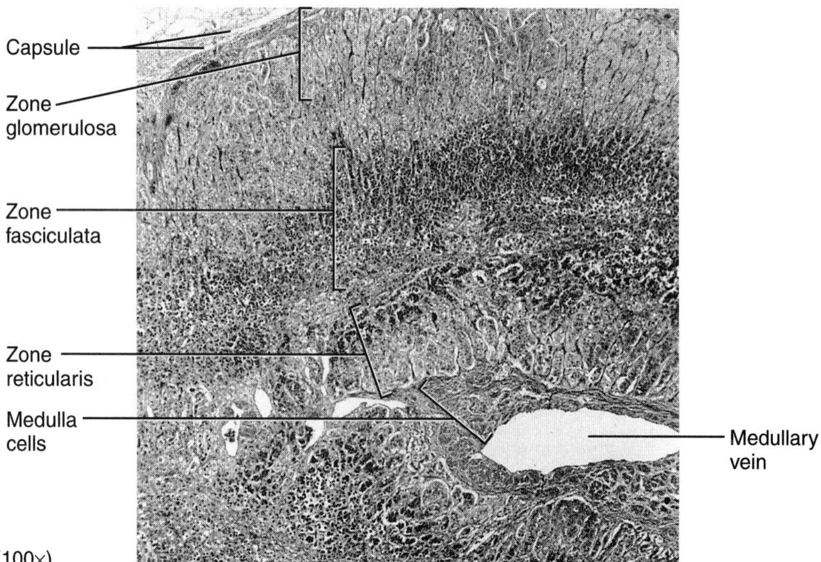

FIGURE 18.11
LM of the adrenal gland at a higher magnification. (100×)

CHAPTER 19

Male Reproductive System

The **male reproductive system** consists of the **testes,** which produce **spermatozoa** and **androgens;** the glands that produce the fluid that facilitates the transfer of sperm; a series of ducts and passageways for the transport of sperm and fluids; and a copulatory organ, the **penis,** which delivers the spermatozoa to the female reproductive tract.

The organs and tubular passageways of the male reproductive system are the **testes, seminiferous tubules, rete testis, ductus epididymis, vas deferens, ampulla of the ductus (vas) deferens, seminal vesicles, ejaculatory duct, corpora cavernosum urethrae, penile urethra, prostate,** and **bulbourethral glands.**

Testes **Spermatozoa** are produced through the process of spermatogenesis, which occurs in the testes. The testes are also the site of the production of androgens, the male sex hormones. Two testes are suspended in a testicular sac, the **scrotum.** The testes are surrounded by a **testicular capsule** composed of a **tunica vaginalis** (outer layer), **tunica albuginea** (middle layer), and **tunica vasculosa** (innermost layer). Internally, each testis is partitioned into approximately 250 pyramid-shaped compartments, the **lobuli testis,** which are partially separated by

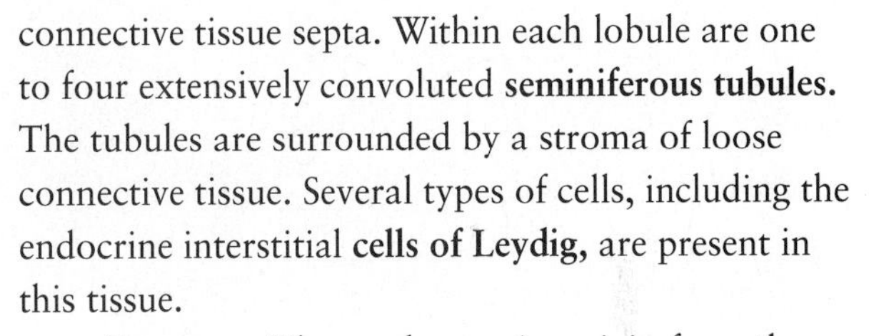

connective tissue septa. Within each lobule are one to four extensively convoluted **seminiferous tubules.** The tubules are surrounded by a stroma of loose connective tissue. Several types of cells, including the endocrine interstitial **cells of Leydig,** are present in this tissue.

Prostate The **urethra,** at its origin from the **urinary bladder,** is surrounded by the prostate gland. Approximately 30 to 50 compound **tuboalveolar** glands make up the prostate. The prostate gland is surrounded by a highly vascularized **fibroelastic capsule** with smooth muscle cells. Prostate secretion is a thin, milky white fluid that is rich in acid phosphates, proteolytic enzymes, and fibrinolysin, which liquifies the semen. **Corpora amylacea,** if present in the prostate, are calcified prostatic secretory concentrations.

Bulbourethral Glands (Glands of Cowper) The bulbourethral glands, or glands of Cowper, are small, paired tuboalveolar bodies in the connective tissue, behind the membranous urethra. The bulbourethral glands secrete a clear viscid mucus that facilitates sperm movement and transportation.

Penis The penis serves as a copulatory organ and a common outlet for seminal fluid and urine. The penis is composed of three cylinders of erectile tissue: two **corpora cavernosa penis** and a single **corpus cavernosum urethrae** or **corpus spongiosum.** The penile urethra is surrounded by corpus spongiosum. The three erectile tissues are surrounded by subcutaneous connective tissue that is rich in smooth muscle and elastic fibers.

The thin overlying skin of the penis is firmly attached to the underlying connective tissue. Distally, the skin folds over the **glans penis** to form an inverted covering, the prepuce. Proximally, small sweat glands are present in the skin with occasional sebaceous glands that are associated with hair follicles.

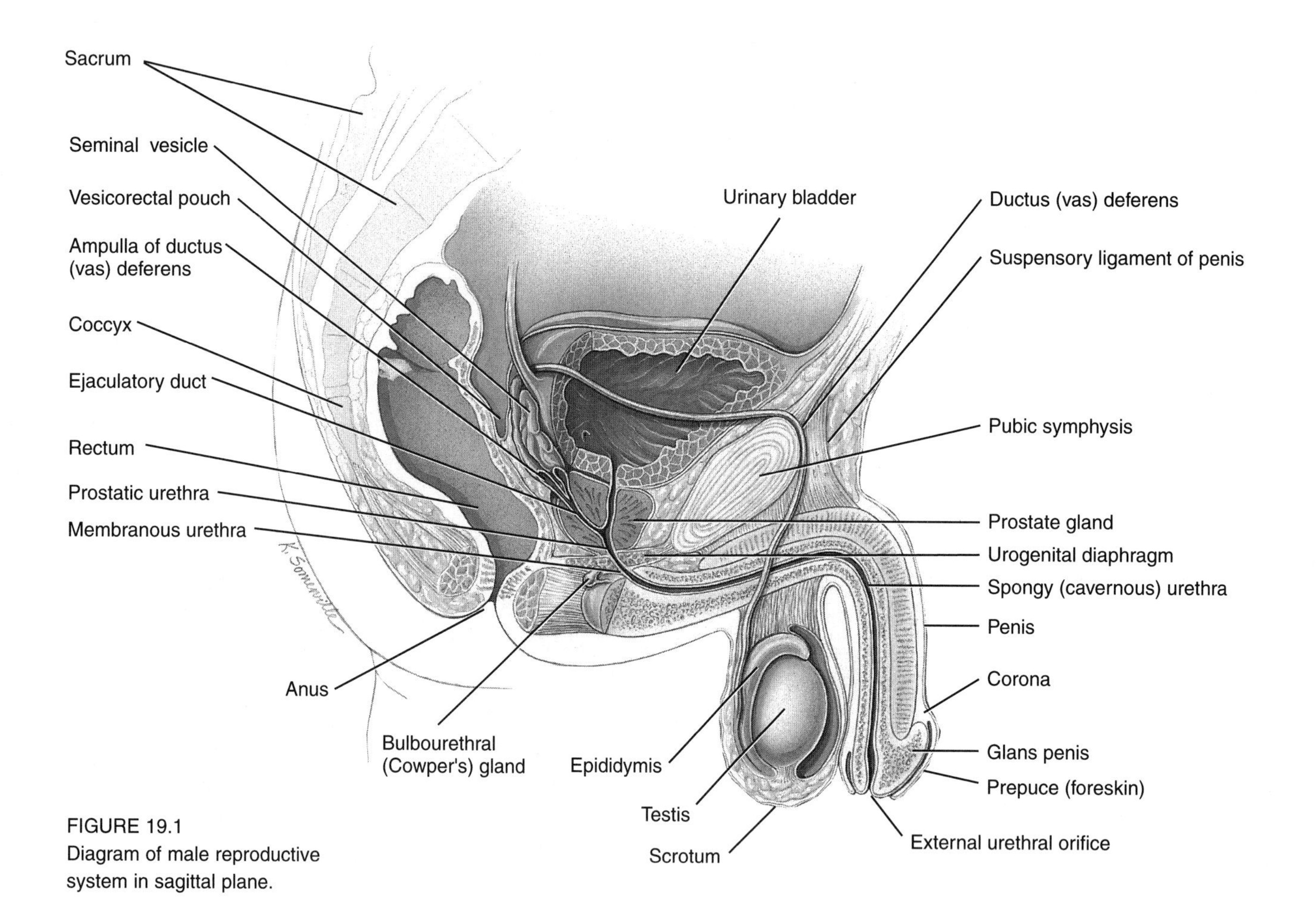

FIGURE 19.1
Diagram of male reproductive system in sagittal plane.

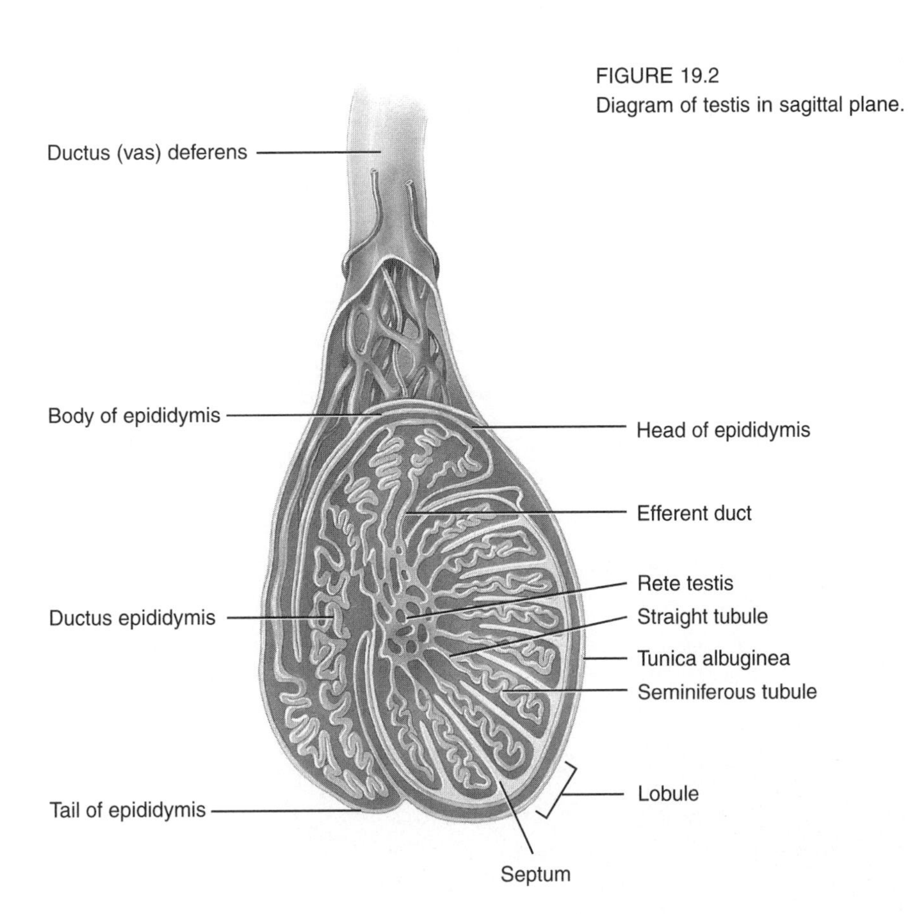

FIGURE 19.2
Diagram of testis in sagittal plane.

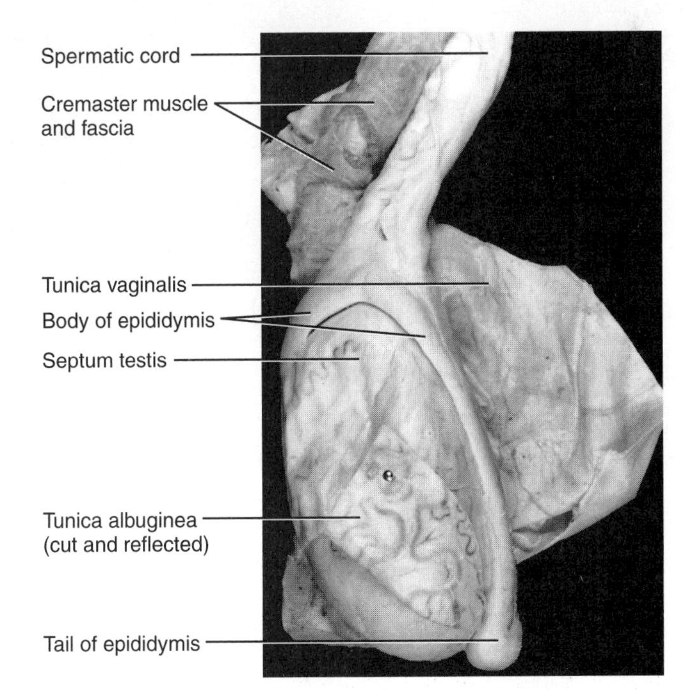

FIGURE 19.3
Bull testis and supporting structures.

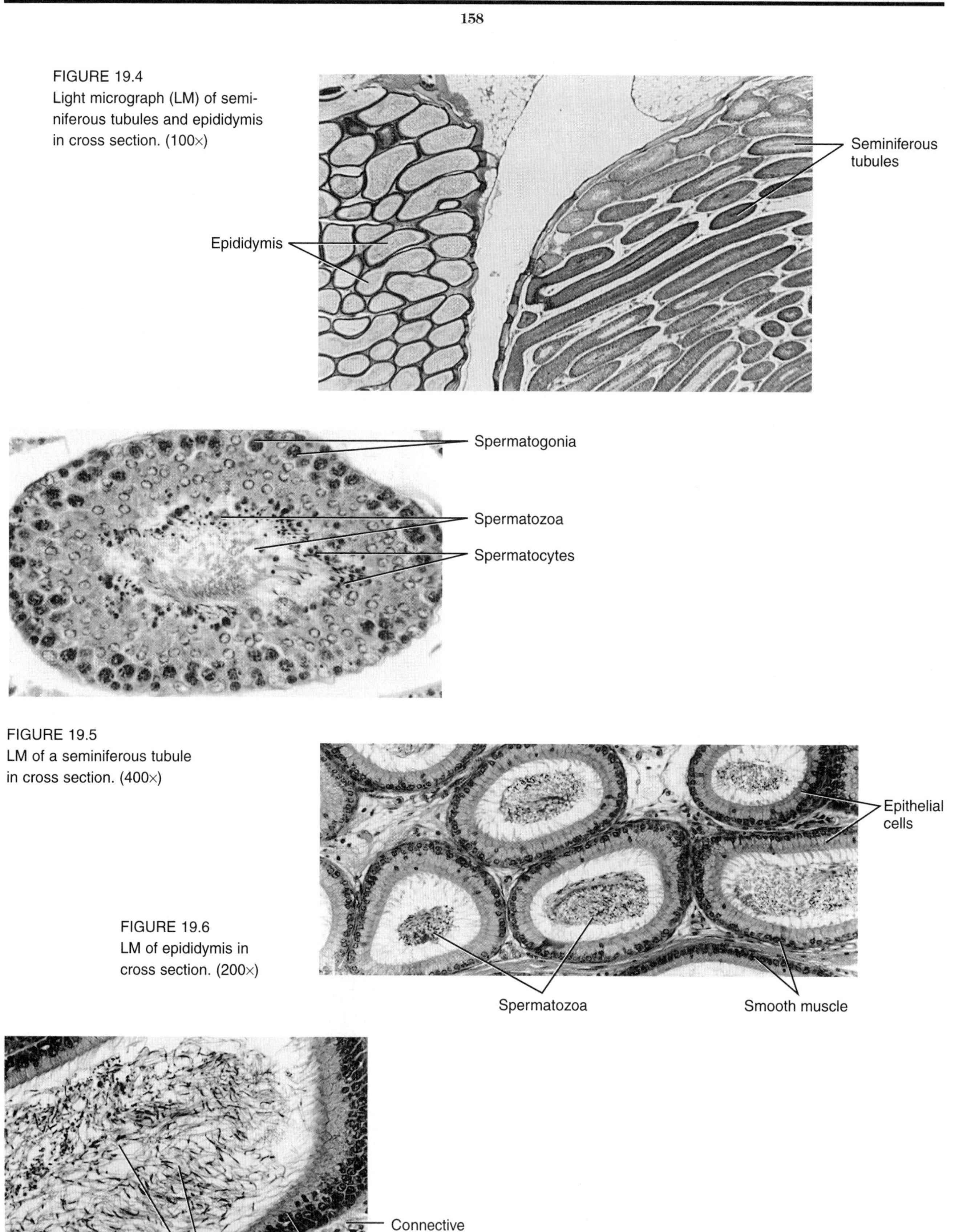

FIGURE 19.4
Light micrograph (LM) of seminiferous tubules and epididymis in cross section. (100×)

FIGURE 19.5
LM of a seminiferous tubule in cross section. (400×)

FIGURE 19.6
LM of epididymis in cross section. (200×)

FIGURE 19.7
LM of epididymis in section, displaying large number of spermatozoa in the lumen. (400×)

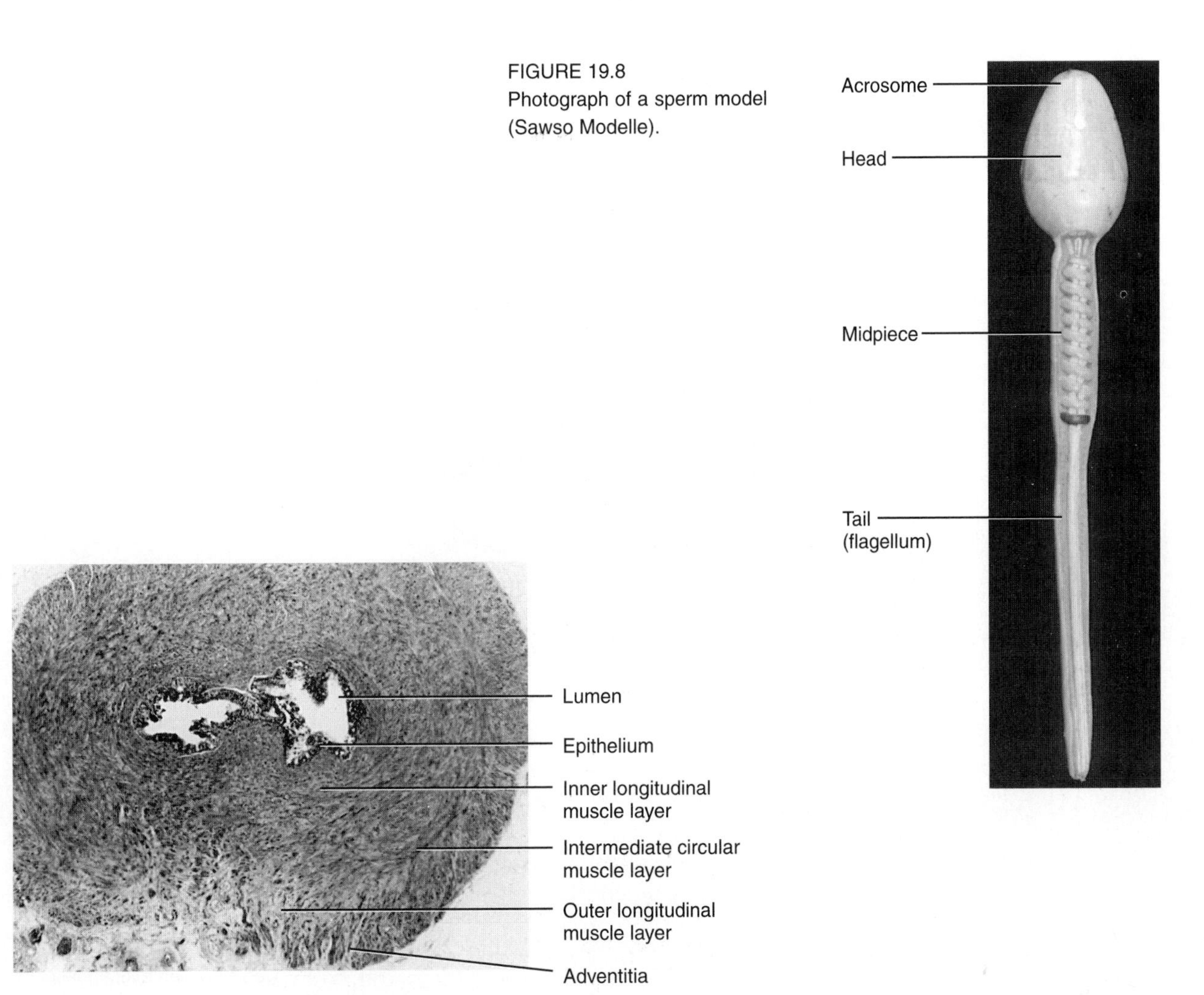

FIGURE 19.8
Photograph of a sperm model (Sawso Modelle).

FIGURE 19.9
LM of a cross section of the ductus (vas) deferens. (40×)

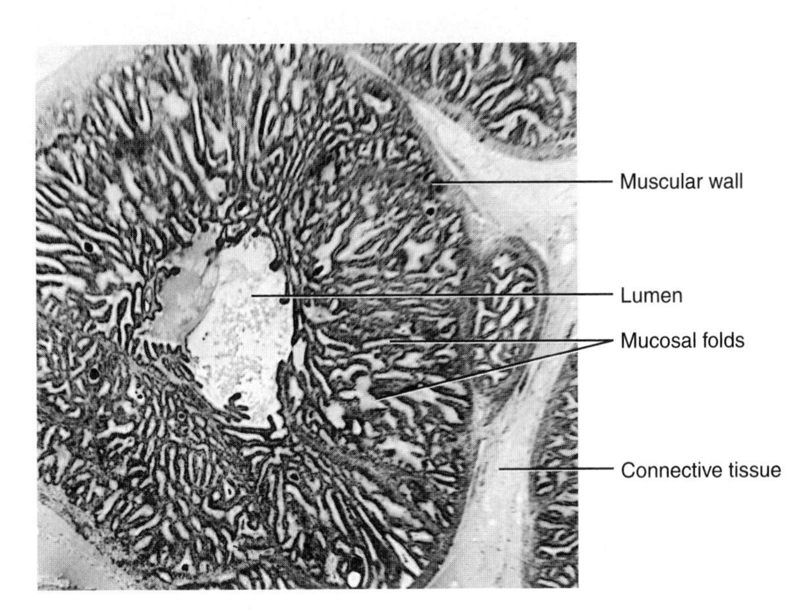

FIGURE 19.10
LM of seminal vesicle in cross section. (40×)

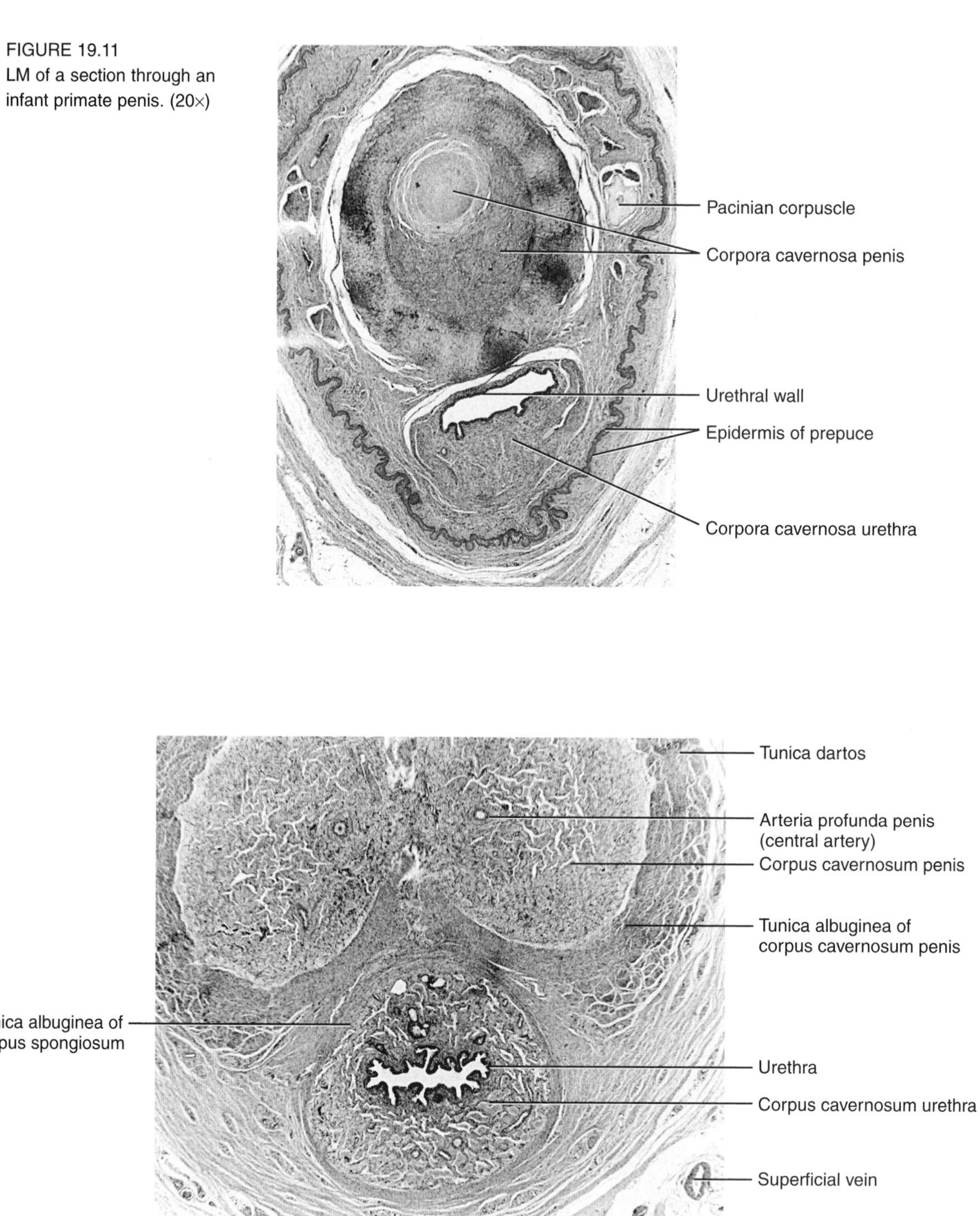

FIGURE 19.11
LM of a section through an infant primate penis. (20×)

FIGURE 19.12
LM of a section through a human penis. (20×)

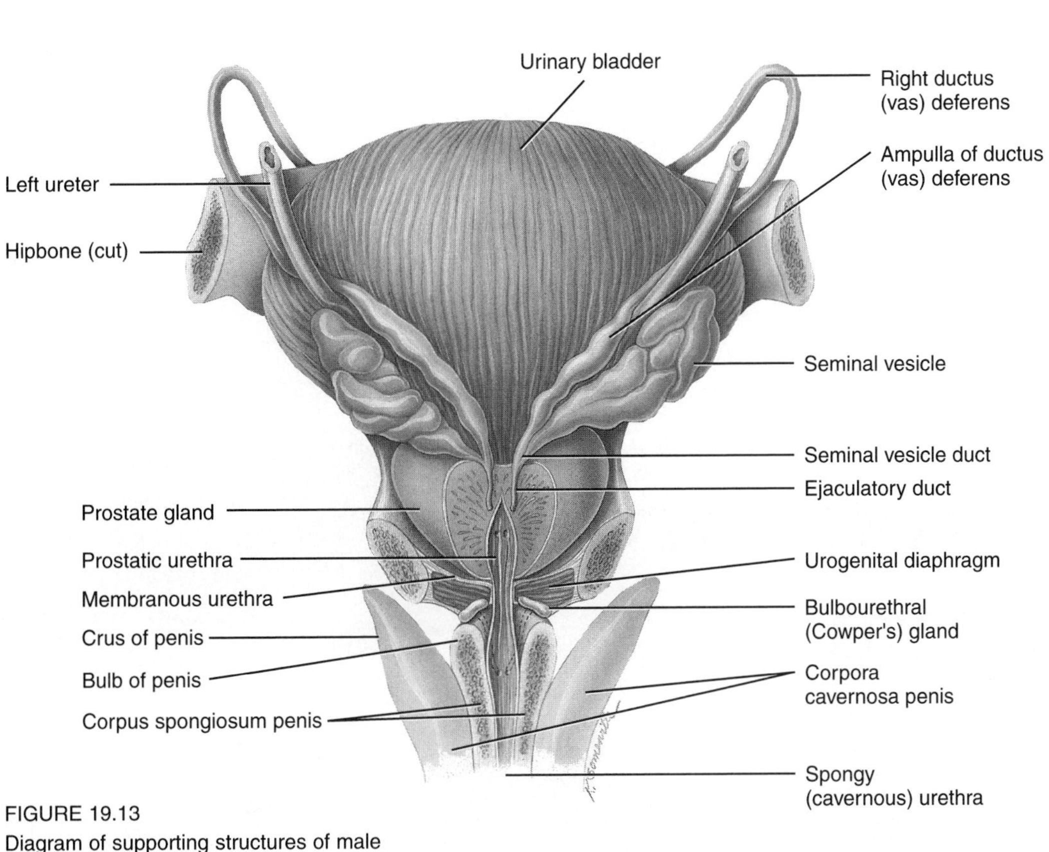

FIGURE 19.13
Diagram of supporting structures of male reproductive system.

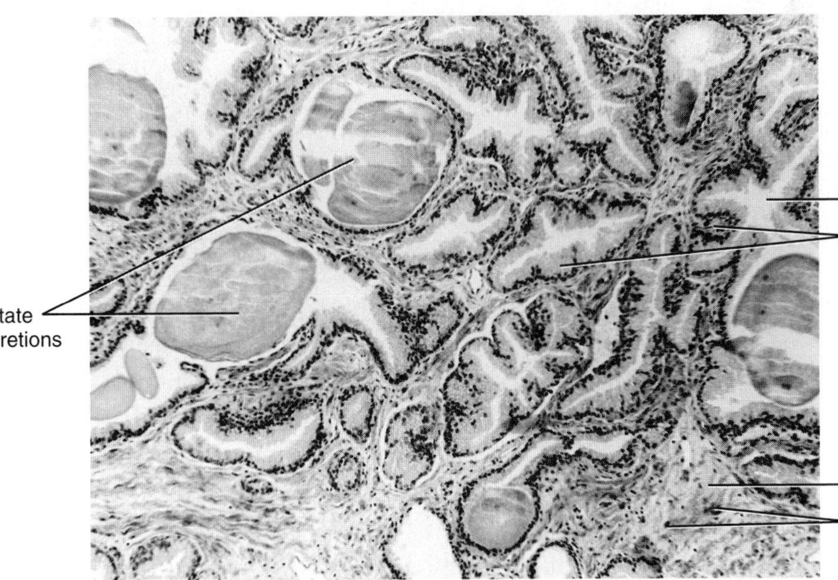

FIGURE 19.14
LM of a cross section through a prostate gland lobule of a young male. (200×)

Female Reproductive System

The female reproductive system consists of the **ovaries, oviducts, uterus,** and **vagina.** The supporting external reproductive structures include the **labia majus, labia minus,** and clitoris. The **mammary glands** are also included in the female reproductive system, although they are not part of the genitalia.

Ovaries The two ovaries are the oval bodies that lie on either side of the uterus, suspended by a mesentery from the broad ligament. Because of their secretions, the ovaries are classified as **endocrine** and **exocrine (cytogenic holocrine)** glands. As endocrine glands, the ovaries are the source of the cyclic secretion of the sex hormones, **estrogen** and **progesterone.** As exocrine glands, the ovaries produce a cyclic secretion of whole cells or **ova.**

Fallopian Tubes or Oviducts The two **fallopian tubes** connect the ovaries to the uterus. At the ovarian end, the oviduct is open and communicates with the peritoneal cavity. The other end of the oviduct opens into the **lumen** of the uterus.

Uterus The uterus is a pear-shaped muscular organ, approximately 7 cm in length and 5 cm in width. The uterus can be divided into the **fundus** (rounded upper end of the body), the **corpus uteri** or **body** (the broad part of the uterus), and the **portio vaginalis,** a narrow cylindrical neck or cervix that projects into the vagina. The fallopian tubes enter the uterus at the **fundus.** The body wall of the uterus can be separated into three zones: the outer **serosa** or **perimetrium,** the middle **myometrium** or **muscularis,** and the inner **endometrium.**

Cervix The cervix is the inferior segment of the uterus and is essentially a dense collagenous connective tissue. The body wall of the cervix lacks smooth muscle. The mucous membrane consists of tall mucus-secreting columnar cells mixed with some ciliated cells.

Vagina The vagina forms the lowermost segment of the female reproductive tract. It is a fibromuscular sheath with a mucous membrane. Ordinarily, it is collapsed with anterior and posterior walls in contact. The body wall of the vagina is composed of the mucous layer, which forms transverse folds, or **rugae,** lined by thick, nonkeratinizing, stratified squamous epithelium.

Mammary Glands Both sexes have mammary glands; however, owing to the lack of female hormones, the glands in the male remain rudimentary throughout life. The mammary glands are modified sweat glands located within the subcutaneous tissue. Each mammary gland is divided into 15 to 20 lobes separated and surrounded by connective tissue composed primarily of adipose cells. Each lobe is an independent gland with its own duct that opens at the apex of the **nipple.**

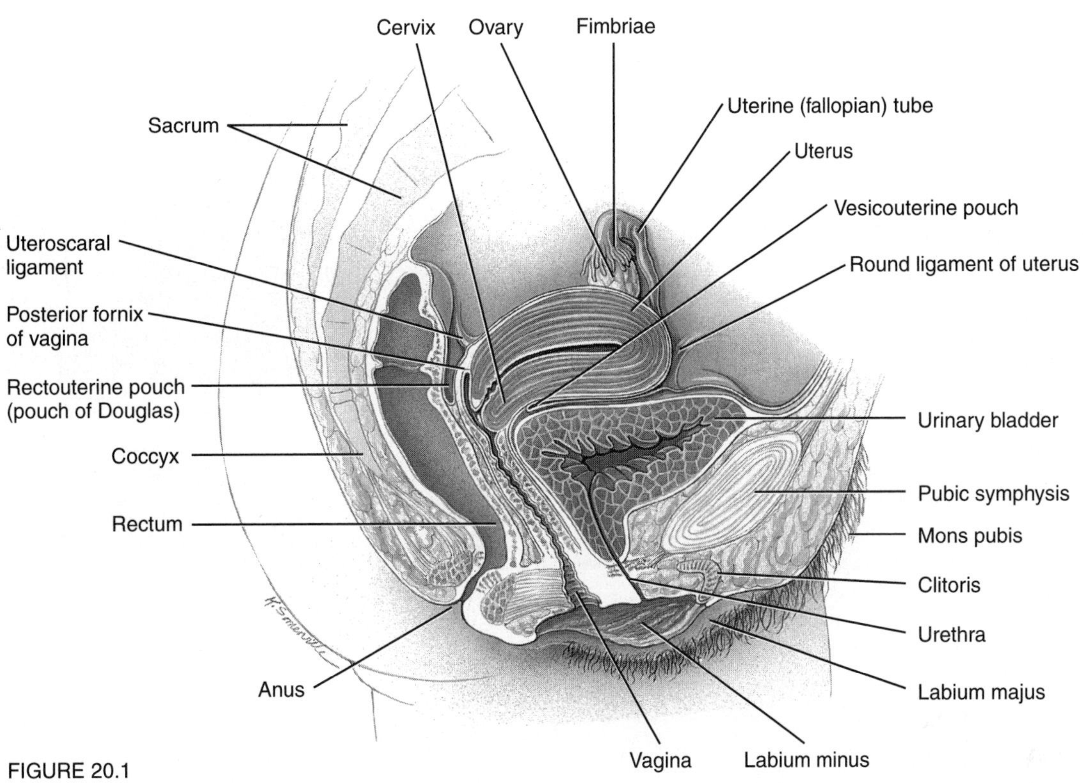

FIGURE 20.1
Diagrammatic presentation of female reproductive organs in sagittal section.

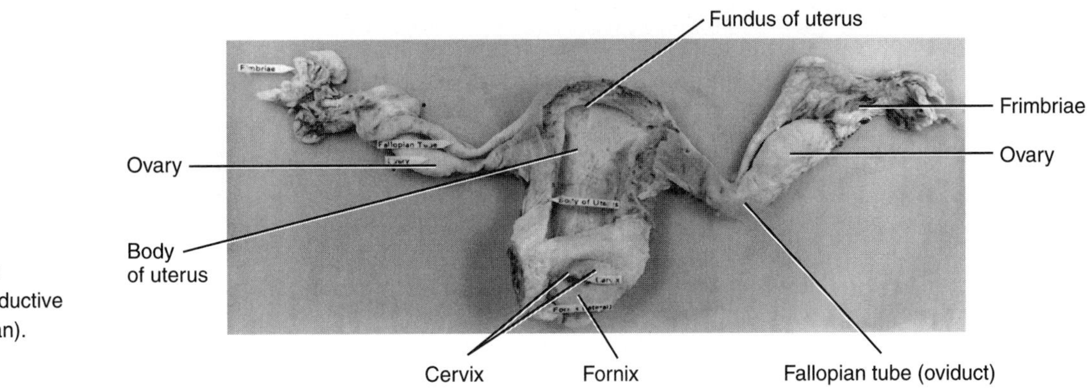

FIGURE 20.2
Female reproductive organs (human).

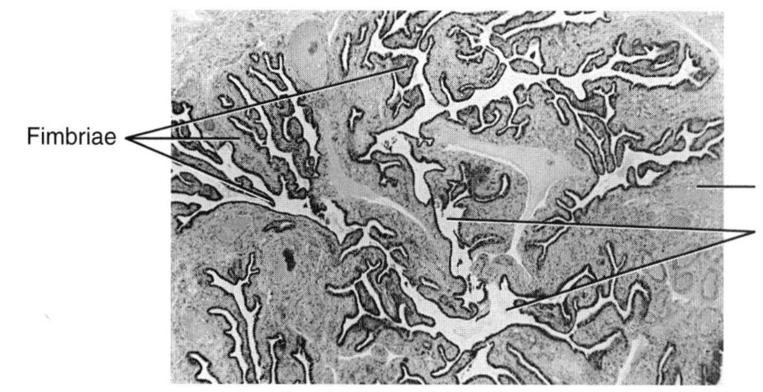

FIGURE 20.3
Light micrograph (LM) of a cross section through the infundibulum of the uterine (fallopian) tube. (200×)

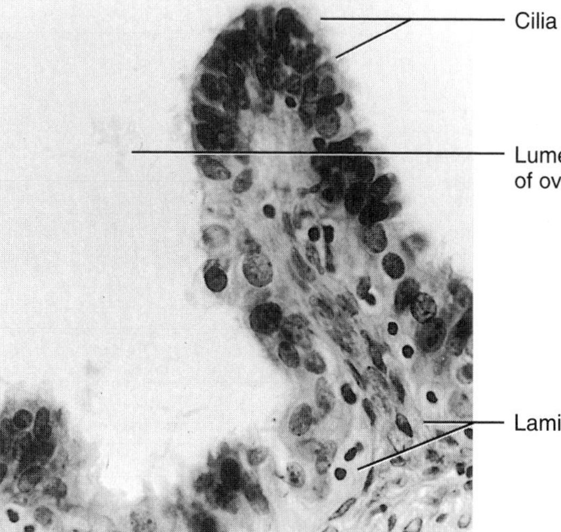

FIGURE 20.4
LM of a cross section through a small portion of the oviduct, displaying ciliated epithelium. (1000×)

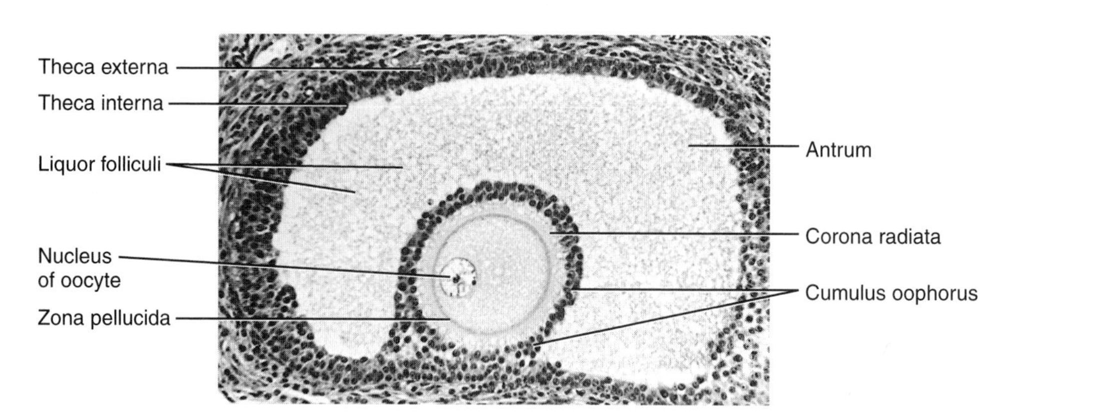

FIGURE 20.5
LM of a graafian follicle in sagittal section. (200×)

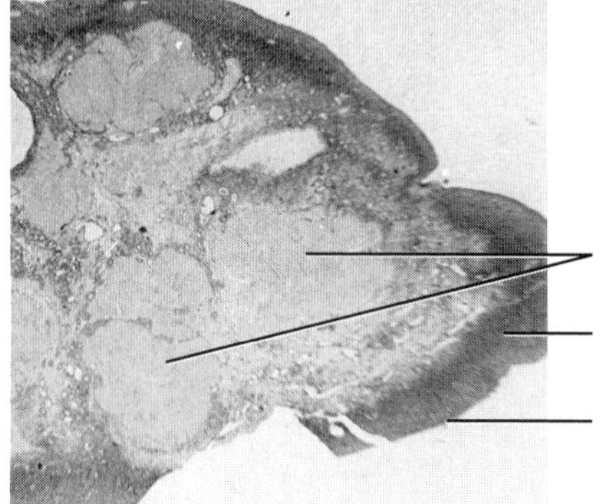

FIGURE 20.6
LM of a sagittal section through the ovary, demonstrating the development of the corpus luteum of pregnancy. (20×)

FIGURE 20.7
Diagrammatic presentation of the female reproductive organs.

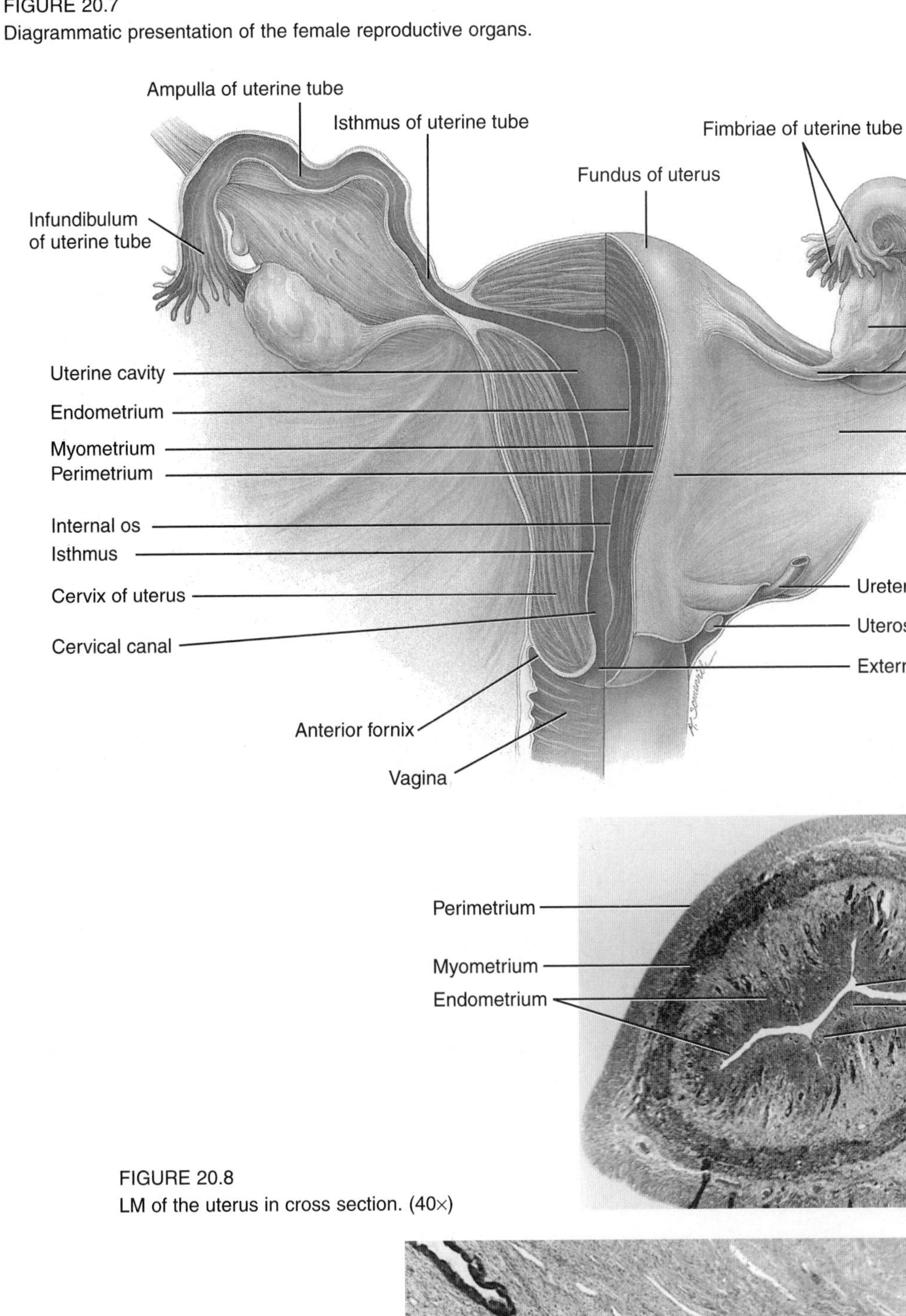

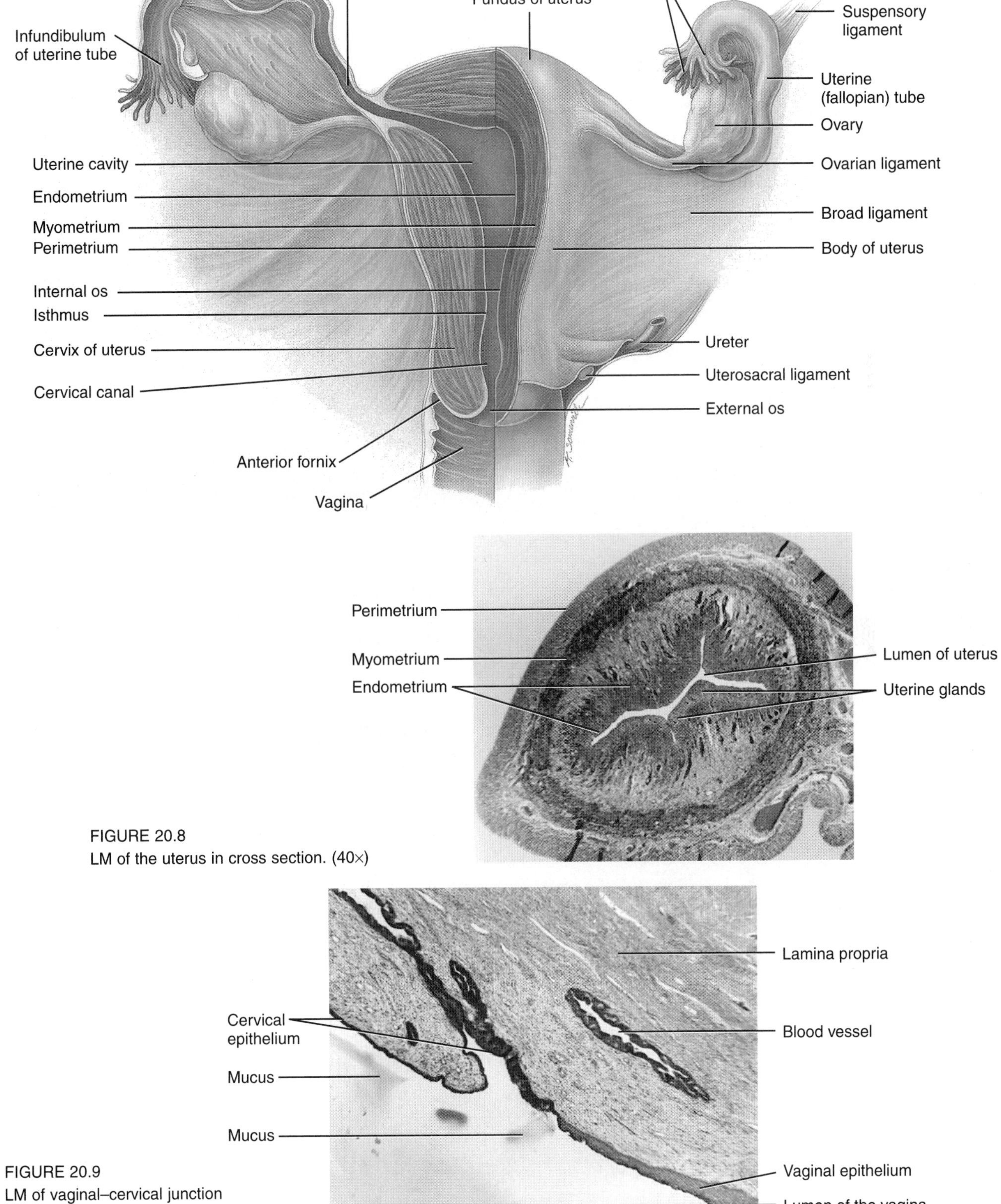

FIGURE 20.8
LM of the uterus in cross section. (40×)

FIGURE 20.9
LM of vaginal–cervical junction (portovaginalis).

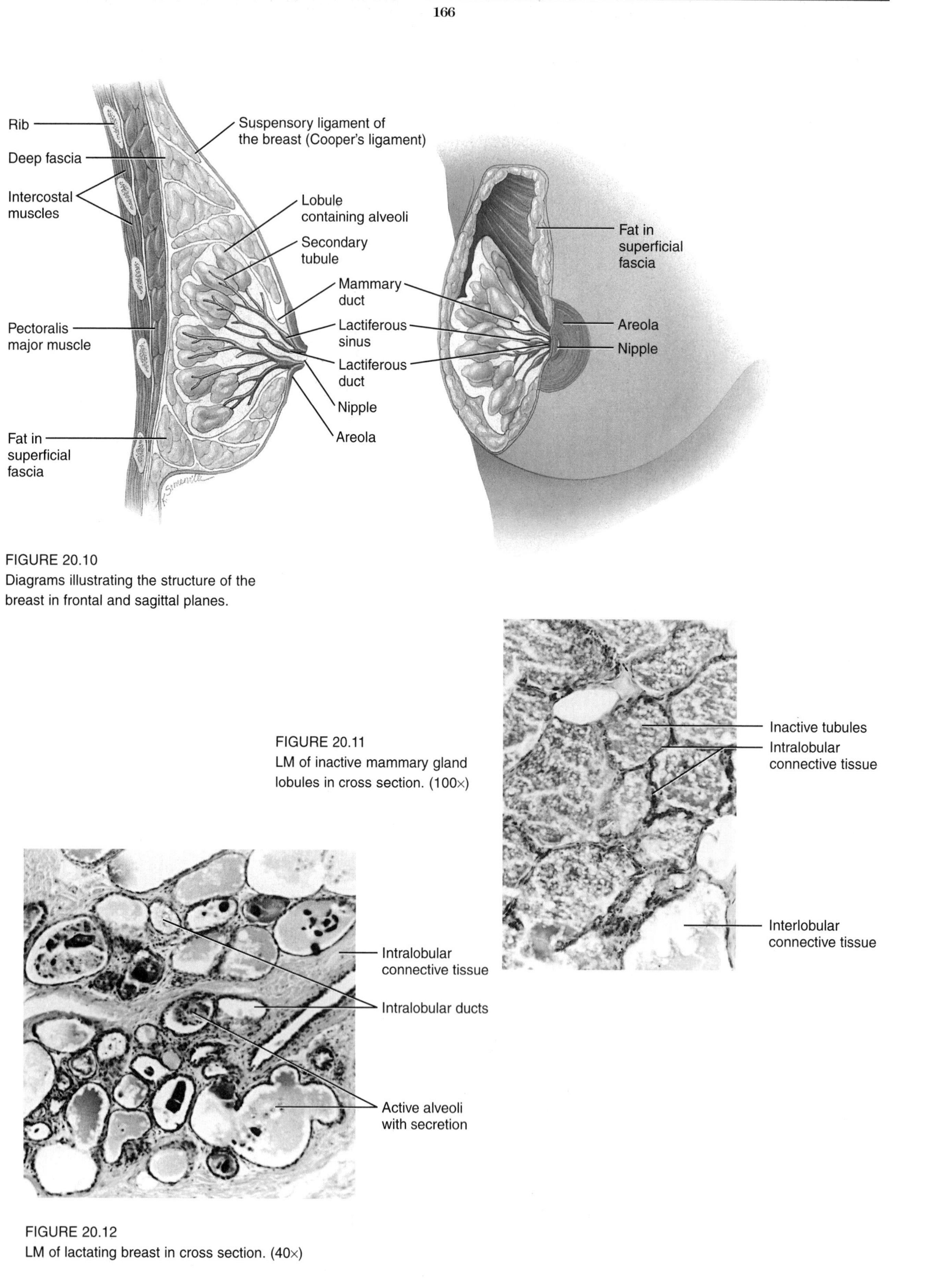

FIGURE 20.10
Diagrams illustrating the structure of the breast in frontal and sagittal planes.

FIGURE 20.11
LM of inactive mammary gland lobules in cross section. (100×)

FIGURE 20.12
LM of lactating breast in cross section. (40×)

Fetal Development

The embryonic development begins from a single fertilized cell, the zygote in higher animals. The zygote is a combination of both male spermatozoa and female ovum. The zygote goes through morphogenesis that includes all development processes beginning with a two-week pre-embryonic stage, a six-week embryonic stage, and a 30-week fetal development stage. The fetal stage leads to the formation of a human infant.

Hormones play an important role in the process of morphogenesis. Human growth hormone (hGH) is associated with general body growth and regulation of cellular functions. The thyroid gland secretes thyroxine, which is associated with metabolism of all cells of the body, and calcitonin hormone, which regulates calcium levels in the blood plasma and bone. Other hormones such as adrenocorticotropic hormone (ACTH) has a profound effect in maintaining homeostasis, whereas melanin stimulating hormone (MSH) is associated with levels of melanin in the skin.

Fetal Development

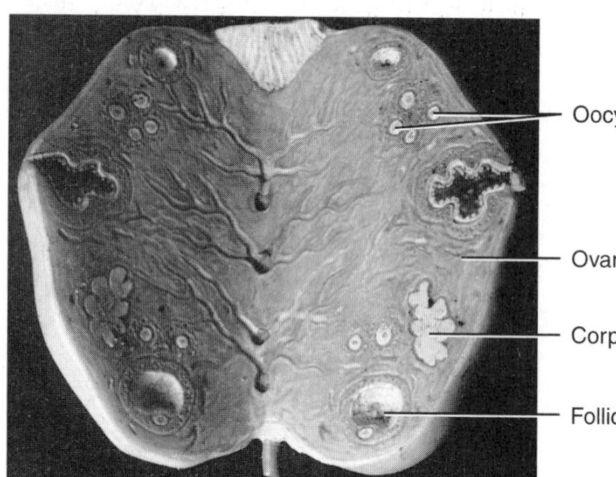

FIGURE 21.1
Model illustrating the general structure of the ovary and stages of follicular development that lead to the formation of an ovum (Sawso Modelle).

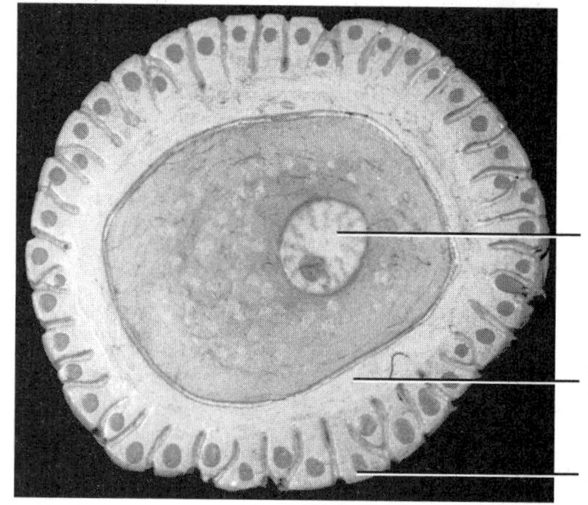

FIGURE 21.2
Model of an ovum prior to fertilization.

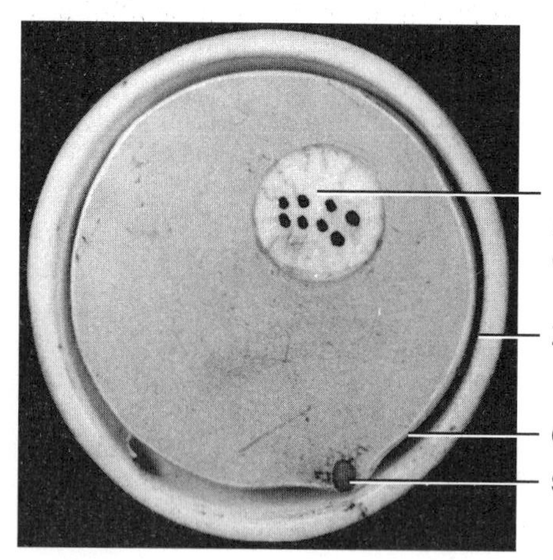

FIGURE 21.3
Model illustrating the penetration of sperm into the egg.

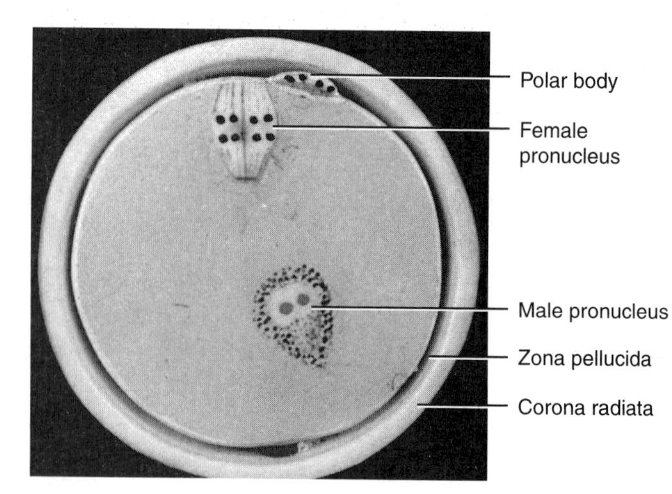

FIGURE 21.4
Formation of a sperm nucleus and a female pronucleus.

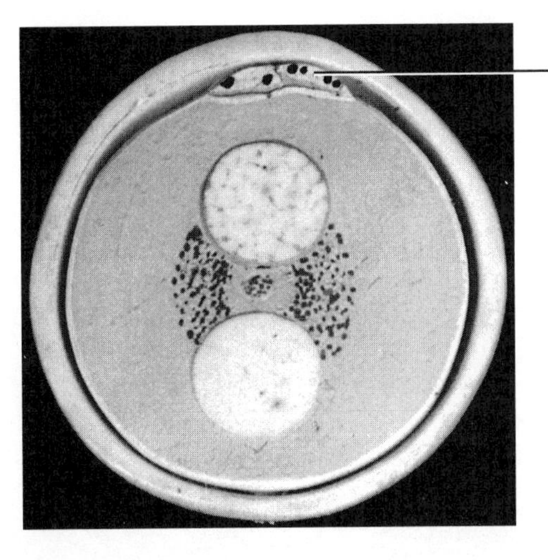
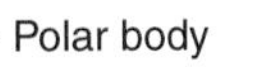

FIGURE 21.5
Union of male and female pronuclei and arrangement of the centrioles.

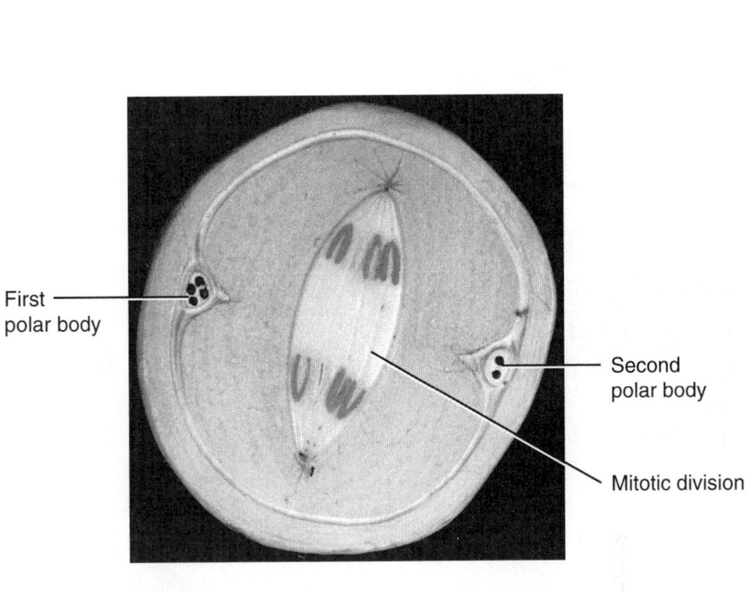

FIGURE 21.6
Mitotic division in the zygote.

Amnion
Neural plate
of forebrain
Neural groove
Somite
Henson's node
Primitive streak

FIGURE 21.7
Model of a germinal disk (embryonic primordium).

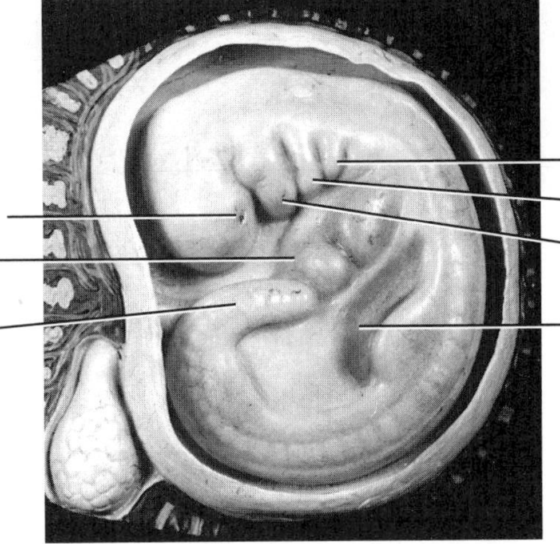

FIGURE 21.8
Embryo at 7-mm development stage.

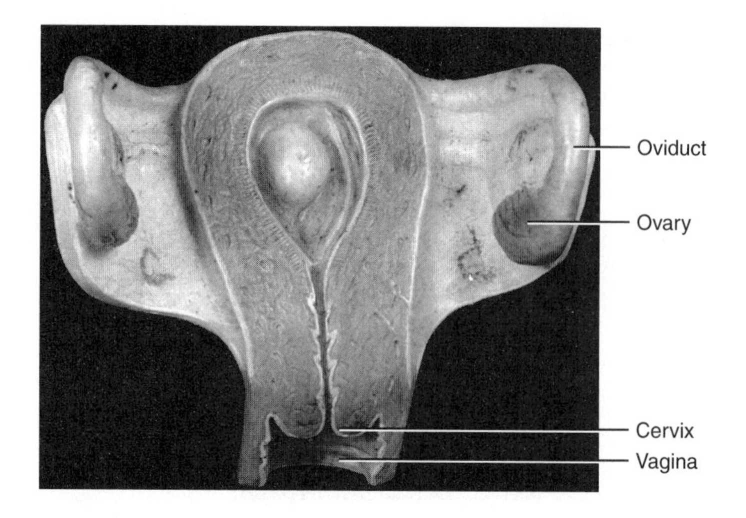

FIGURE 21.9
Exposed uterus illustrating embryo development at the end of fourth week.

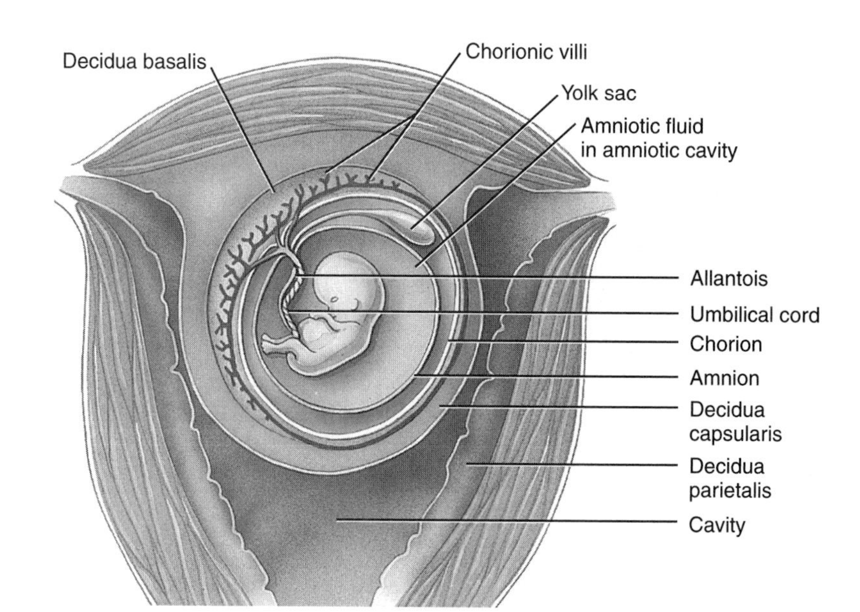

FIGURE 21.10
Diagram of a developing embryo and supporting structures.

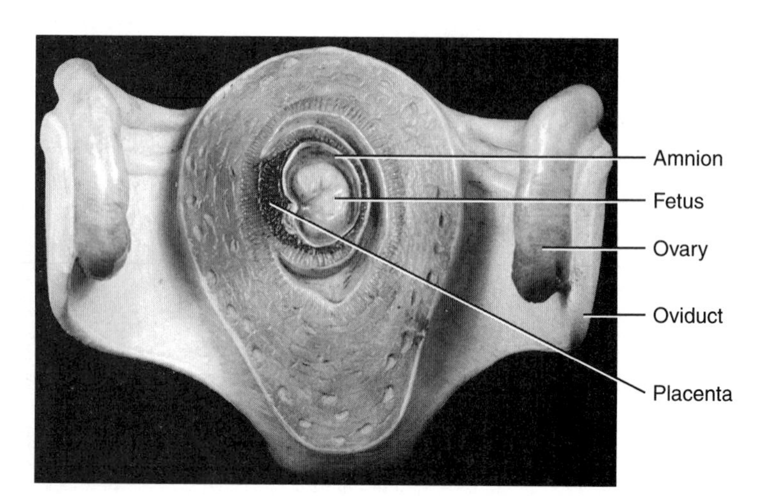

FIGURE 21.11
Embryo development at the end of eighth week.

FIGURE 21.12
Fetal development at the end of tenth week.

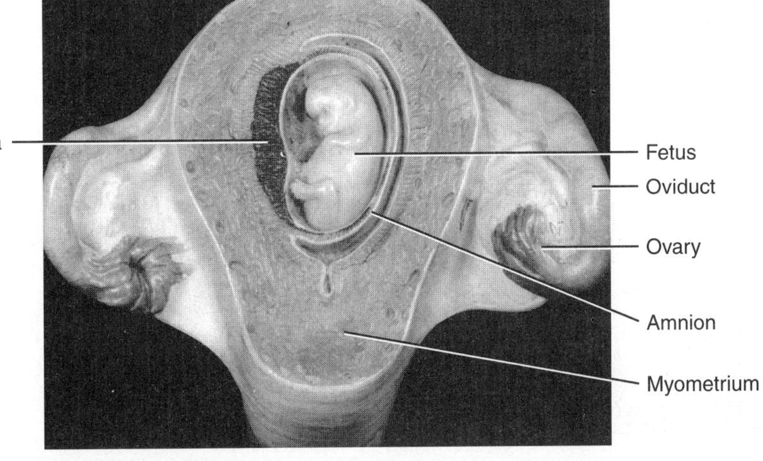

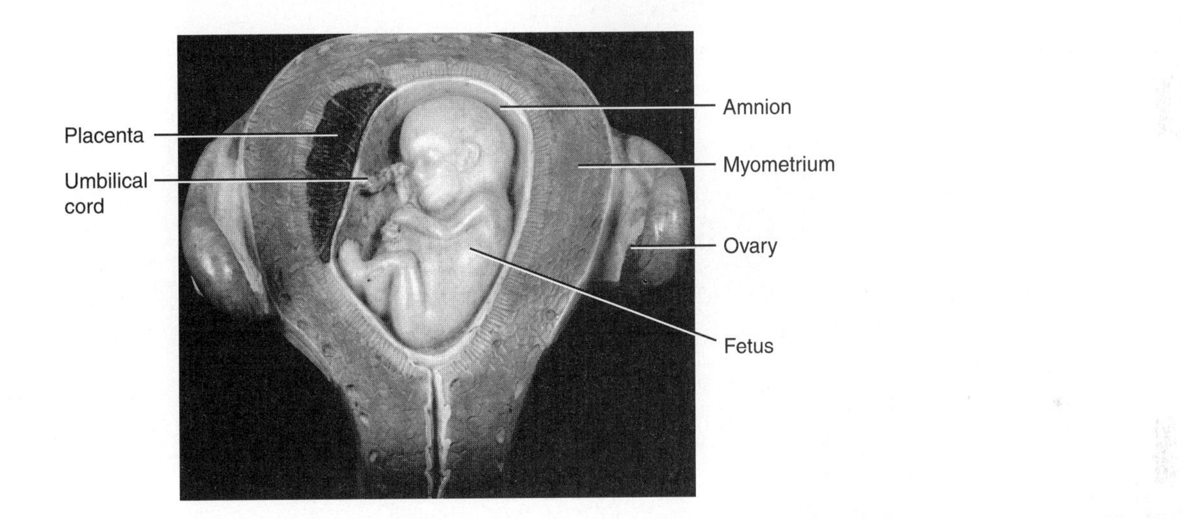

FIGURE 21.13
Fetal development at the beginning of four months.

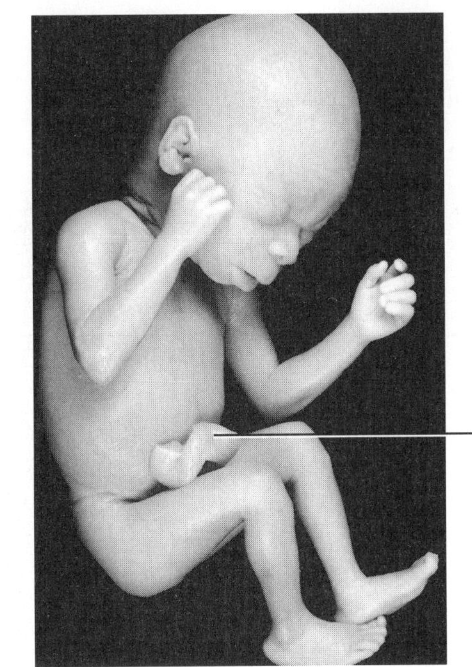

FIGURE 21.14
Fetal development at six months.

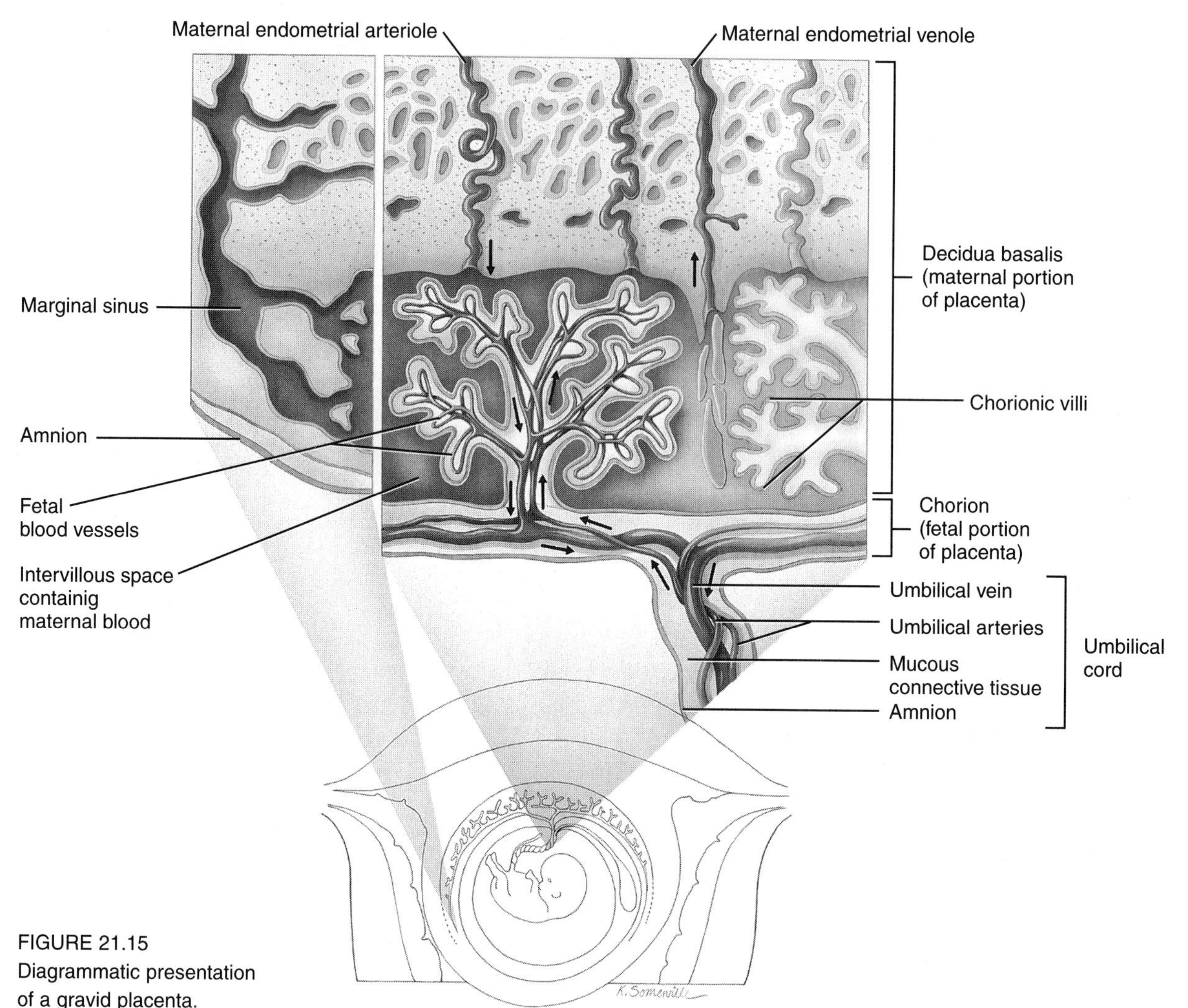

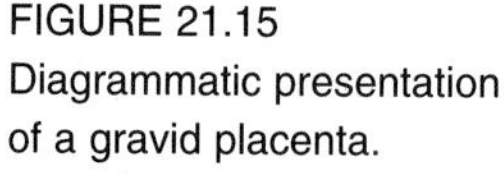

FIGURE 21.15
Diagrammatic presentation of a gravid placenta.

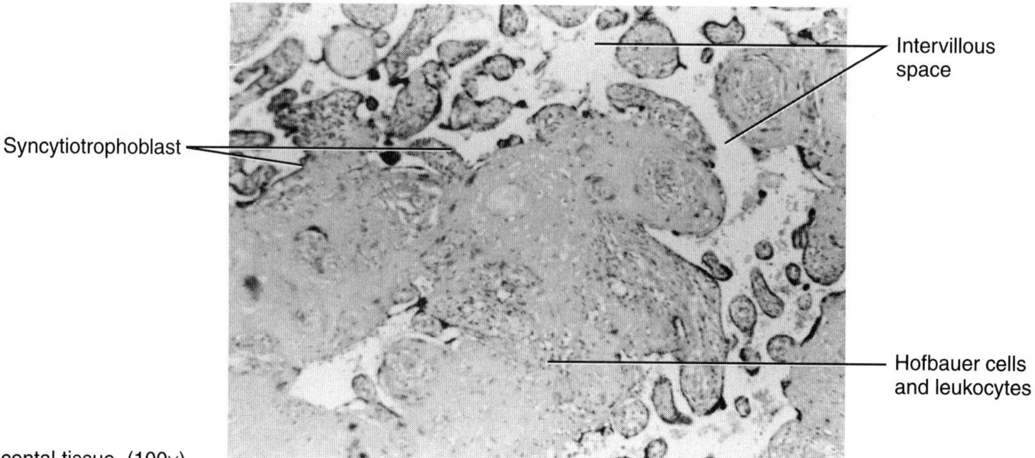

FIGURE 21.16
Light micrograph (LM) of placental tissue. (100×)

Organs of Special Sense

The organs of **sense** are distributed extensively in the epithelium, connective tissues, muscles, and tendons; however, special receptors associated with taste, sight, hearing, and smell are localized in the **tongue,** the **eyes,** the **ears,** and the **nose.** The taste buds of the tongue are described in Chapter 15 with the digestive system, and the nose is discussed in Chapter 16 with the respiratory system.

The Eye

The functioning of the eye has been compared to that of a camera; however, no such analogy can do justice to the relationship between the eye and the nervous system. The eye functions as a camera in that light from the external environment enters the eye through a lens, which focuses an image on the photosensitive cells of the **retina.**

The wall of the eyeball is a complex tissue structure. There are three identifiable layers: the outer **corneoscleral coat,** the middle **uveal layer,** and the inner **retina.** The uveal layer is the nutrient-rich middle vascular coat; it can be divided further into a posterior **choroid,** a **ciliary body,** and an **anterior iris.** The **retina**—the inner coat, which is considerably more complex than the other two coats—consists of photopigmented photoreceptors that convert light energy into chemical energy, which is then transformed into electrical energy by means of nerve impulses.

The Ear

The three divisions of the ear are the **external ear, middle ear,** and **inner ear.**

External Ear

The external ear includes the **auricle (pinna), external auditory meatus, ceruminous glands,** and **cerumen.** The auricle is an irregular elastic cartilage covered with a **perichondrium.** The skin covers the perichondrium of the elastic cartilage. In humans, hair follicles, sweat glands, and sebaceous glands may be present in the dermis. The external auditory meatus is a long, narrow canal approximately 2.5 cm in length, with an irregular path from the auricle to the tympanic membrane. The skin lining the meatus contains sebaceous and ceruminous glands that secrete cerumen, a bitter, protective secretion.

Middle Ear

The middle ear consists of the **tympanic cavity, auditory ossicle, tensor tympani, stapedius, auditory (eustachian) tube,** and **tympanic membrane.** Also present in the middle ear are a series of canals and cavities in the endolymph-filled membranous labyrinths.

Tympanic Membrane The tympanic membrane is an obliquely oriented, oval, fibrous membrane at the end of the meatus, separating the external ear from the middle ear. The outer fibers are radial, and the inner fibers are circular.

Tympanic Cavity The tympanic cavity lies behind the tympanic membrane. The epithelial lining in the cavity is squamous or low cuboidal, shifting to columnar and ciliated at the opening of the **auditory** or **eustachian tube** in the middle ear. Present in the tympanic cavity are three bony ossicles: the **malleus, incus,** and **stapes.** Associated with the three ossicles are the **tensor tympani** and the **stapedius muscle.**

Auditory or Eustachian Tube The auditory tube is about 3.5 cm long and connects the **nasopharynx** with the tympanic cavity. The epithelium is ciliated columnar at the opening into the tympanic cavity and pseudostratified ciliated mixed with goblet cells near the nasopharynx. Present near the nasopharynx opening are **seromucous glands** in the underlying lamina propria.

Inner Ear

The inner ear includes the **bony labyrinth, membranous labyrinth, perilymph,** and **endolymph.** The inner ear consists of canals and cavities within the fluid-filled **perilymph** in the **petrous region** of the temporal bone, the **osseous labyrinth.** The bony and membranous labyrinths have two functionally different components: the **vestibular labyrinth,** which controls sensory equilibrium, and a **cochlea,** which consists of sensory neurons and special cells for hearing.

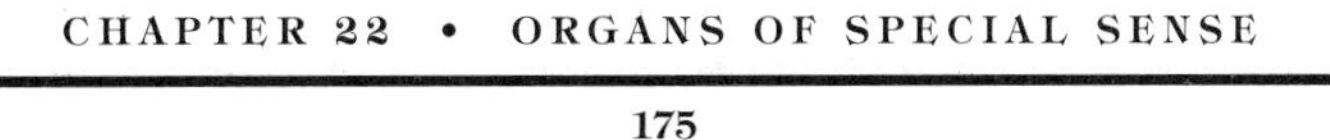

The Eye

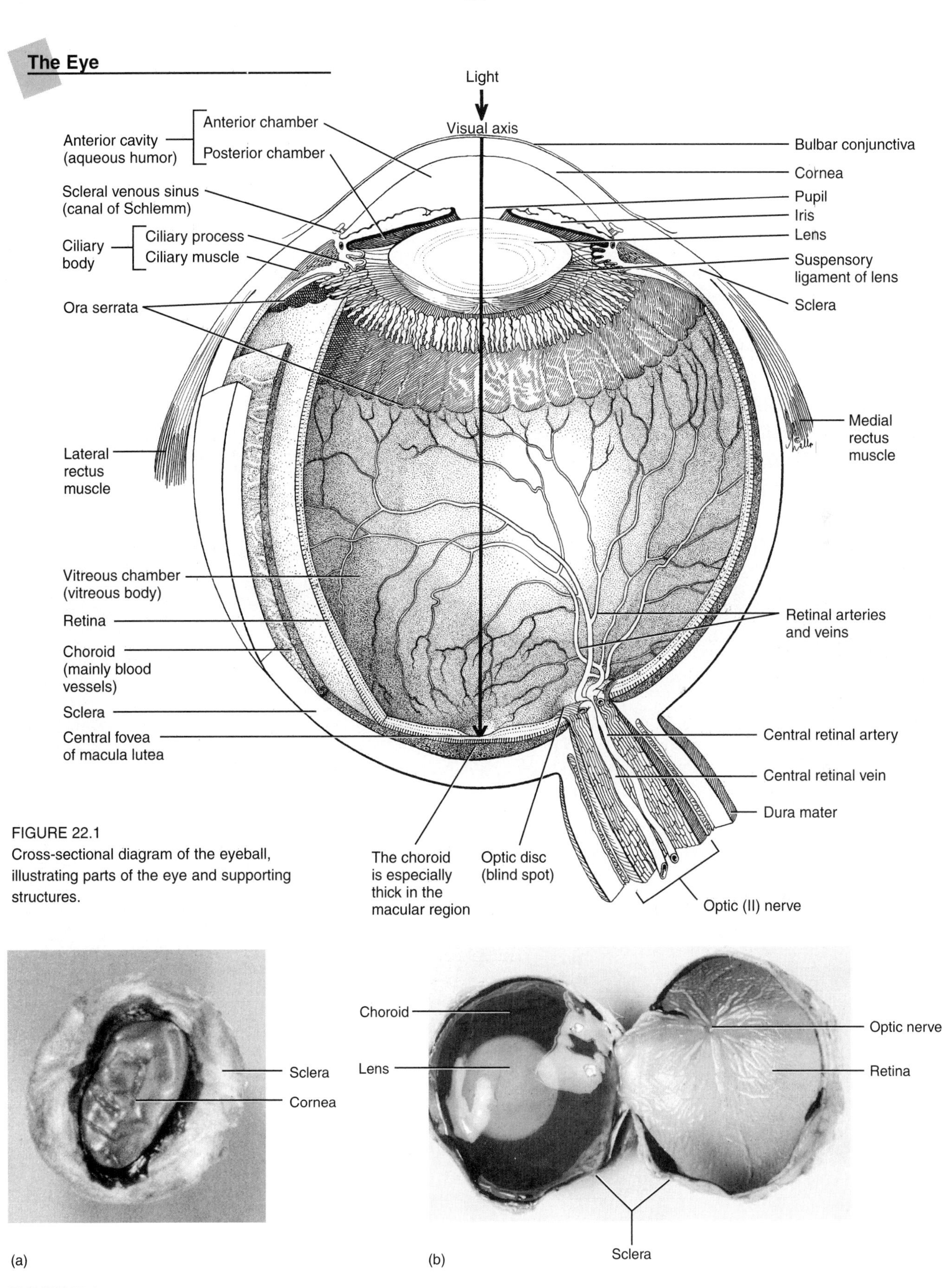

FIGURE 22.1
Cross-sectional diagram of the eyeball, illustrating parts of the eye and supporting structures.

(a)

(b)

FIGURE 22.2
Sheep eyeball: (a) whole and (b) transversely sectioned to show the internal anatomy.

FIGURE 22.3
Diagram of the eye and supporting structures.

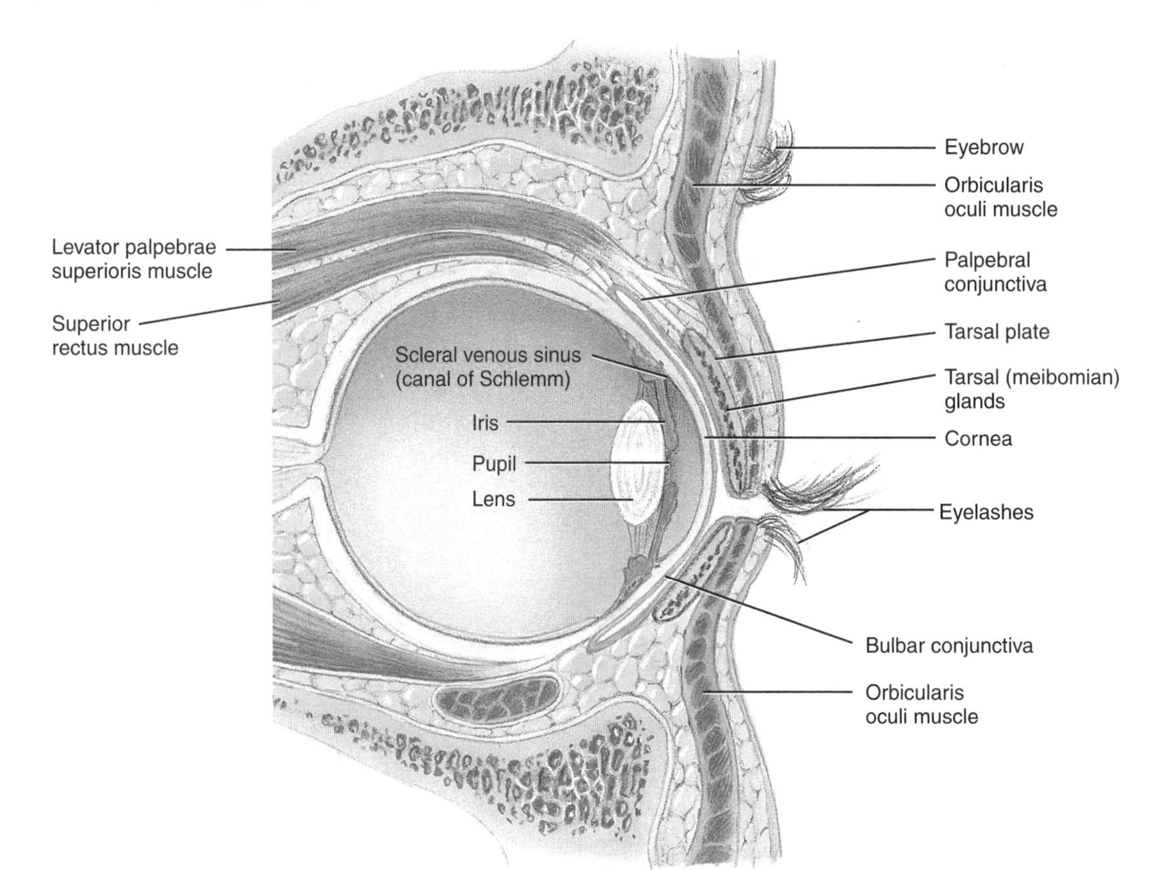

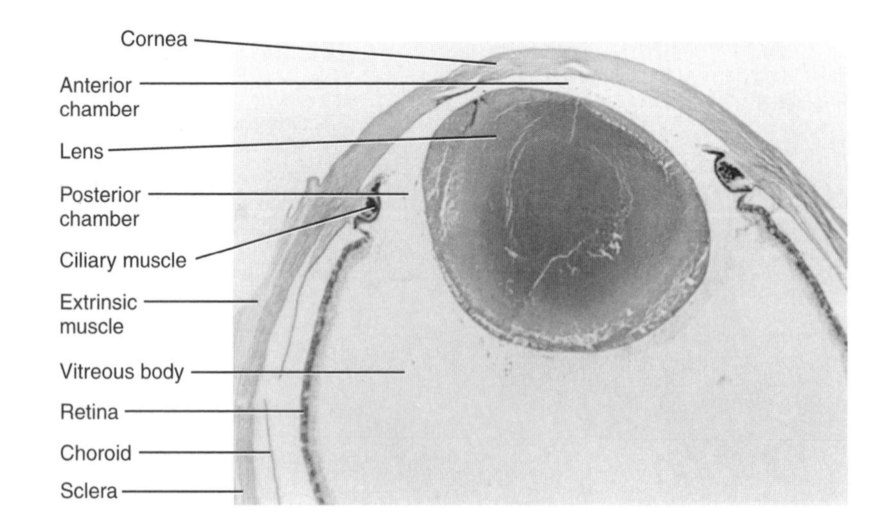

FIGURE 22.4
Light micrograph (LM) of a sagittal section through the eye. (2×)

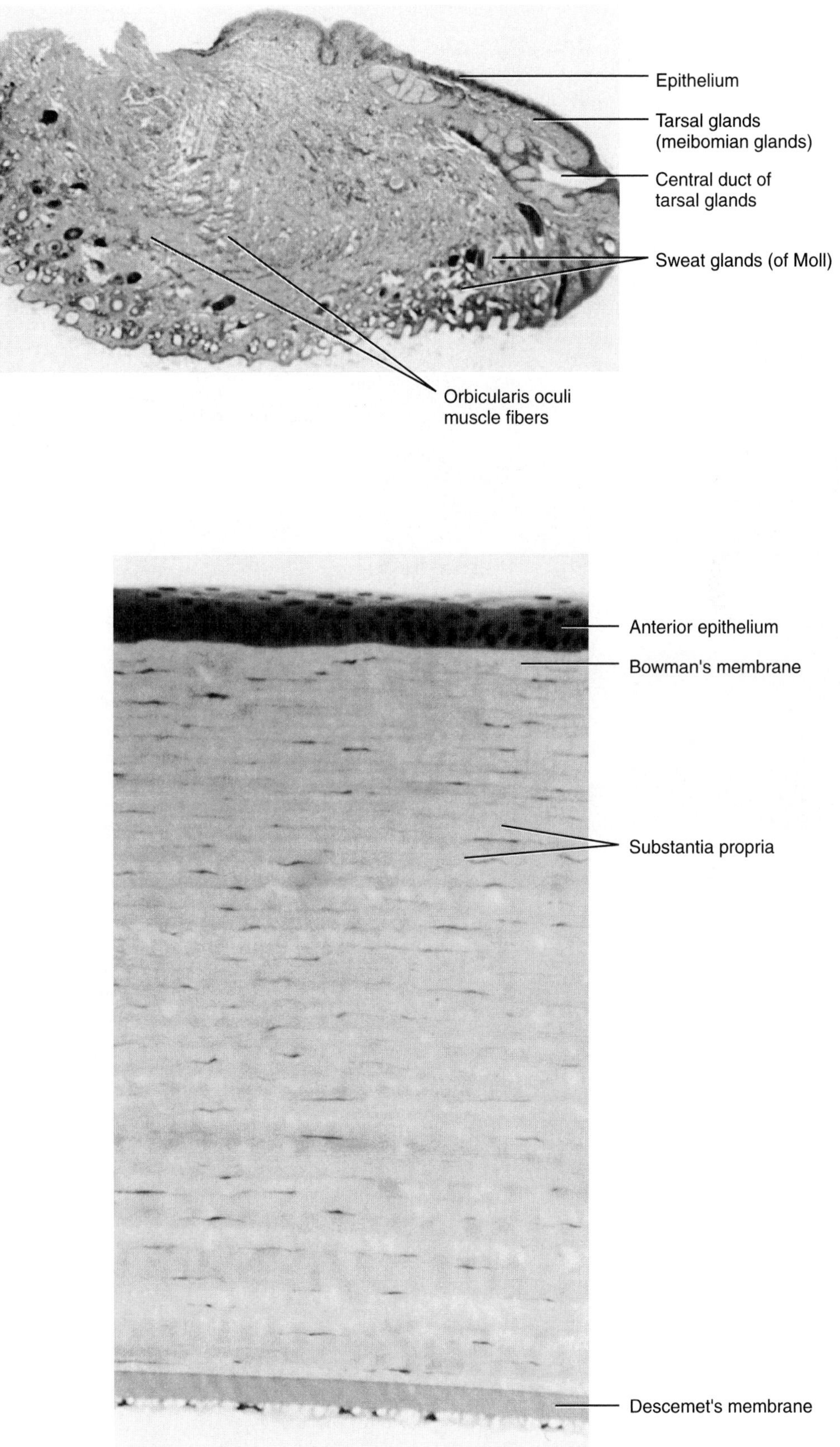

FIGURE 22.5
LM of a cross section through the eyelid. (40×)

FIGURE 22.6
LM of a section through the cornea. (200×)

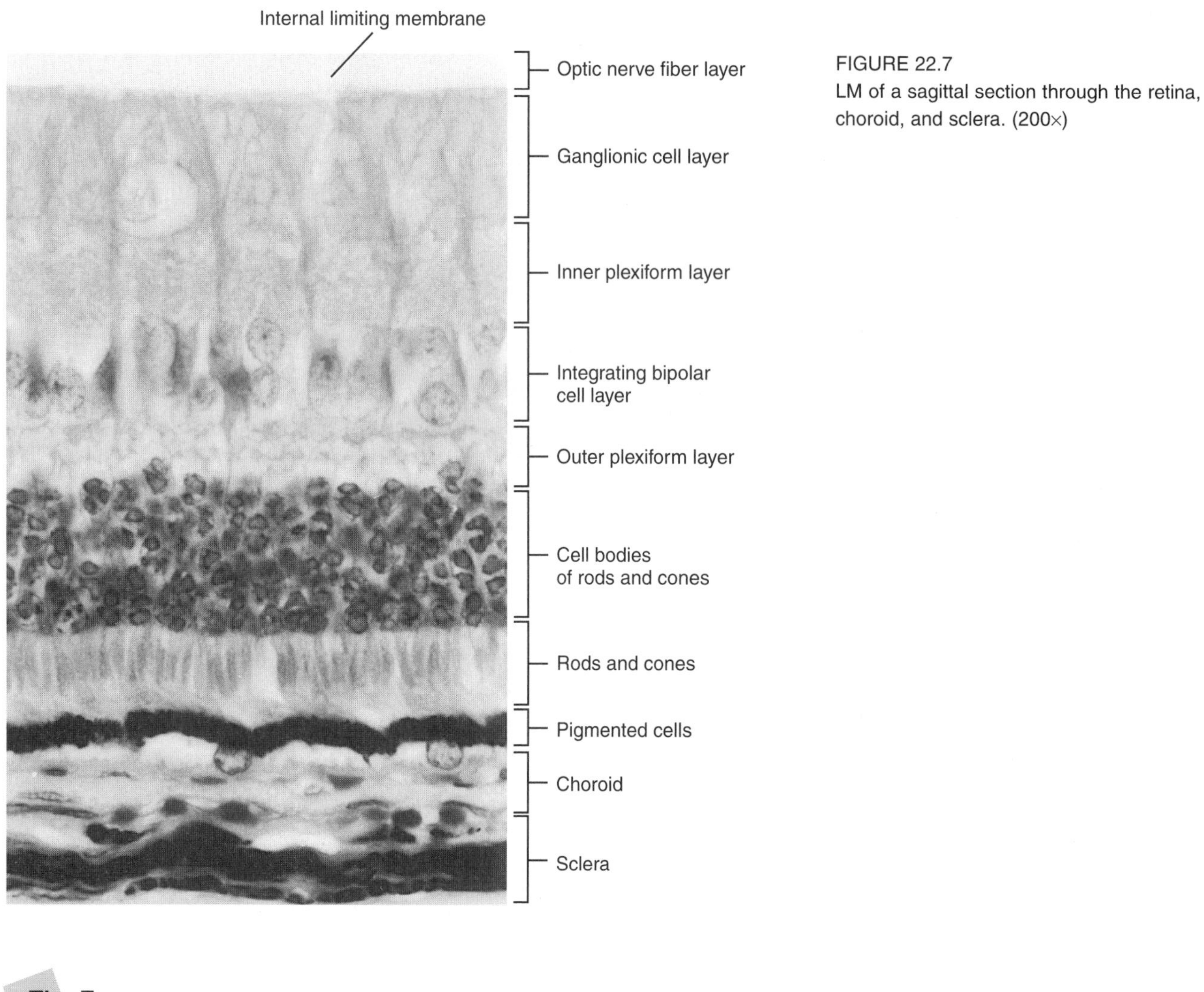

FIGURE 22.7
LM of a sagittal section through the retina, choroid, and sclera. (200×)

The Ear

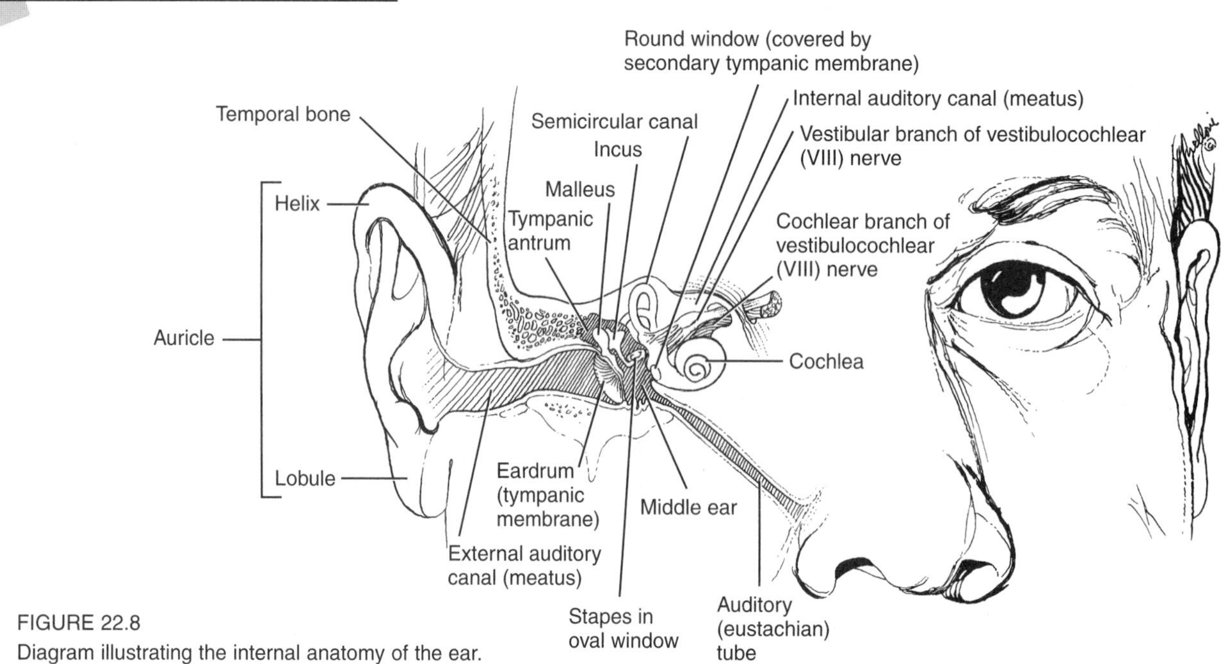

FIGURE 22.8
Diagram illustrating the internal anatomy of the ear.

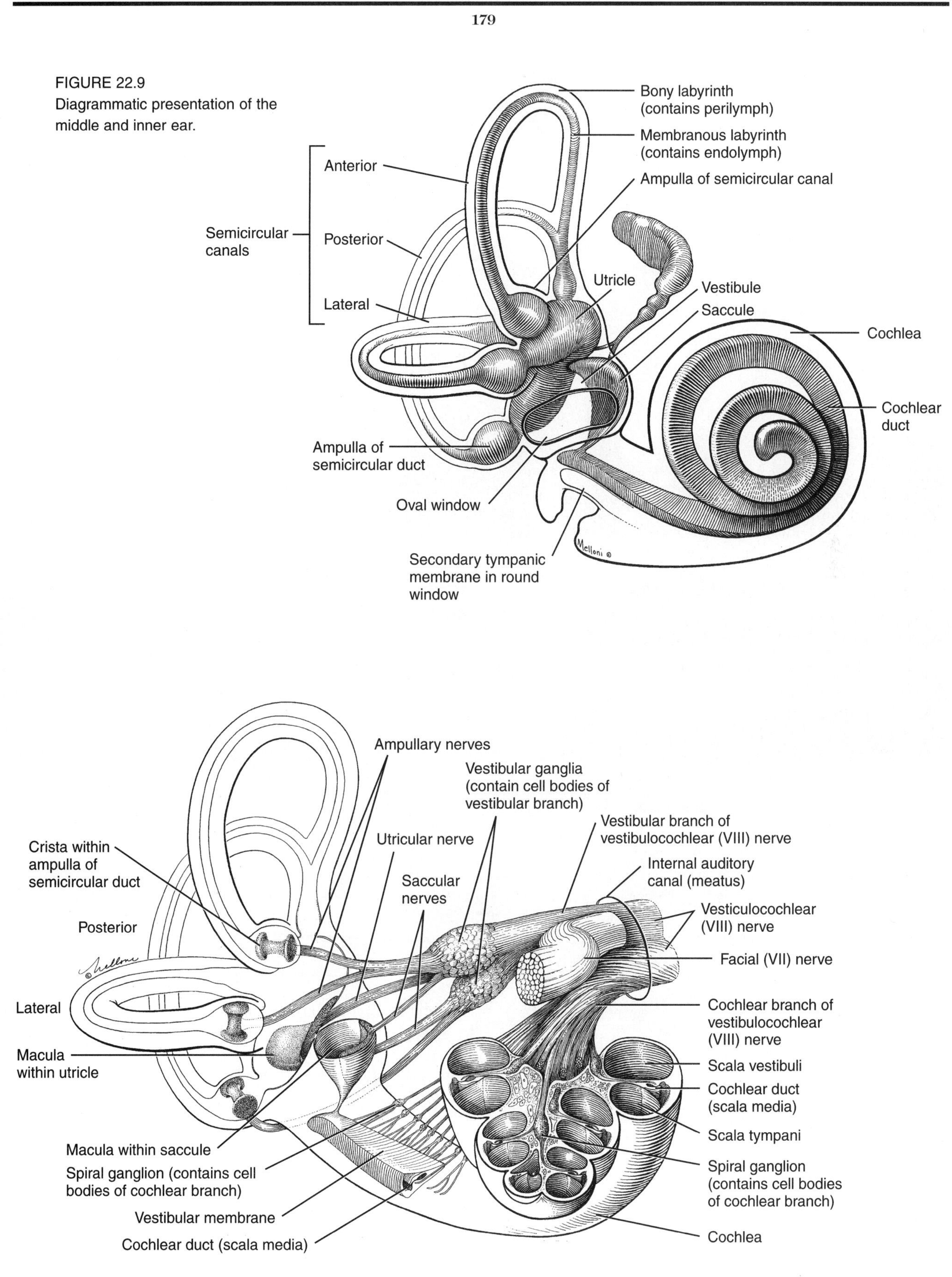

FIGURE 22.9
Diagrammatic presentation of the middle and inner ear.

FIGURE 22.10
Sectional presentation of the semicircular canals, the cochlea, and the distribution of the vestibulocochlear (VIII) nerve.

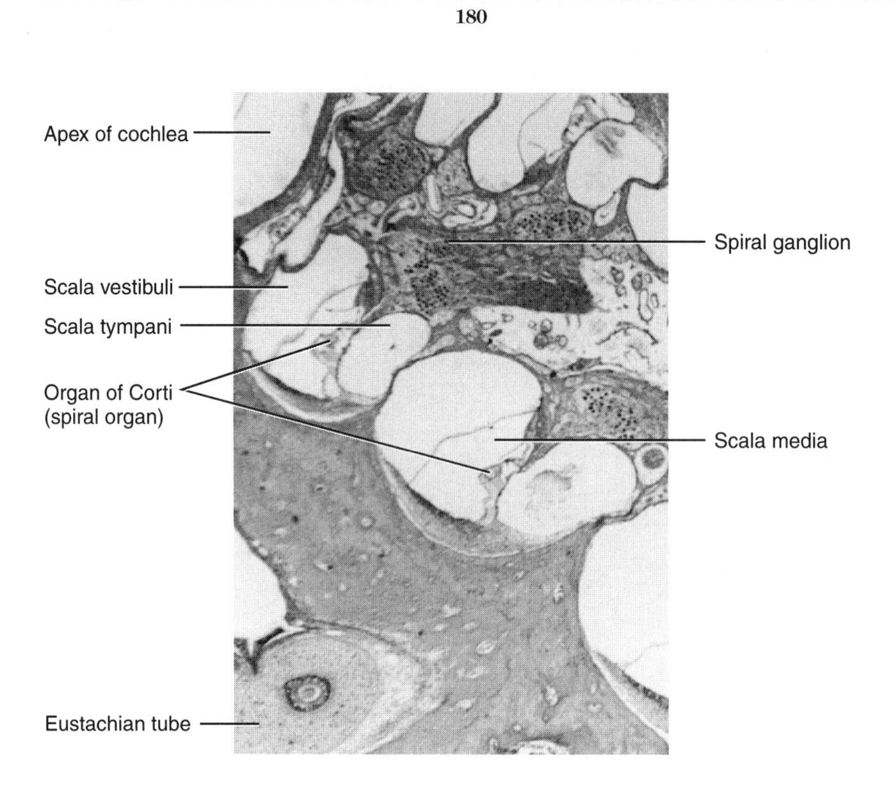

FIGURE 22.11
LM of a section through the cochlea. (40×)

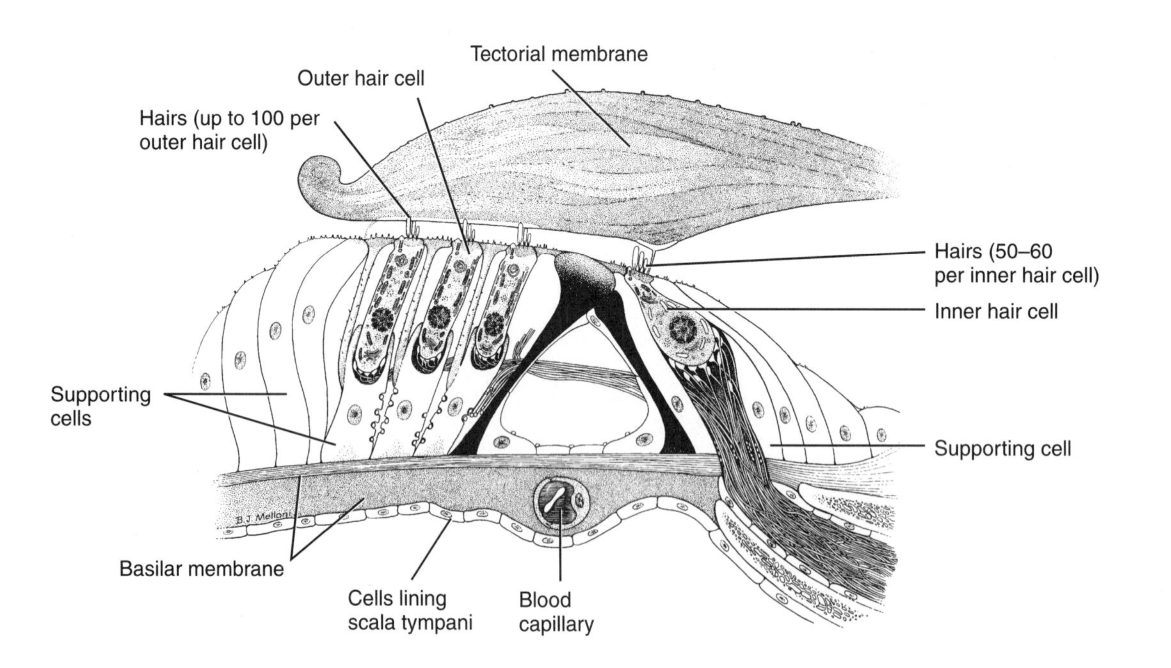

FIGURE 22.12
Diagram of organ of Corti (spiral organ).

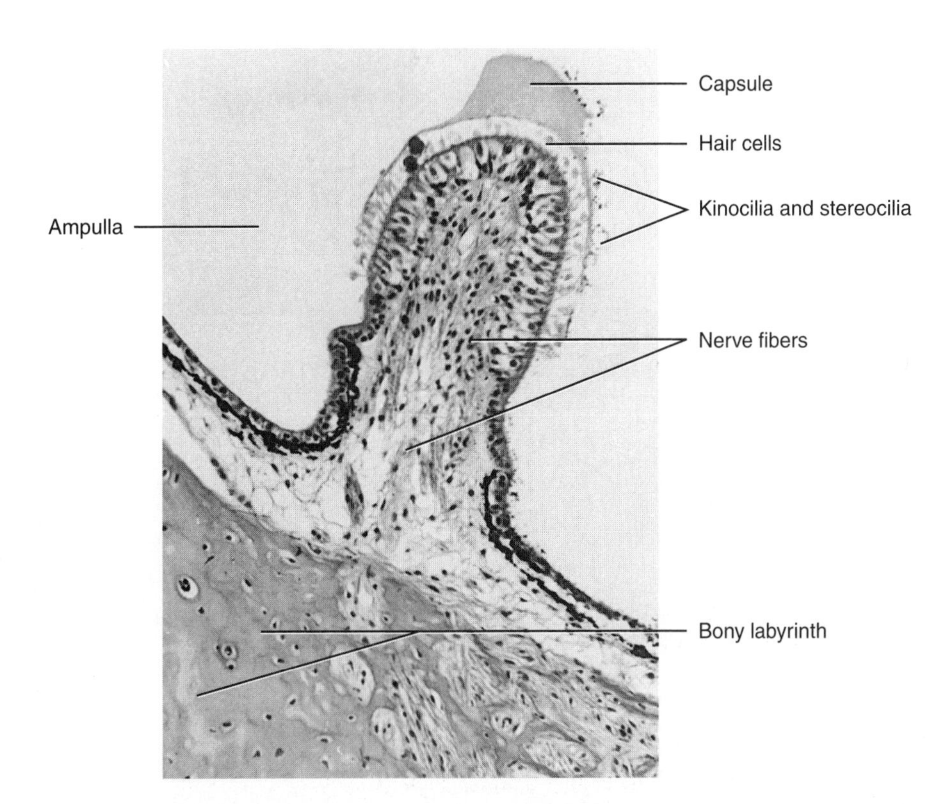

FIGURE 22.13
LM of a section through the crista ampullaris. (400×)

Cat Dissection

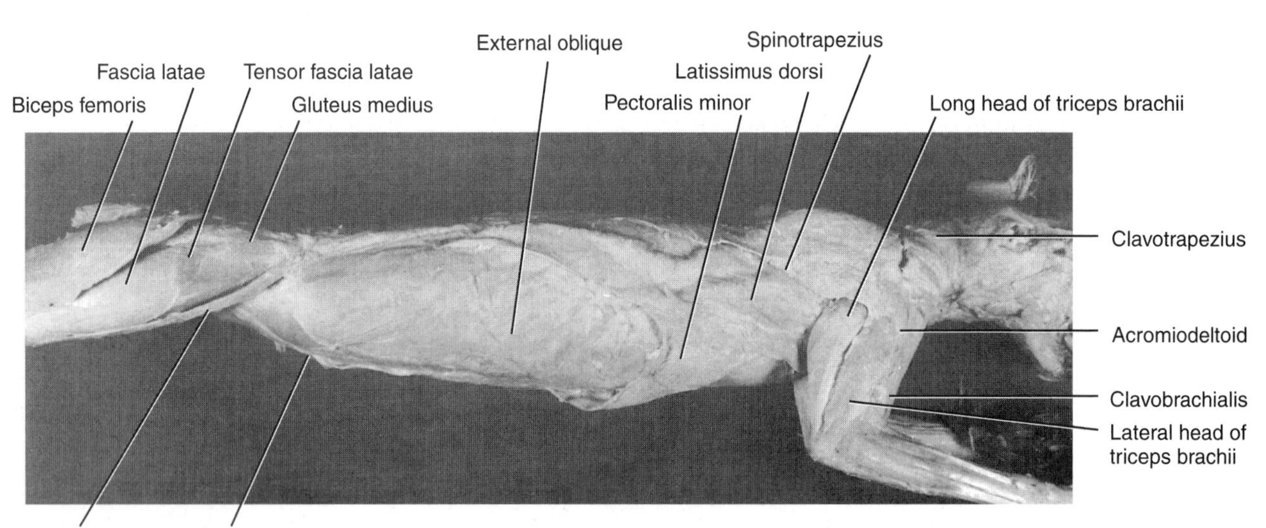

FIGURE 23.1
Superficial muscles of the cat, dorsal and lateral surfaces.

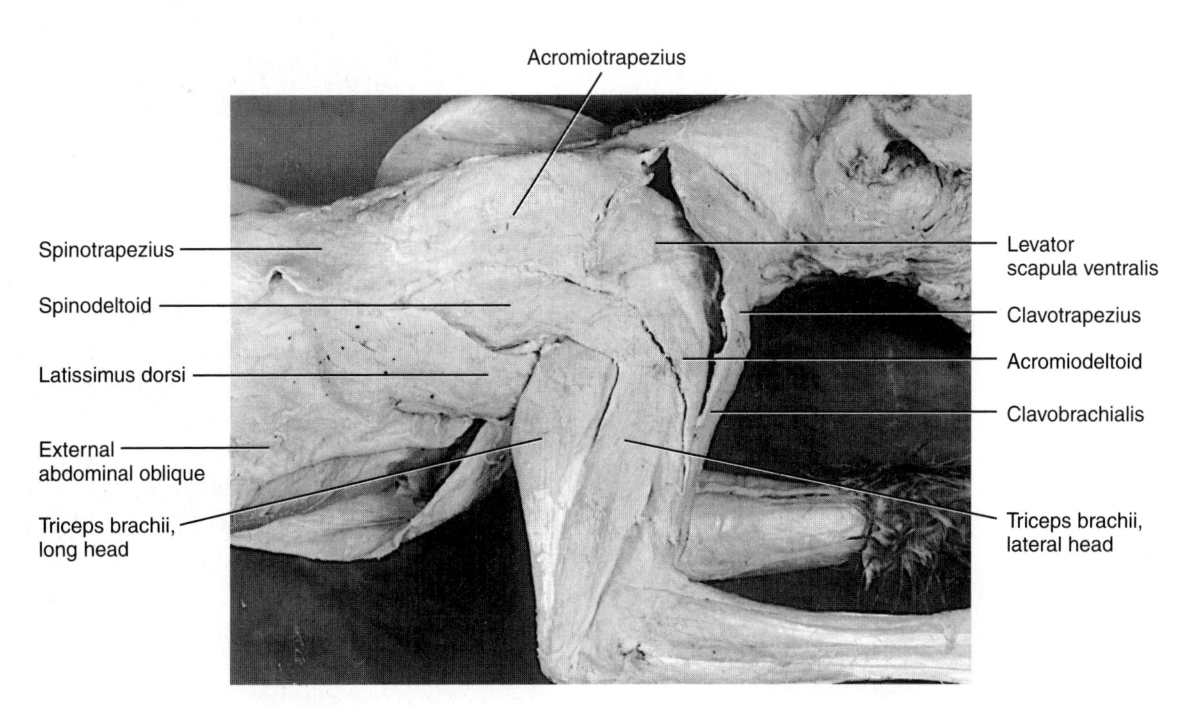

FIGURE 23.2
Dorsal muscles of the shoulder and forearm.

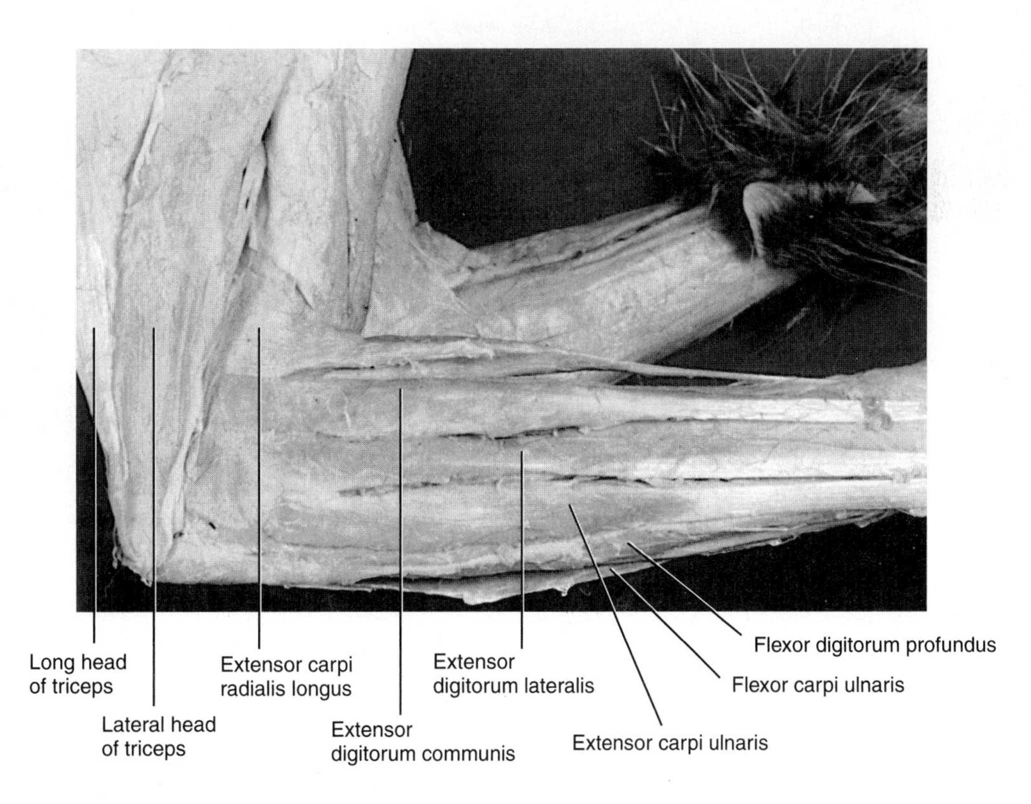

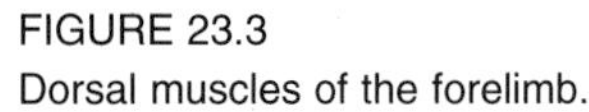
FIGURE 23.3
Dorsal muscles of the forelimb.

FIGURE 23.4
Abdominal muscle.

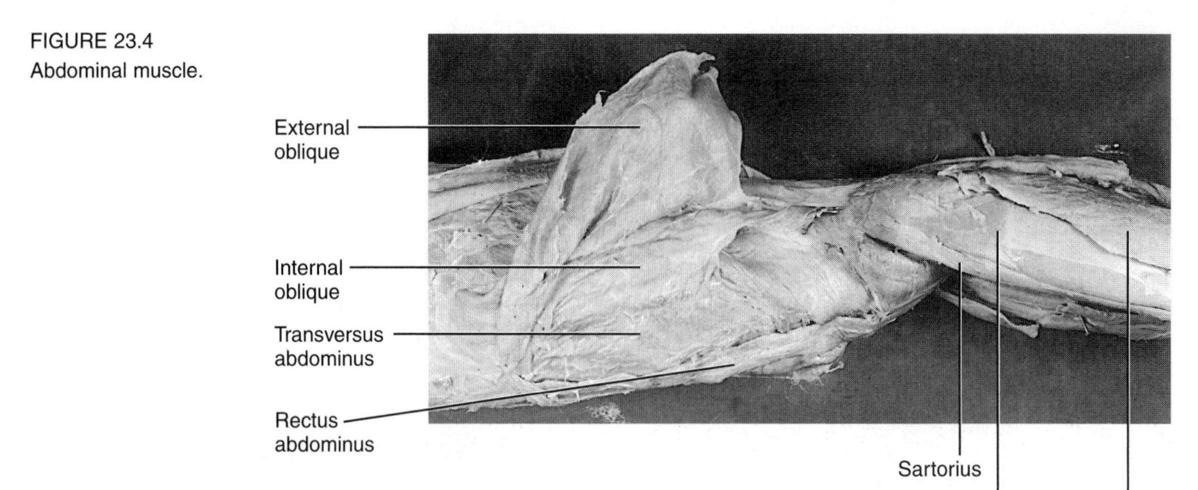

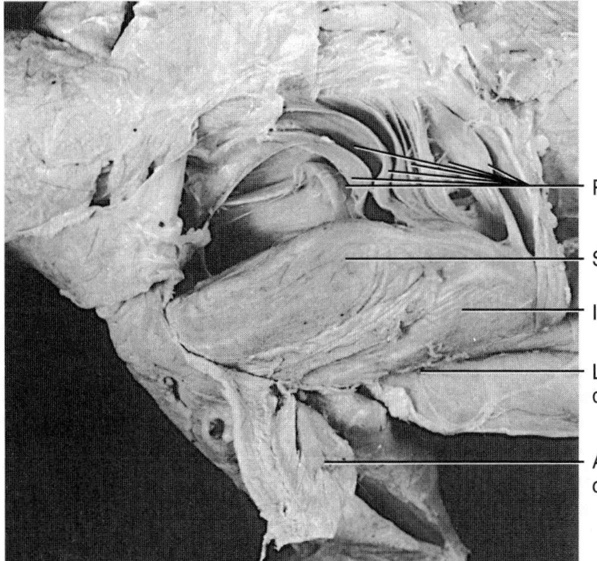

FIGURE 23.5
Deep muscles of the shoulder, dorsal view.

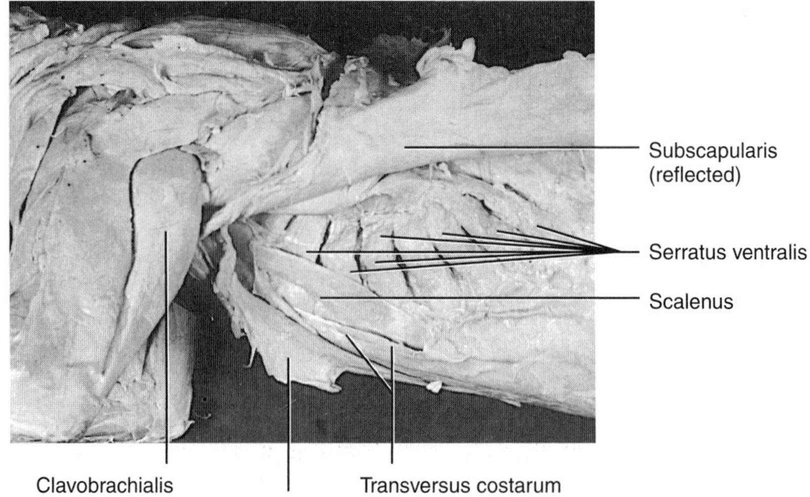

FIGURE 23.6
Deep muscles of the axillary region.

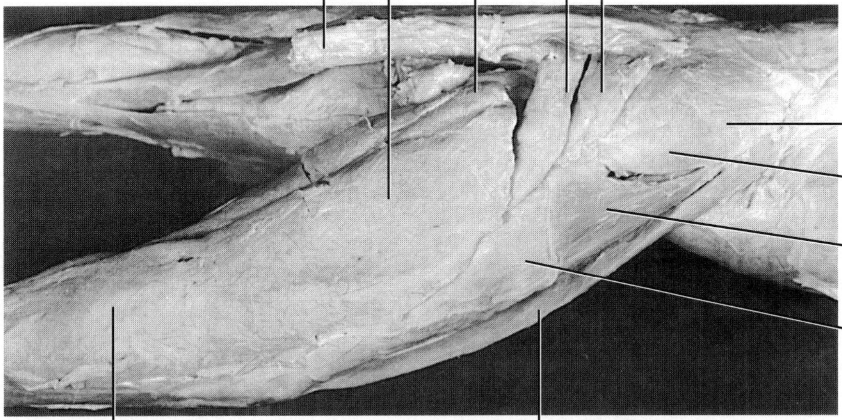

Biceps femoris
Caudofemoralis
Tail
Semitendinosus
Gluteus maximus
Thoracolumbar fascia
Gluteus medius
Tensor fasciae latae
Fascia lata
Gastrocnemius
Sartorius

FIGURE 23.7
Superficial muscles of the thigh, lateral view.

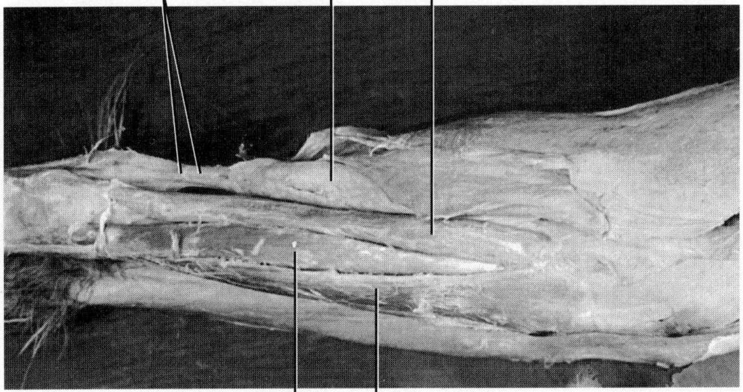

Tendon of Achilles
Gastrocnemius
Soleus
Peroneus longus
Extensor digitorum longus

FIGURE 23.8
Superficial lower leg muscles.

Gastrocnemius (cut and reflected)
Tenuissimus
Biceps femoris reflected
Peroneus longus
Extensor digitorum longus
Tibialis anterior
Sciatic nerve
Fascia lata
Tensor fasciae latae
Sartorius

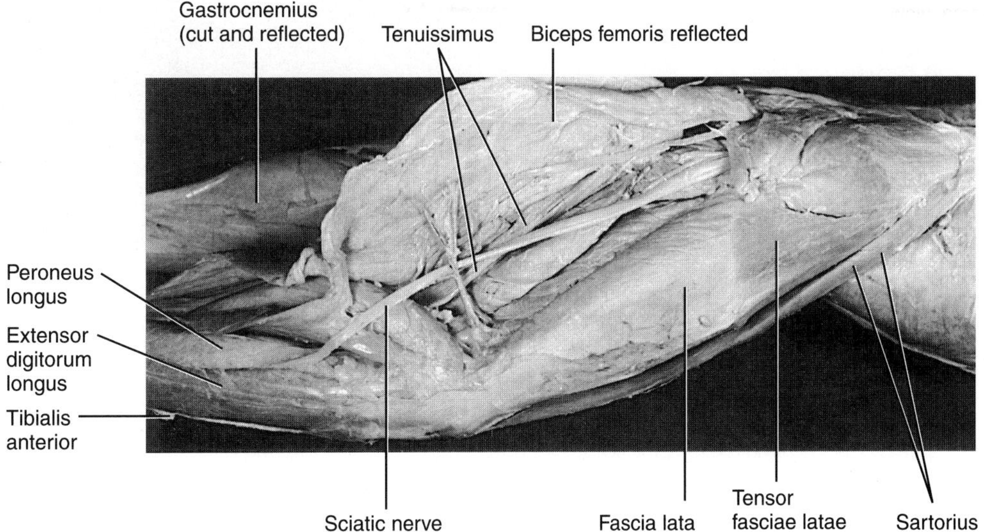

FIGURE 23.9
Deep muscles of the shank region, lateral view.

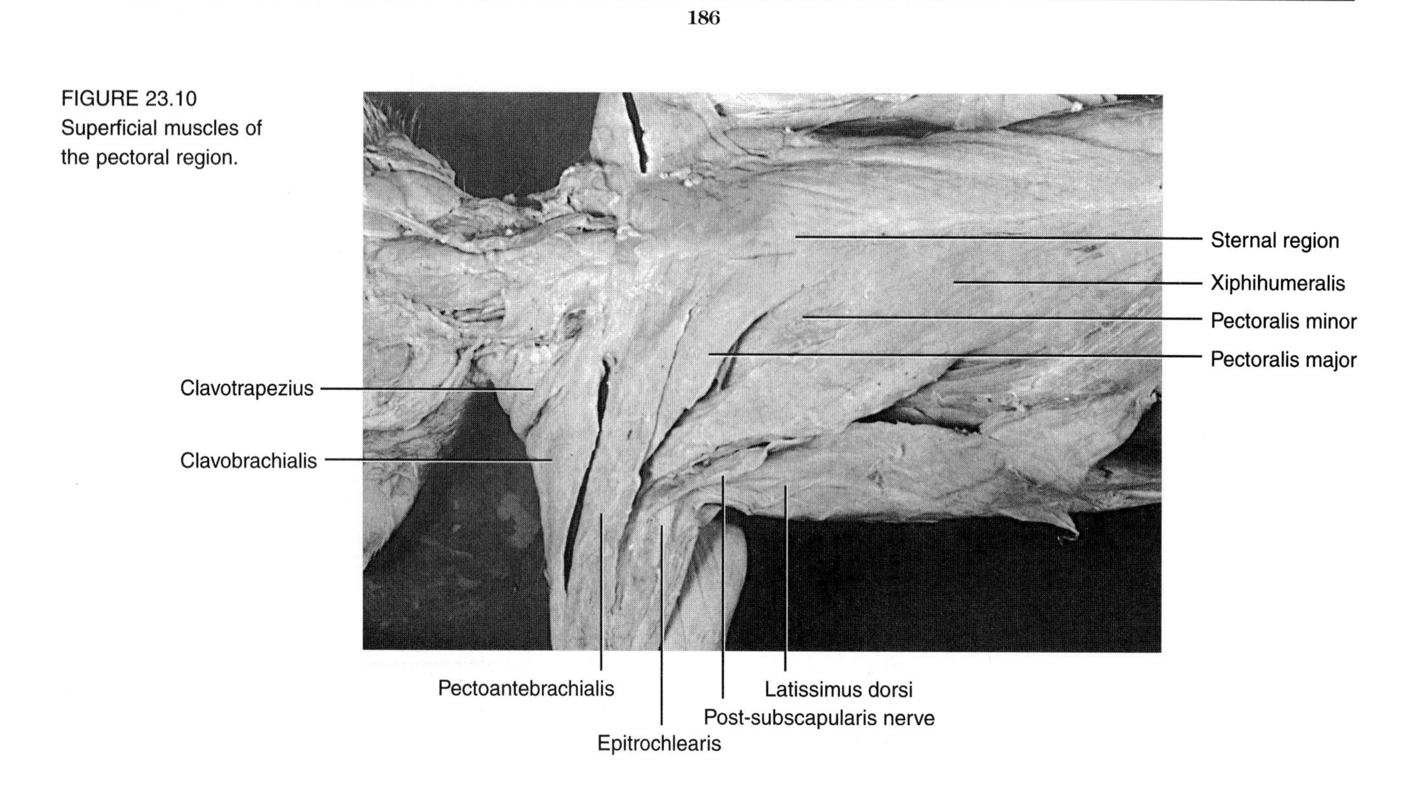

FIGURE 23.10
Superficial muscles of the pectoral region.

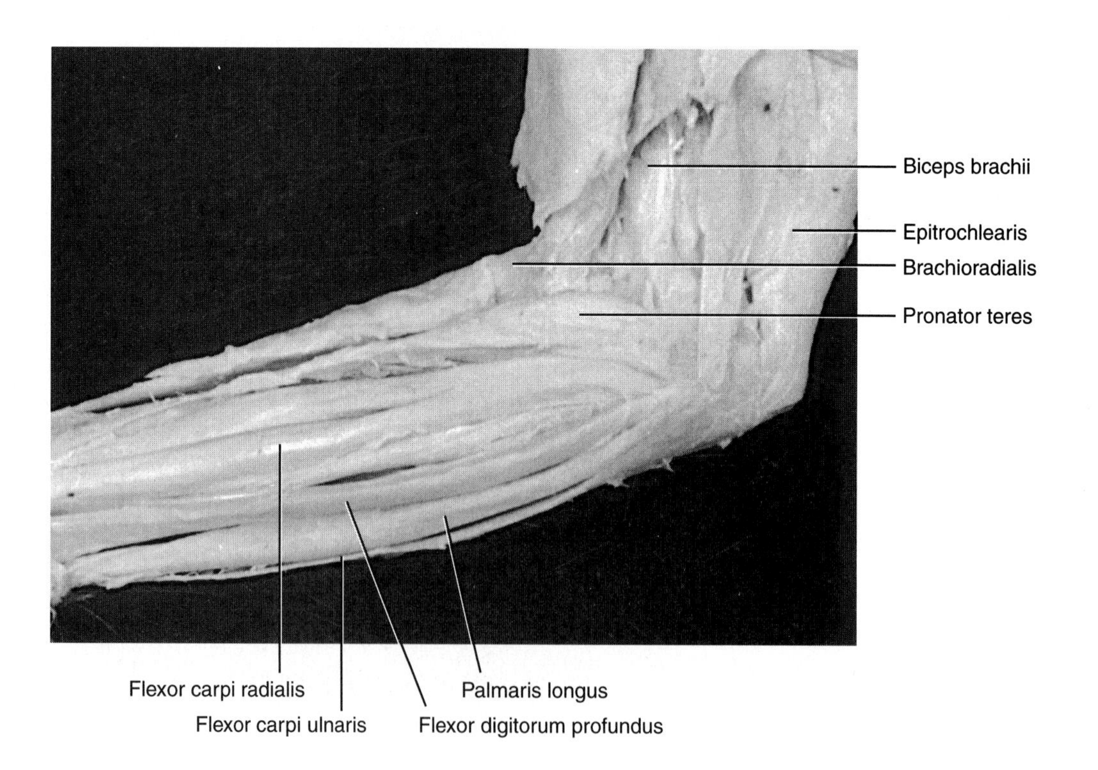

FIGURE 23.11
Muscles of the forelimb, medial view.

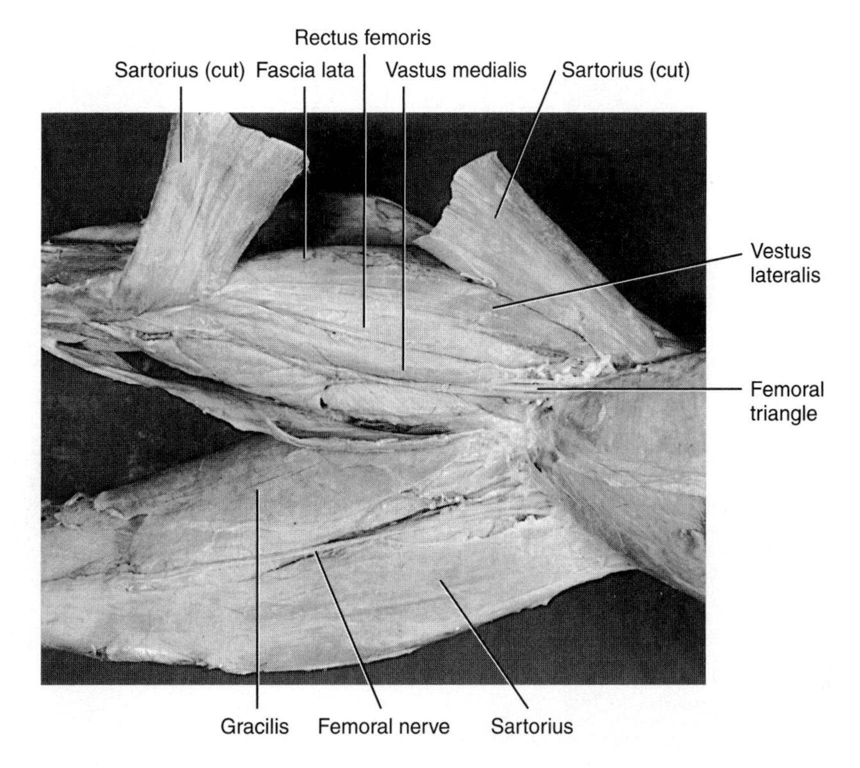

FIGURE 23.12
Thigh muscles, medial view.

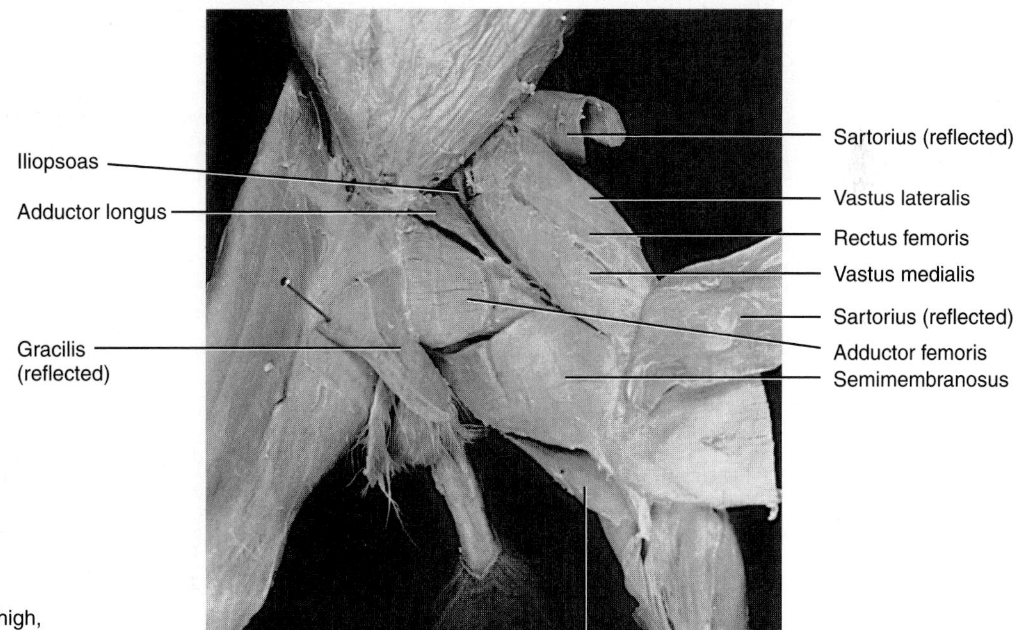

FIGURE 23.13
Deep muscles of the thigh, medial view.

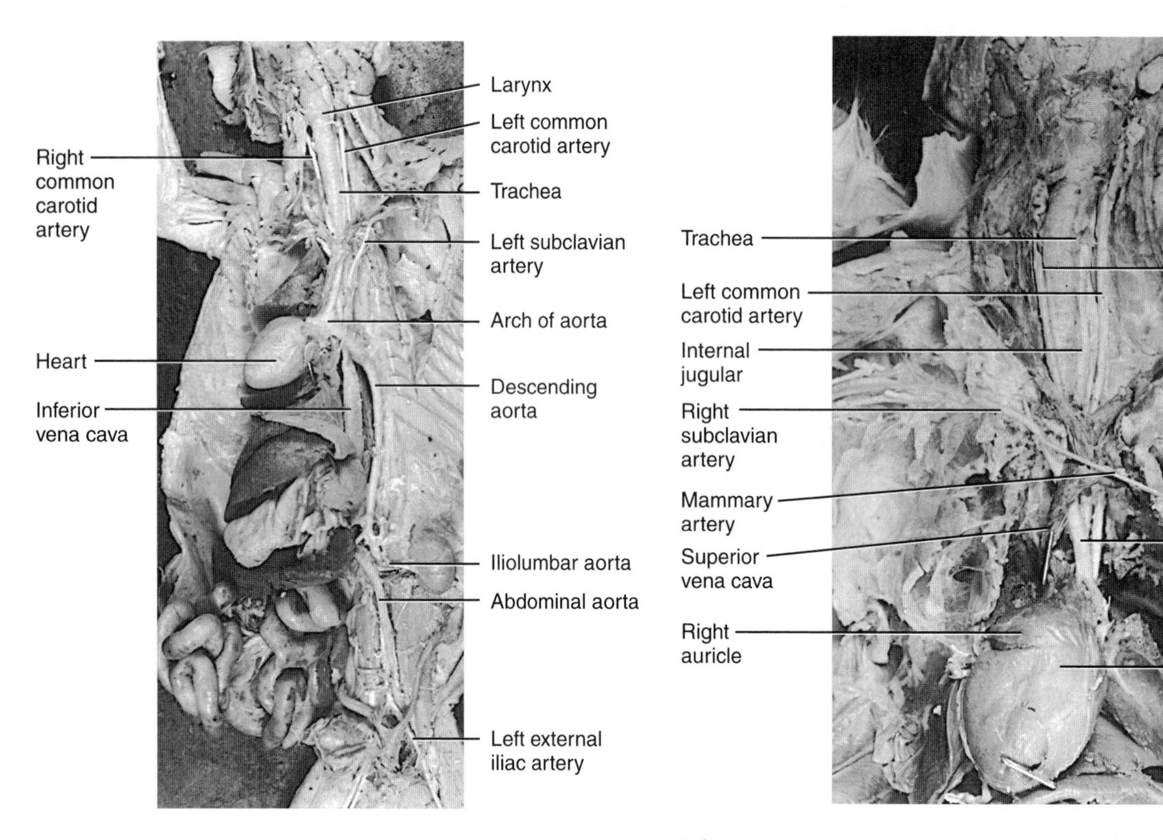

FIGURE 23.14
Major blood vessels of the cat.

FIGURE 23.15
Major blood vessels of the cat, anterior region.

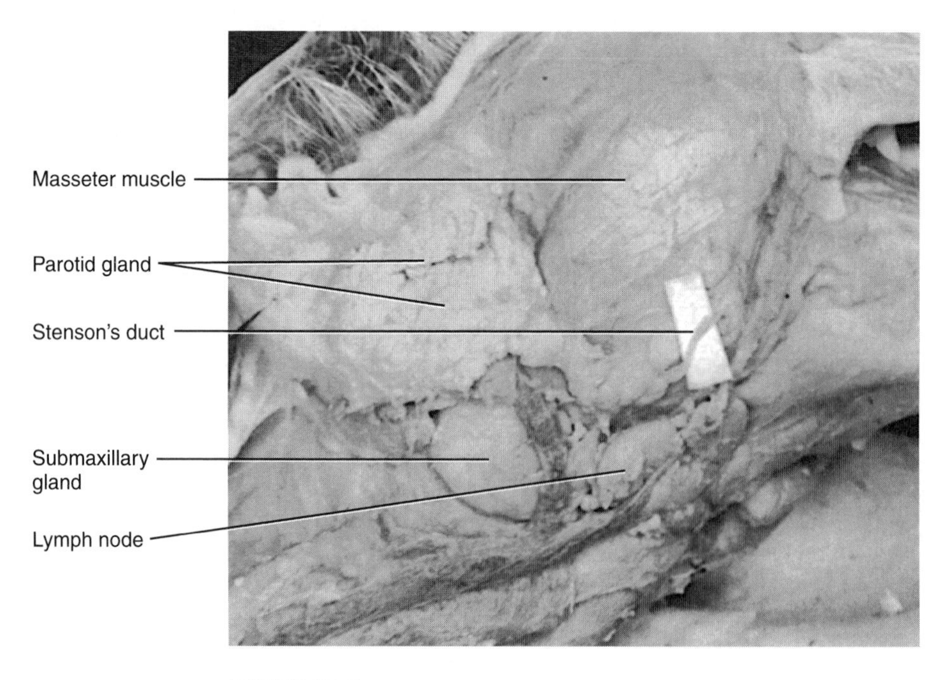

FIGURE 23.16
Salivary glands and lymph nodes.

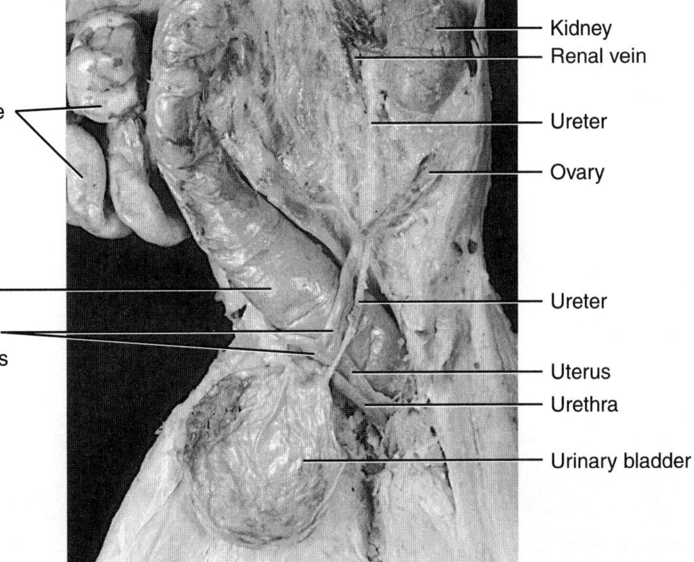

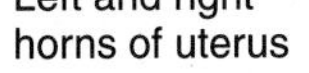

FIGURE 23.17
Urogenital system in a nonpregnant cat.

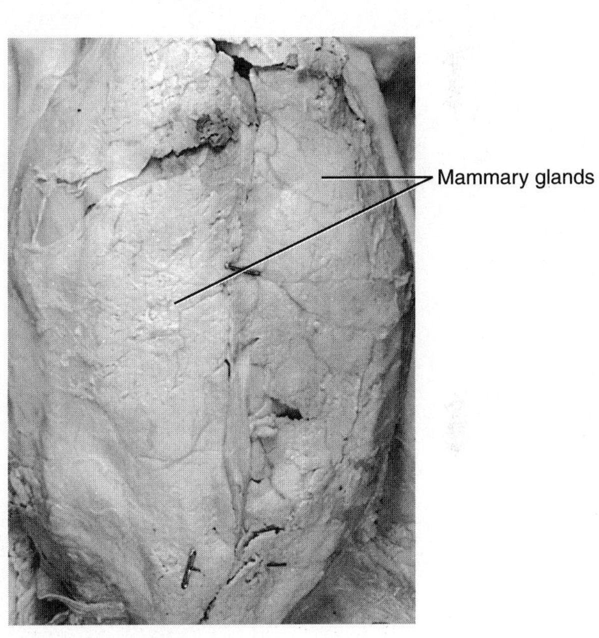

FIGURE 23.18
Mammary glands in a pregnant cat.

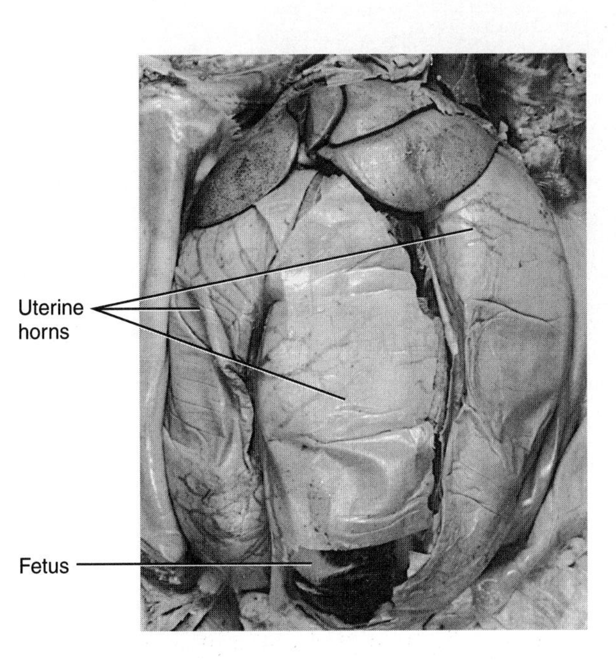

FIGURE 23.19
Uterine horns in a pregnant cat.

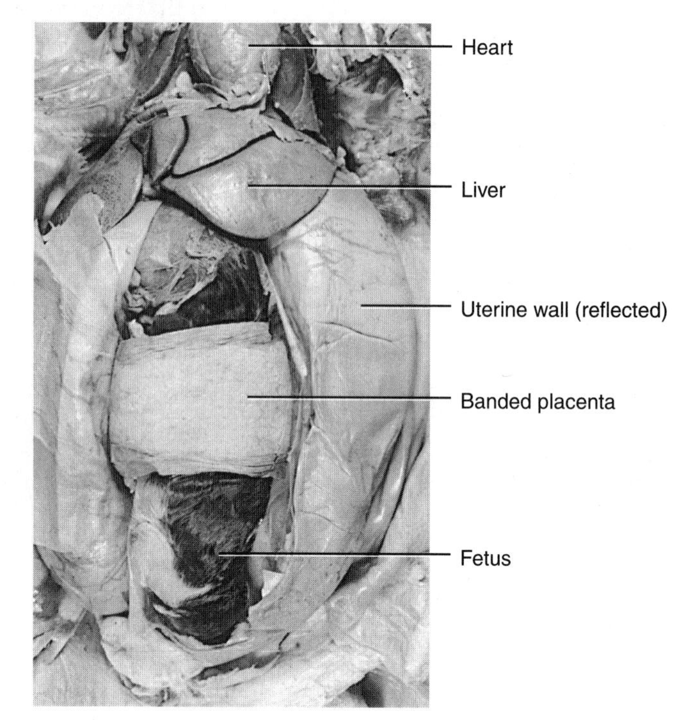

FIGURE 23.20
Uterine horn cut and reflected to display fetus and the banded placenta.

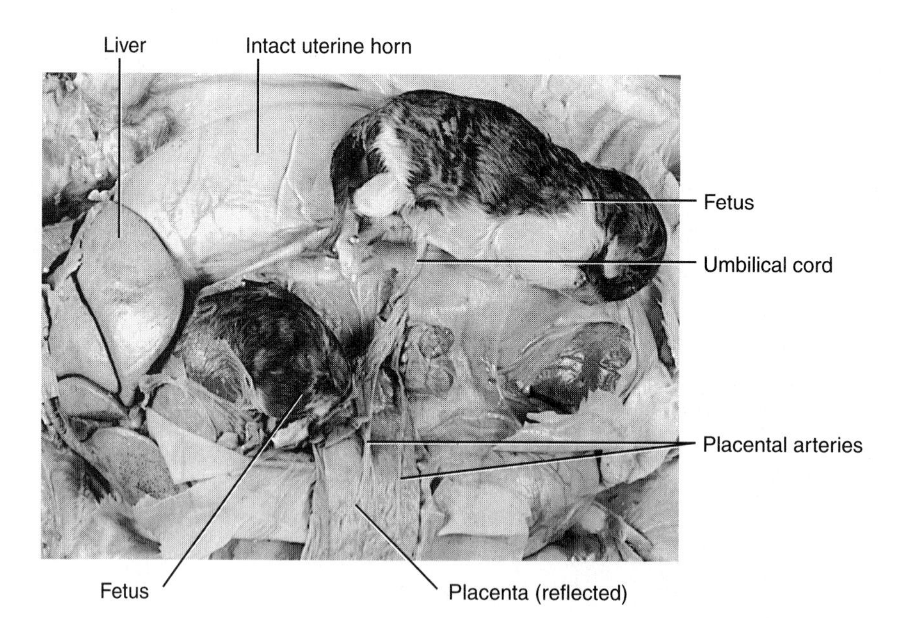

FIGURE 23.21
Banded placenta cut and reflected to show the blood vessels, fetus, and umbilical cord.

Fetal Pig Dissection

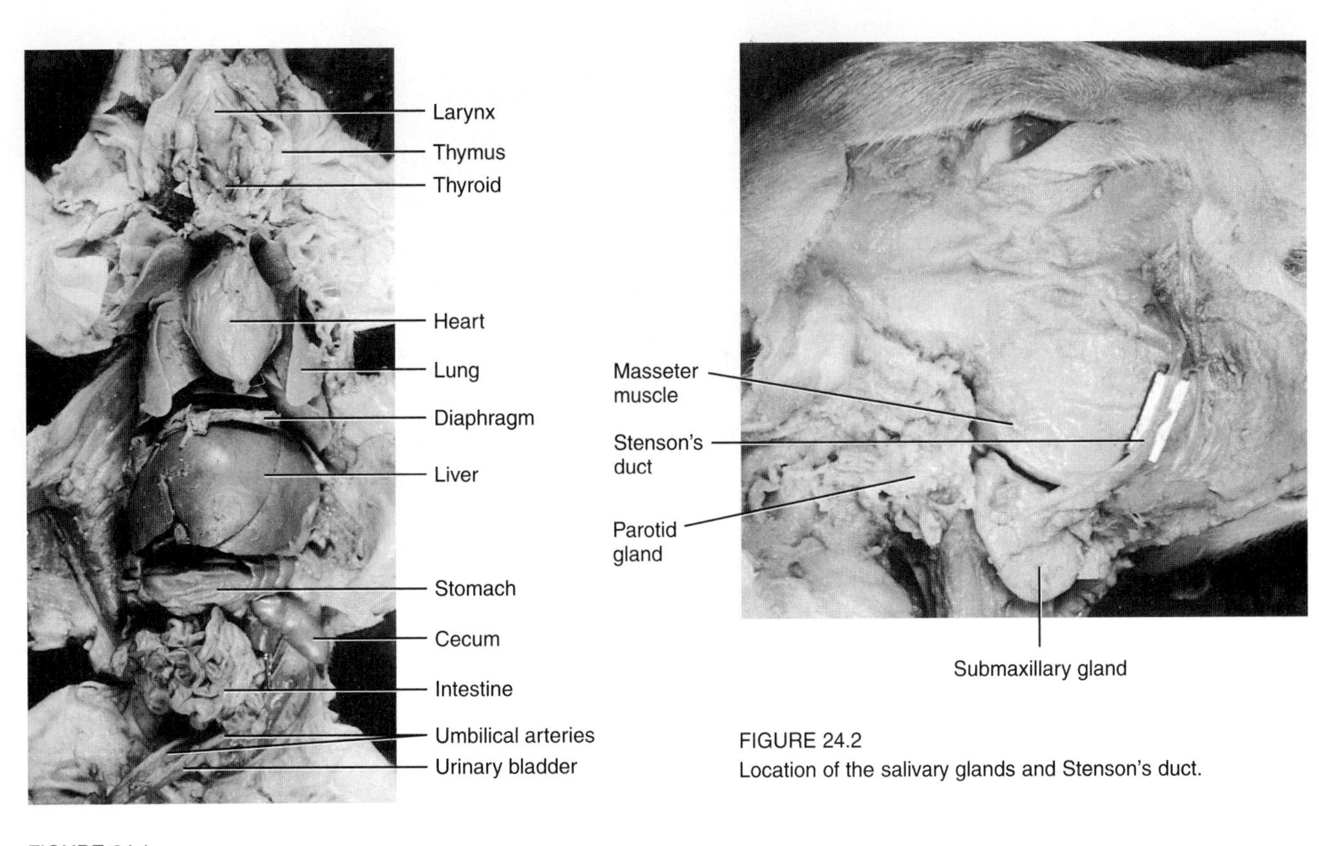

FIGURE 24.1
General internal anatomy of the fetal pig.

FIGURE 24.2
Location of the salivary glands and Stenson's duct.

Renal artery
Abdominal aorta
Kidney
Renal vein
Ureter
Umbilical arteries
Urinary bladder
Horns of uterus
Vagina

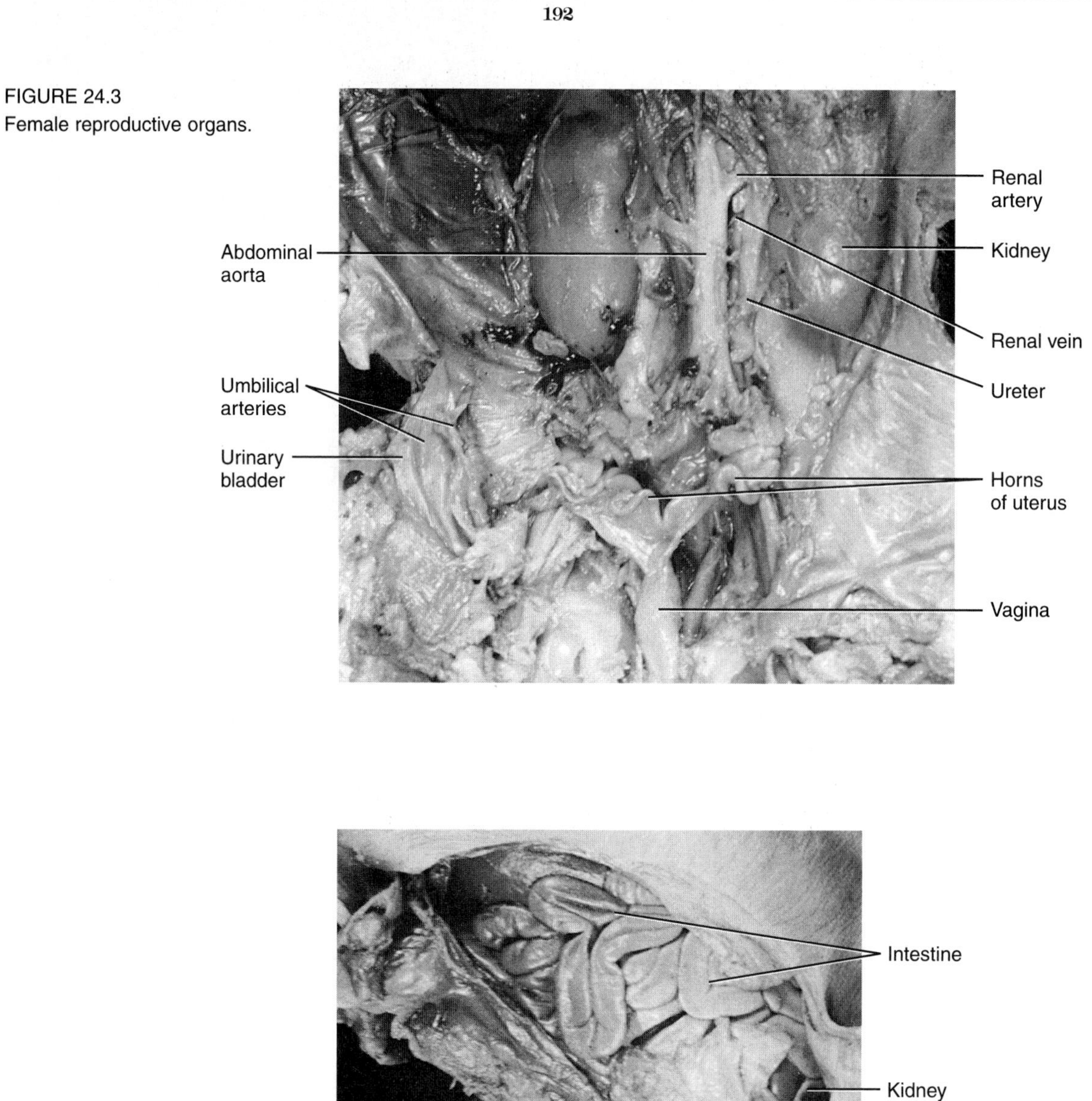

FIGURE 24.3
Female reproductive organs.

Intestine
Kidney
Urinary bladder
Penis
Glans penis
Spermatic cord
Testis

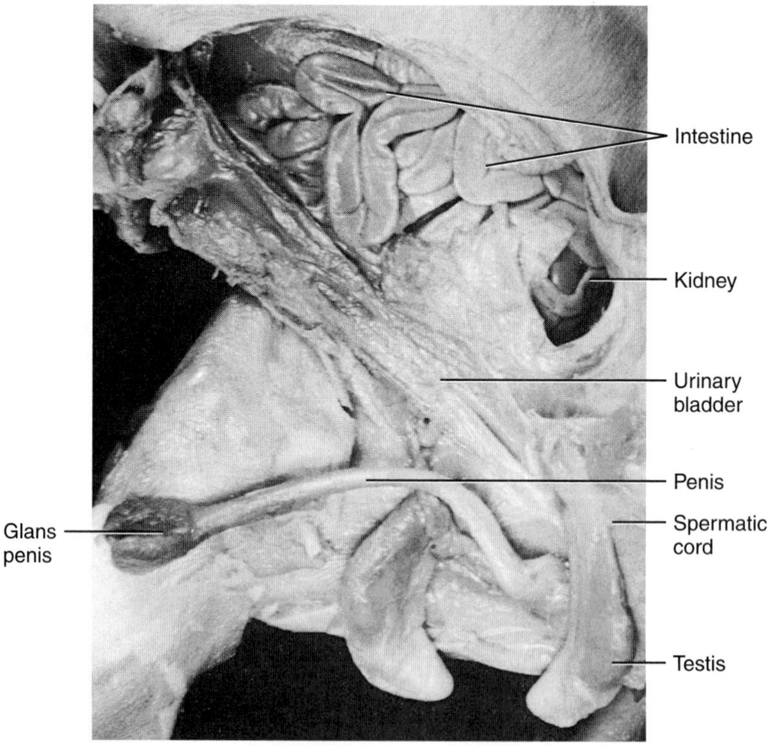

FIGURE 24.4
Male reproductive organs.

Credits

Photos

All photographs courtesy of Dennis Strete, Ph.D., except for the following:
7.63b courtesy of Lester Bergman and Associates
11.1 inset © 1991 Jan Leesma, M.D./Custom Medical Stock Photo.

Line Art

Leonard Dank

7.1a,b, 7.10a,b, 7.63a, 10.9, 10.10, 10.11, 10.12, 10.13, 10.14a,b,c,d, 10.15, 10.16, 10.17, 10.18, 10.19a,b, 10.20a,b.

Sharon Ellis

11.1a,b, 11.18, 11.27, 11.28, 11.29, 14.1, 14.2, 22.3.

Lauren Keswick

2.1, 2.2, 2.3, 3.4, 6.1 inset, 7.4 inset, 12.1.

Biagio John Melloni, Ph.D.

22.1, 22.8, 22.9, 22.10, 22.12.

Hilda Muinos

13.1, 13.8, 13.9, 15.13, 15.17.

Lynn O'Kelley

Chap. 1 opener, 16.1, 16.5, 18.1.

Jared Schneidman Design

3.1.

Nadine Sokol

15.5a,b, 17.1.

Kevin Somerville

1.5a,b, 1.6, 1.7, 1.8, 19.1, 19.2, 19.13, 20.1, 20.8, 20.10, 21.10, 21.15.

Index

A bands, in skeletal muscle, 82
Abdomen, 1, *5f*
  muscles of, *90f*, *184f*
  skin from, *109f*
  transverse plane of, *90f*
Abdominal, definition, *4f*
Abdominal cavity, 126
Abdominal muscle, of cat, *184f*
Abdominal oblique muscle, external, of cat, *183f*
Abdominopelvic regions, *7f*
Abducens (VI) nerve, *99f*
Abduction
  of arm, *73f*
  of carpometacarpal joint, *73f*
  of hand, *73f*
  of joints, 66
  of metacarpophalangeal joints, *73f*
  of phalanges, *73f*
  of synovial joints, 65
Abductor pollicis longus muscle, *85f–86f*, *88f–89f*
Accessory (XI) nerve, *99f*, *102f*
Acervilli, of pineal gland, *151f*
Acetabulum, *55f*, *68f*, *70f*
  labrum of, *77f*
  notch of, *56f*
Achilles tendon, *85f–86f*, *93f*
  of cat, *184f*
Acoustic meatus, internal, *44f*
Acromial, definition, *4f–5f*
Acromiodeltoid muscle, of cat, *182f–183f*
Acromion, *3f*, *56f–57f*, *76f*
  process of, *3f*, *56f–57f*
Acromiotrapezius muscle, of cat, *184f*
Acrosome, of sperm, *159f*
ACTH. *See* Adrenocorticotropic hormone
Actin, of skeletal muscle, *84f*
Adam's apple, *87f*
Adduction
  of arm, *73f*
  away from axis, *73f*
  of carpometacarpal joint, *73f*
  of hand, *73f*
  of joints, 66
  of metacarpophalangeal joints, *73f*
  of phalanges, *73f*
  of synovial joints, 65
  towards axis, *72f*
Adductor femoris muscle, of cat, *187f*
Adductor longus muscle, *85f*, *91f*
  of cat, *187f*
Adductor magnus muscle, *85f–86f*, *91f*
Adenosine triphosphate (ATP), 150
Adherens junction, of cells, *13f*
Adipocytes, *30f*, 79
Adipose cell(s), 27
  of bone marrow, *81f*
  in brown adipose tissue, *30f*
Adipose tissue, 26, *97f*, *108f*, *110f–111f*, *119f*
  brown, *30f*
  LM of, *30f*
  in neurovascular bundle, *119f*
  at tendon-skeletal muscle junction, *84f*
  of thymus, *124f*
Adrenal cortex, 150
Adrenal (suprarenal) gland(s), 150, *151f*, *154f*
  capsule of, *154f*
  cortex of, *154f*
  medulla of, *154f*
Adrenal medulla, 150
Adrenocorticotropic hormone (ACTH), 167
Adventitia
  arterial, 112, 121
  of bronchioles, 139
  of capillaries, 113
  of ductus (vas) deferens, *159f*
  intestinal, 127
  of ureter, *149f*
Afferent arteriole(s)
  of glomeruli, 144
  of nephron, *146f*
Afferent lymphatic vessel, *123f*
Aggregated lymphatic follicle. *See* Peyer's patch(es)
Agranulocytes, 79
Albumins, 78
Alimentary tract, 126
  mucosal glands of, 126
Allantois, *169f*
Alveolar air, 138
Alveolar bone, *67f*, *130f*
  angle of, *67f*
  body of, *67f*
  gomphosis joint of, *67f*
Alveolar duct(s), 138–139
Alveolar margin(s), *42f–43f*, *46f*, *49f*
Alveolar sacs. *See* Alveoli
Alveolar wall, *143f*
Alveoli, 138–139
  in bronchiole, *143f*
  in lobule, *166f*
  in lung tissue, *143f*
  in parenchymal epithelium, 108
Amino acids, 150
Amnion, *169f–171f*
Amniotic cavity, *169f*
Amniotic fluid, *170f*
Amphiarthrosis, 65–66
Ampullary nerves, *179f*
Anal canal, *136f*
Anal sphincter, external, *92f*
Anaphase
  in meiosis, *19f*
  in mitosis, *18f*
    micrograph of, *19f*
Anatomical positions, 1–11
Anconeus muscle, *86f*, *89f*
Androgens, 155
Angular movement, of synovial joints, 65–66
Ankle, *5f*
  joint, *64f*
Annular ligament, *76f*
Anococcygeal raphe muscle, *92f*
Anterior, definition, 2
Anthelix, *3f*
Antrum, *164f*
Anus, *92f*, 126–127, *156f*, *163f*
  epithelium of, *136f*
Aorta, *115f*, 139
  abdominal, *117f*
  of cat, *187f*
  of fetal pig, *192f*
  arch of, *114f–115f*, *117f*
  ascending, *114f–115f*, *117f*
  descending, *114f*
  of cat, *187f*
  elastic connective tissue in, *29f*
  iliolumbar, of cat, *188f*
  thoracic, *117f*
Appendix, *7f*, 121, *122f*, 127
  vermiform, *135f*
Aqueous humor, of eye, *175f*
Arachnoid membrane, 94
Arcuate artery, 144, *146f*
Arcuate vein, 144, *146f*
Areola, *166f*
Areolar tissue, 26
  cells of, 27
  in omentum, *28f*
Arm(s), 1, *5f*. *See also* Upper extremity
  abduction of, *73f*
  adduction of, *73f*
Armpit, *5f*
Arrector pili muscle, *108f*, *110f*
Arterial adventitia, 112, 121
Arteria profunda penis, *160f*
Arteriole(s), 112–113
  afferent, 144, *146f*
  endometrial, maternal, *172f*
  of small intestine, *133f*
Artery(ies), 112–113, *117f*. *See also specific artery*
  adventitia of, 121
  body wall of, *119f*
  bronchial, 139
  cross section, *119f*
  elastic, 112
  in neurovascular bundle, *119f*
  in skin, *108f*
  small, in bronchiole, *143f*
Arthrodial joints, 65–66
  in carpals, *69f*
  in wrists, *69f*
Articular cartilage, *58f*, *77f*
  fetal, *38f*
Articulation, 65–77
  of atlas, *54f*

Articulation *(continued)*
of axis, 54*f*
cartilaginous, 57*f*
cervical, 72*f*
of fibula, 62*f*
of humerus, 59*f*
of radius, 59*f*
of ulna, 59*f*
Arytenoid cartilage, 139
Aster microtubule, in prophase, 18*f*
Astrocytes, 94
fibrous, 96*f*
Atlantoaxial cervical joint, rotation at, 73*f*
Atlas
anterior arch of, 54*f*
anterior tubercle of, 53*f*–54*f*
articulation of, 54*f*
inferior view of, 53*f*
posterior arch of, 53*f*
posterior tubercle of, 53*f*
spinous process of, 69*f*
superior articular facet of, 53*f*
superior articular surface of, 54*f*
superior view of, 53*f*
ATP. *See* Adenosine triphosphate
Atrium, cardiac, 114*f*
Auditory canal (meatus)
external, 3*f*, 43*f*, 48*f*, 178*f*
internal, 178*f*–179*f*
Auditory ossicle, 174
Auditory (eustachian) tube(s), 139, 174, 178*f*, 180*f*
cartilage of, 32
orifice of, 140*f*
Auricle
of ear, 173, 178*f*
helix of, 3*f*
of heart, 114*f*–115*f*, 139
of cat, 188*f*
Auricular muscle, posterior, 87*f*
Autonomic nerve, 108*f*
Axilla, 8*f*–9*f*
definition, 4*f*
dense irregular connective tissue in, 28*f*
Axillary, definition, 5*f*
Axillary artery, 117*f*
Axillary node, 122*f*
Axillary region, muscles of, cat, 184*f*
Axillary skin, 109*f*
Axillary vein, 118*f*
Axis
articulation of, 54*f*
dens of, 54*f*
facet for, 53*f*
inferior view of, 53*f*
lamina of, 54*f*
pedicle of, 54*f*
superior articular surface of, 54*f*
superior view of, 53*f*–54*f*
Axolemma, 95*f*
Axon(s), 95*f*. *See also* Nerve fibers
of motor neuron, 103*f*
of peripheral nerve, 97*f*
of skeletal muscle fiber, 97*f*
Axon collateral, 95*f*
Axon hillock, 95*f*
Axon terminal, 95*f*

Back, 5*f*
Ball and socket joints, 65–66, 70*f*
Barr body, of neutrophil, 81*f*
Basal lamina, of capillaries, 113
Basement lamina, 21
in trachea, 25*f*
Basement membrane, of ureters, 145
Basilar membrane, of ear, 180*f*
Basilic vein, 118*f*
Basophil(s), 79, 80*f*–81*f*
granules of, 80*f*
Basophilic erythroblast, 81*f*
Bertz's cell, 97*f*
Biceps brachii muscle, 8*f*, 85*f*–86*f*, 88*f*, 90*f*
of cat, 186*f*
long head, of cat, 182*f*
tendon of, 76*f*, 88*f*
Biceps femoris muscle, 11*f*, 86*f*, 93*f*
of cat, 182*f*, 184*f*–185*f*
tendon of, 93*f*
Biceps muscle, 9*f*
Bicuspid teeth, 129*f*
Bicuspid (mitral) valve, 115*f*
Bile, 127
Bile duct, 136*f*
Billroth's cords, 121
Bilobed nucleus, 80*f*
Blind spot, 175*f*
Blood, 78–81
functions of, 78
Blood cells, 78
in neurovascular bundle, 119*f*
Blood plasma. *See* Plasma
Blood tissue, 26
Blood vessel(s), 22*f*, 114*f*, 123*f*. *See also* Artery(ies); Vein(s)
body wall of, 22*f*
bronchial, 142*f*
in bronchiole, 143*f*
in brown adipose tissue, 30*f*
of cat, 187*f*
in colon, 135*f*–136*f*
in compact bone tissue, 35*f*
endothelial cells of, 22*f*
fetal, 172*f*
in gallbladder, 137*f*
in jejunum, 134*f*
in lingual tonsil, 124*f*
lumen of, 22*f*
in lungs, 139
in lymph node, 29*f*
in nephron, 146*f*
in pancreas, 153*f*
in pineal gland, 151*f*
in pituitary gland, 152*f*
in respiratory tract, 138
in compact bone tissue, 35*f*
in trachea, 142*f*
in ureter, 149*f*
vaginal, 165*f*
B lymphocytes, 120
Body
central axis of, 1
organization of, 1–11
by subcutaneous muscles, 8*f*
planes of, 8*f*
regions of, 1–11, 5*f*, 7*f*
Bone(s), 34–64. *See also specific bone*
canaliculi of, 34, 35*f*–36*f*
cells, 34
of foot, superior view of, 64*f*
of hand, 60*f*
of skull, disarticulated, 46*f*
of wrist, 60*f*
Bone joints, cartilage of, 31
Bone marrow, 78–81, 111*f*, 122*f*
basophilic erythroblast of, 81*f*
cartilage and, 32*f*–33*f*
cells, 81*f*
in developing bone, 37*f*
in fetal bone, 38*f*
functions of, 79
red, 58*f*, 81*f*
in spongy bone trabeculae, 35*f*
yellow, 58*f*
Bone tissue
classification of, 34–35
components of, 34
nourishment of, 34
remodeling of, 34
of skull, developing, 36*f*
Bony labyrinth
of crista ampullaris, 181*f*
of ear, 174, 179*f*
Bowel, gas in, 56*f*
Bowman's capsule, 22*f*, 144–145, 146*f*
Bowman's membrane, of cornea, 177*f*
Brachial, definition, 4*f*–5*f*
Brachial artery, 88*f*, 117*f*
Brachialis muscle, 85*f*–86*f*, 88*f*
Brachial vein, 118*f*
Brachiocephalic trunk, 115*f*, 117*f*
Brachiocephalic vein, 114*f*
of cat, 188*f*
left, 114*f*
right, 114*f*, 118*f*
Brachioradialis muscle, 9*f*, 85*f*–86*f*, 88*f*–89*f*
of cat, 186*f*
Brain, 94
dorsal surface of, 105*f*
lateral view of, 99*f*
midsagittal section of, 100*f*, 106*f*
plains, MRI of, 104*f*
sagittal plane of, 105*f*
sagittal section of, 100*f*
sheep's, 105*f*–106*f*
transverse plane of, 105*f*
ventral aspect of, 100*f*
ventral surface of, 99*f*, 106*f*
Brain sand, 151*f*
Brain stem, 101*f*
anterior view, 99*f*, 103*f*
posterior view, 102*f*
Breast, 5*f*
frontal plane, 166*f*
lactating, 166*f*
sagittal plane, 166*f*
suspensory ligament of, 166*f*
Broad ligament, 165*f*
Bronchiole(s), 138–139, 141*f*
cross section, 143*f*
lumen of, 143*f*
mucosa of, longitudinal folds of, 143*f*
terminal, 141*f*
Bronchus/bronchi, 138–139, 141*f*
arterial system of, 139
cartilage of, 31
primary, 139, 141*f*–142*f*
secondary, 139, 141*f*
tertiary, 139, 141*f*
Brunner's glands, 134*f*
Buccinator muscle, 87*f*
Bulbar conjunctiva, 175*f*–176*f*
Bulbocavernosus muscle, 92*f*
Bulbospongiosus muscle, 92*f*
Bulbourethral glands, 155, 156*f*, 161*f*
Buttock, 5*f*

Calcaneal, definition, 4*f*–5*f*
Calcaneal muscle, 85*f*
Calcaneal tendon, 86*f*, 93*f*
Calcaneus, 64*f*, 70*f*
anterior tubercle of, 64*f*
Calcitonin hormone, 167
Calf, 5*f*
Calyx/calyces, renal
major, 146*f*
minor, 146*f*
cAMP. *See* Cyclic adenosine monophosphate
Canal(s)
anal, 136*f*
auditory. *See* Auditory canal
carotid, 43*f*, 47*f*
central, of spinal cord, 103*f*–104*f*
cervical, 165*f*
condylar, 45*f*, 48*f*
fascial, 3*f*
haversian, 35, 35*f*–36*f*
hypoglossal, 44*f*
incisive, 44*f*
lacrimal, 46*f*
optic, 47*f*
palatonasal, 43*f*
pterygoid, 47*f*
of Schlemm, 175*f*–176*f*
semicircular, 178*f*–179*f*
Volkmann's, 36*f*
Canaliculi, of bone, 34, 35*f*–36*f*
Cancellous bone tissue. *See* Spongy bone
Canine teeth, 42*f*–43*f*, 46*f*, 129*f*
Capillary(ies), 113, 121
in adipose connective tissue, 30*f*
adventitia of, 113
basal lamina of, 113
of ear, 180*f*
peritubular, 144, 146*f*
of small intestine, 133*f*
Capillary plexus, 139
Capitate bone, 60*f*
Capitulum, 58*f*–59*f*
Capsular space
of Bowman's corpuscle, 22*f*
of renal corpuscle, 148*f*
Capsule, 164*f*
of lymph node, 123*f*
of pituitary gland, 152*f*
Carbon dioxide, 138
Cardiac muscle, 82, 84*f*, 116*f*
cross section of, 84*f*
endomysium of, 84*f*
fibers, 116*f*
intercalacted disk in, 84*f*
nucleus of, 84*f*, 116*f*
perimysium of, 84*f*
Cardiac prominence, embryonic, 169*f*
Cardiac sphincter, 127
Cardiac stomach, 132*f*
Cardiac vein, great, 118*f*
Cardiovascular system, 112–119
functions of, 112
vasculature of, 112–113

layers of, 112
Carina, 141*f*
Carotid artery
of cat, 188*f*
common, 114*f*, 117*f*
of cat, 188*f*
external, 117*f*
internal, 117*f*
Carotid canal, 43*f*
groove of, 47*f*
Carotid groove, 45*f*
Carpal, definition, 4*f*–5*f*
Carpals, 6*f*, 40*f*, 69*f*, 89*f*
joints of, 65
Carpometacarpal joint
abduction of, 73*f*
adduction of, 73*f*
Cartilage, 31–33
articular, 58*f*, 77*f*
fetal, 38*f*
arytenoid, 139
calcifying, 38*f*
chemical composition of, 31
corniculate, 139
costal, 56*f*
cricoid, 139, 140*f*
cuneiform, 139
elastic, 31–32, 33*f*, 141*f*
epiphyseal, 37*f*
fibrous, 31–32, 33*f*
hyaline, 31, 32*f*, 142*f*
of larynx, 139, 140*f*
nasal, 31, 87*f*
proliferating, 38*f*
reserve, 38*f*
thyroid, 87*f*, 139, 140*f*, 142*f*
Cartilage matrix, 31
in fetal bone, 38*f*
Cartilaginous joints, 65–66
Cat
digestive system of, 128*f*
dissection of, 182–190
Caudofemoralis muscle, of cat, 185*f*
Cavernous tissue, 138
Cecum, 7*f*, 127
in fetal pig, 191*f*
glandular epithelium of, 135*f*
Celiac trunk artery, 117*f*
Cell(s), 12–16. *See also specific cells*
components of, 12
division of. *See* Cytokinesis; Mitosis
functions of, 12
sectional view of, 13*f*
Cell cycle, 17*f*–18*f*
Cell junctions
locations of, 13*f*
types of, 13*f*
Cell membrane, 12, 13*f*
in interphase, 18*f*
structures of, 13*f*
Cells of Leydig, 155
Cementum, of teeth, 126
Central canal, of spinal cord, 103*f*–104*f*
Central nervous system, 94
membranes of, 94
Central vein, of liver lobule, 136*f*
Centriole(s), 12, 13*f*–14*f*
in meiosis, 19*f*
Centromere(s)
in meiosis, 19*f*
in mitosis, 18*f*
Centrosome(s), 13*f*
Cephalic, definition, 2, 4*f*–5*f*
Cephalic vein, 118*f*
Cerebellar cortex, 95
granular layer, 102*f*
molecular layer, 102*f*
Purkinje cells in, 96*f*, 102*f*
sagittal section, 102*f*
white layer, 102*f*
Cerebellar peduncle
inferior, 102*f*
middle, 99*f*, 102*f*
superior, 102*f*
Cerebellum, 95, 101*f*, 106*f*
midsagittal section, 100*f*
anterior aspect of, 99*f*
hemispheres of, 95
sagittal plane, 105*f*
sagittal section, 100*f*
vermis of, 95, 105*f*
Cerebral cortex, 94–95
Bertz's cell in, 97*f*
external granular layer, 101*f*
external pyramidal layer, 101*f*
internal granular layer, 101*f*
internal pyramidal (ganglionic) layer, 101*f*
molecular layer, 101*f*
multiform (polymorphic) layer, 101*f*
Cerebral hemisphere, 105*f*
Cerebral peduncle, 99*f*–100*f*, 106*f*
Cerebrum, 99*f*
ventral surface, 106*f*
Cerumen, of external ear, 173
Ceruminous glands, 173
Cervical, definition, 4*f*–5*f*
Cervical articulation, 72*f*–73*f*
Cervical canal, 165*f*
Cervical epithelium, 165*f*
Cervical nerve, first, dorsal root of, 102*f*
Cervical node, 122*f*
Cervical spine, 45*f*
curvature of, 52*f*
Cervical vertebra(e), 52*f*
arch of, 54*f*
body of, 54*f*
first, 103*f*
inferior view, 53*f*
lamina of, 54*f*
pedicle of, 54*f*
superior view, 53*f*–54*f*
Cervix, 162, 163*f*, 170*f*
Channels (pores), of plasma membrane, 13*f*
Chest, 1, 5*f*
Chiolar cells, 143*f*
Choanae, 43*f*
Cholesterol
derivatives of, 150
of plasma membrane, 13*f*
Chondroblasts, 31
Chondrocytes, 31, 32*f*
in developing bone lacunae, 37*f*
in elastic cartilage, 33*f*
in fetal bone, 38*f*
in fibrous cartilage, 33*f*
Chondromucoids, 31
Chondronectin, 31
Chordae tendineae, 115*f*
Chorion, 170*f*, 172*f*
Chorionic villi, 170*f*, 172*f*
Choroid, 173, 175*f*–176*f*, 178*f*
Chromaffin bodies, 150
Chromatids, in meiosis, 19*f*
Chromatin, 13*f*
in interphase, 18*f*
Chromatophilic substance, 95*f*
Chromosome(s)
female, karyotype of, 20*f*
male, karyotype of, 20*f*
in meiosis, 19*f*
in mitosis, 18*f*
metaphase of, 20*f*
onset of, 20*f*
paired homologous, 19*f*
Cilia, 12, 13*f*
of epithelium, 25*f*
TEM of, 15*f*
Ciliary body, 173, 175*f*
Ciliary muscle, 175*f*–176*f*
Ciliary process, 175*f*
Circular muscle(s)
of ductus deferens, 159*f*
of gastrointestinal tract, 132*f*
of small intestine, 133*f*
of ureter, 149*f*
Circulatory system, 120
Circumduction, of synovial joint, 65–66
Circumvallate papillae, of tongue, 126, 130*f*
Cisterna chyli, 122*f*
Clara cells, 143*f*
Clavicle, 3*f*, 6*f*–7*f*, 9*f*, 40*f*, 56*f*, 76*f*, 90*f*
acromial (lateral) end, 57*f*
conoid tuberosity of, 57*f*
costal tuberosity of, 57*f*
inferior view of, 57*f*
joints of, 65
medial (sternal) end, 57*f*
and sternum, gliding movement between, 71*f*
Clavobrachialis muscle, of cat, 182*f*–184*f*, 186*f*
Clavotrapezius muscle, of cat, 182*f*–183*f*, 186*f*
Cleavage furrow
in meiosis, 19*f*
in mitosis, 18*f*
Clitoris, 92*f*, 163*f*
Clivus, 45*f*
Coccygeus muscle, 92*f*
Coccyx, 52*f*, 55*f*–56*f*, 68*f*, 92*f*, 156*f*, 163*f*
Cochlea, 174, 178*f*–180*f*
apex of, 180*f*
Cochlear duct, 179*f*
Collagen fibers, 27
in areolar connective tissue, 28*f*
in axilla, 28*f*
of nerve, 16*f*
TEM of, 16*f*
Collagenous tissue, 26
of dermis, 107
Collateral ligament(s)
of fibula, 77*f*
of radius, 76*f*
of tibial, 77*f*
of ulna, 76*f*
Collecting duct(s), renal, 146*f*–147*f*, 149*f*
Collecting tubules, 148*f*
Collecting vessels, of lymphatic system, 121
Colliculus, cerebral
inferior, 102*f*
superior, 102*f*
Colloid
of pituitary gland, 152*f*
of thyroid gland, 153*f*
Colon, 126–127, 135*f*–136*f*
ascending, 6*f*–7*f*, 127
of cat, 128*f*
descending, 6*f*–7*f*, 127
sigmoid, 127
transverse, 6*f*–7*f*, 127
Columnar epithelium, 107
pseudostratified. *See* Pseudostratified columnar epithelium
simple. *See* Simple columnar epithelium
stratified, 21
Common tendinous ring, 88*f*
Compact bone, 26, 34–35, 35*f*–36*f*, 58*f*, 68*f*
of fetal finger, 38*f*
Concentric lamellae, in compact bone tissue, 35*f*
Conchae. *See* Nasal concha(e)
Condylar canal, 45*f*, 48*f*
Condyle(s)
femoral, 63*f*
lateral, 61*f*, 63*f*
medial, 61*f*
humeral, internal, 76*f*
mandibular, 43*f*, 49*f*, 67*f*
occipital, 44*f*, 48*f*
superior articular facet for, 69*f*
tibial, 63*f*
lateral, 62*f*, 68*f*
medial, 62*f*, 68*f*
Condyloid joint(s), 65–66, 70*f*
Cones, 178*f*
Conjunctiva
bulbar, 175*f*–176*f*
palpebral, 176*f*
Connective tissue, 12, 21, 26–30
adipose. *See* Adipose tissue
adult, 26
areolar, 26–27, 28*f*
bone, 34
cartilage, 26, 31–33
cells, 27
collagenous, 26, 107
composition of, 31
dense. *See* Dense connective tissue
dermal, 23*f*–24*f*, 107, 110*f*
elastic, 26, 107
of epididymis, 158*f*
fetal, 26
of fetal finger, 38*f*
fibers, 27
functions of, 26
ground substance of, 27
of hair follicle, 108
interlobular, 166*f*
intralobular, 166*f*
mesenchymal, 26, 28*f*
mucous, 171*f*
of prostate gland, 161*f*
reticular, 29*f*
of seminal vesicle, 159*f*
of skin, 109*f*, 111*f*
specialized, 26, 78–81. *See also* Blood; Bone(s); Cartilage

Connective tissue *(continued)*
of stratified squamous epithelium, 23*f*
tracheal, 141*f*
of urinary bladder wall, 149*f*
Connective tissue sheath, of hair follicles, 108, 108*f*
Cooper's ligament, 166*f*
Coracoacromial ligament, 76*f*
Coracobrachial muscle, bursa of, 76*f*
Coracohumeral ligament, 76*f*
Cornea, 88*f*, 175*f*–177*f*
Corneosacral coat, 173
Corniculate cartilage, 139
Cornified epithelium, 107
Coronal plane, 2, 8*f*
Coronal suture, 67*f*
Corona radiata, 164*f*, 168*f*
Coronary artery, 114*f*, 117*f*
Coronary sinus, 118*f*
Coronary sulcus, 114*f*
Corpora amylacea, 155
Corpora arenacea, 151*f*
Corpora cavernosa penis, 156, 160*f*–161*f*
tunica albuginea of, 160*f*
Corpora cavernosa urethrae, 155–156, 160*f*
Corpora quadrigemina, 100*f*
Corpus callosum, 100*f*, 105*f*–106*f*
splenium of, 100*f*, 105*f*
Corpus luteum, 150, 164*f*, 168*f*
Corpus spongiosum. *See* Corpora cavernosa urethrae
Corpus spongiosum penis, 161*f*
Corpus uteri, 162
Corrugator supercilii muscle, 87*f*
Cortical cells
pyramidal, 95
spindle-shaped, 95
stellate, 95
Cortical sinus, 123*f*
Cortical stroma, 164*f*
Corticomedullary border area, 144
Corticomedullary junction, 146*f*
cortex of, 146*f*
medulla of, 146*f*
Corticospinal tract fibers
lateral, 103*f*
ventral, 103*f*
Costal cartilage, 56*f*
Cowper's glands. *See* Bulbourethral glands
Coxae, lateral view of, 55*f*
Coxal, definition, 4*f*–5*f*
Coxal joint, 77*f*
Cranial, definition, 2, 5*f*
Cranial nerves, 99*f*, 102*f*–103*f*. *See also specific nerve*
Cranial vault, superior view, 45*f*
Cranium, 94
Cremaster muscle, 157*f*
Cricoid cartilage, 139, 140*f*
Cricoid membrane, 139
Crista/cristae
within ampulla of semicircular duct, 179*f*
in cellular cytosol, 15*f*
Crista ampullaris, 180*f*
capsule of, 180*f*
Crista galli, 45*f*, 50*f*
Cross-sectional plane, 8*f*
Crural, definition, 4*f*–5*f*
Crypt of Lieberkühn, 133*f*
Cubital vein, 118*f*
Cuboid, 64*f*, 70*f*
Cuboidal cells, 23*f*
Cuboidal epithelium
simple, 21, 23*f*
stratified, 21
of submaxillary gland, 24*f*
of sweat glands, 107
Cumulus oophorus, 164*f*
Cuneiform bone(s)
first (medial), 64*f*, 70*f*
second (intermediate), 64*f*, 70*f*
third (lateral), 64*f*, 70*f*
Cuneiform cartilage, 139
Cuspid, 129*f*
Cyclic adenosine monophosphate (cAMP), 150
Cystic duct, 127
Cytogenic holocrine gland. *See* Exocrine gland
Cytokinesis, 17*f*–18*f*
Cytoplasm, 12, 13*f*
components of, 13*f*
of neurolemmocyte, 95*f*
of neuron, 95*f*
Cytosol, 13*f*
in interphase, 18*f*
TEM of, 15*f*

Daughter cells
formation of, 18*f*
micrograph of, 19*f*
Decidua
basalis, 170*f*, 172*f*
capsularis, 170*f*
parietalis, 170*f*
Deciduous teeth, development of, 41*f*
Deltoid muscle, 3*f*, 8*f*–9*f*, 85*f*–86*f*, 90*f*
Demifacet, 54*f*
Dendrites, 95*f*
Dense bone tissue. *See* Compact bone
Dense connective tissue, 119*f*
in axilla, 28*f*
irregular, 26, 28*f*
regular, 28*f*
of ligamentum nuchae, 29*f*
of skin, 109*f*
of tooth, 130*f*
Dentin, 126, 130*f*
Dentitions
deciduous, 129*f*
permanent, 129*f*
Depression
of joints, 66
of mandible, 74*f*
Depressor anguli oris muscle, 85*f*, 87*f*
Depressor labii inferioris muscle, 87*f*
Dermal connective tissue, 23*f*–24*f*, 107, 110*f*
Dermal papilla(e), 108*f*–109*f*, 111*f*
of stratified squamous epithelium, 23*f*–24*f*
Dermal sheath, of hair follicle, 110*f*
Dermis, 23*f*, 107, 108*f*–111*f*
layers of, 107
Descemet's membrane, 177*f*
Desmosome, 13*f*
Diaphragm, 6*f*–7*f*, 117*f*, 127, 141*f*
of cat, 128*f*
of fetal pig, 191*f*
urogenital, 156*f*, 161*f*
inferior fascia of, 92*f*
Diaphysis, of long bones, 35, 58*f*, 68*f*
Diarthrosis, 65–66
Digastric fossa, 49*f*
Digestive enzymes, 127
Digestive system, 126–137
of cat, 128*f*
proper, 126
Digital, definition, 4*f*–5*f*
Digitalis artery, 117*f*
Digitalis vein, 118*f*
Diploë, 45*f*
Directional terms, 6*f*
Distal, definition, 2, 6*f*
Distal convoluted tubule, 146*f*–148*f*
Dorsal, definition, 2, 4*f*–5*f*
Dorsal arch artery, 117*f*
Dorsal interosseus muscle, 89*f*
Dorsalis pedis artery, 117*f*
Dorsal muscles, of cat, 183*f*
Dorsiflexion, 66
of foot, 75*f*
Dorsum, definition, 2
Dorsum sellae, 45*f*, 47*f*
Duct(s)
alveolar, 138–139
bile, 136*f*
cochlear, 178*f*
collecting, 146*f*, 147*f*, 149*f*
cystic, 127
efferent, of testis, 157*f*
ejaculatory, 155, 156*f*, 161*f*
excretory. *See* Excretory duct(s)
gastrointestinal, 132*f*
interlobular, 137*f*
intralobular, 166*f*
lactiferous, 166*f*
lymphatic, 122*f*
mammary, 166*f*
papillary, 146*f*
semicircular, 178*f*
of seminal vesicle, 161*f*
Stenson's
of cat, 188*f*
of fetal pig, 191*f*
thoracic, 122*f*
Ductus (vas) deferens, 156*f*–157*f*, 159*f*
ampulla of, 155, 156*f*, 161*f*
right, 161*f*
Ductus epididymis. *See* Epididymis
Duodenum, 126–127, 134*f*
glands of, 134*f*
Dura mater, 94, 106*f*, 175*f*
Dust cells. *See* Macrophage(s)

Ear, 5*f*, 173–174
basilar membrane of, 180*f*
external, 173–174
cartilage of, 32
inner, 174, 179*f*
internal anatomy, 178*f*
middle, 174, 178*f*–179*f*
spiral ganglia of, 179*f*–180*f*
Eardrum, 178*f*
Earlobe, 3*f*, 178*f*
Eccrine glands, 107
Ectoderm, 150
Efferent arteriole, of nephron, 146*f*
Efferent duct, of testis, 157*f*
Efferent lymphatic vessel, 123*f*
Egg. *See* Ovum/ova
Ejaculatory duct, 155, 156*f*, 161*f*
Elastic arteries, 112
Elastic cartilage, 31–32, 33*f*, 141*f*
Elastic connective tissue, 26, 107
Elastic fiber(s), 27
in areolar connective tissue, 28*f*
in lung tissue, 143*f*
Elastic lamina
of elastic arteries, 112
of veins, 113
Elbow
extension of, 72*f*
flexion of, 72*f*
joints of, 59*f*, 65, 76*f*
Electrolytes, elimination of, 144
Elevation
of joints, 66
of mandible, 74*f*
orbital, of frontal bone, 45*f*
Ellipsoidal joints, 65–66, 70*f*
Embryo
developing, 169*f*
at end of 4th week, 170*f*
at end of 8th week, 170*f*
Embryonic primordium, 169*f*
Enamel, 126, 130*f*
Endocardium, 113
Endochondral ossification, 32*f*
in developing bone, 37*f*
in fetal bone, 38*f*
Endocrine gland(s), 127, 151*f*, 162
Endocrine system, 150–154
Endoderm, 150
Endolymph, 174, 179*f*
Endometrial arteriole, maternal, 172*f*
Endometrial venole, maternal, 172*f*
Endometrium, 165*f*
Endomysium, of cardiac muscle, 84*f*
Endoneurium, of peripheral nerve, 97*f*
Endoplasmic reticulum, 12
of freeze-fractured cell, 14*f*
infiltrated with virus, 16*f*
rough (granular), 12, 13*f*
in cellular cytosol, 15*f*
smooth (agranular), 12
Endosteum, 58*f*
Endothelial cells, of blood vessel, 22*f*
Endothelium
of arterioles, 113
of elastic arteries, 112
Eosinophil(s), 79, 80*f*
granules of, 80*f*
nucleus of, 80*f*
Ependymal cells, 94
ciliated, 96*f*
Epicardium, 113
Epicondyle(s)
of femur
lateral, 61*f*
medial, 61*f*
of humerus
lateral, 58*f*, 89*f*
medial, 8*f*, 9*f*, 58*f*, 88*f*–89*f*
Epicranial aponeurosis muscle, 87*f*

Epidermis, 107, 108f–109f, 111f
cells of, 107
fetal, 28f, 38f
layers of, 109f
of prepuce, 160f
Epididymis, 155, 156f–158f
body of, 157f
epithelial cells of, 25f, 158f
head of, 157f
tail of, 157f
Epigastric region, 7f
Epiglottis, 139, 140f
cartilage of, 32, 33f
mucosa of anterior surface, 141f
sagittal section, 141f
Epineurium, of peripheral nerve, 97f
Epiphyseal cartilage, 37f
Epiphyseal growth plate, 37f
Epiphyseal plates, 63f, 65
of long bones, 68f
Epiphysis, 32f
developing, 37f
distal, 58f
of long bones, 35, 68f
proximal, 58f
Epiphysis cerebri. See Pineal gland
Epithelial tissue/epithelium, 12, 21–25, 32f, 111f
anal, 136f
cervical, 165f
ciliated, 141f, 164f
columnar, 25f, 107
pseudostratified. See Pseudostratified columnar epithelium
simple. See Simple columnar epithelium
stratified, 21
of cornea, 176f
cornified, 107
cuboidal, 107
simple, 21, 23f
stratified, 21, 24f
of ductus (vas) deferens, 159f
of eyelid, 176f
of fallopian tubes, 164f
gastric, 131f–132f
glandular, 135f
keratinized, 21, 23f
lingual, 130f
nonkeratinized, 21, 24f
olfactory, 140f
parenchymal, 108
of prostate gland, 161f
pseudostratified, 158f
rectal, 136f
of small intestine, 133f
squamous, 139
simple. See Simple squamous epithelium
stratified. See Stratified squamous epithelium
of sweat glands, 107
transitional, 21, 24f, 145, 149f
of ureters, 24f, 145, 149f
of urinary bladder, 145, 149f
vaginal, 165f
Epitrochlearis muscle, of cat, 186f
Eponychium, 107, 111f
fetal, 38f
Erectile tissue, 138
Erosion, zone of, in fetal bone, 38f
Erythrocytes, 78–79, 80f
Esophageal glands, mucous acini of, 131f
Esophagus, 6f, 126–127, 131f–132f, 139, 140f
cross section, 142f
epithelial lining, 131f, 142f
junction with stomach, 131f
lumen of, 142f
upper region of, 24f
Estrogen, 162
Ethmoidal nerve, groove for, 45f
Ethmoidal sinus, 50f, 138
Ethmoid bone, 44f
cribriform plate of, 50f
disarticulated, 46f
horizontal plate of, 50f
inferior view of, 50f
lateral mass of, 50f
lateral view of, 50f
orbital plate of, 50f
perpendicular plate of, 44f, 50f
superior view of, 50f
Euchromatin, 14f–15f
Eustachian tube(s). See Auditory (Eustachian) tube(s)
Eversion
of joints, 66
of sole of foot, 74f
Excretory duct(s), 137f
lumen of, 24f
of nephron, 144
of pancreatic lobules, 137f
of parotid gland, 129f
of submaxillary gland, 128f
of tongue, 130f
Exocrine (cytogenic holocrine) gland, 127, 162
Extension
angular, 71f
of elbow, 72f
of head, 72f
of hip, 72f
of joints, 66
of knee, 72f
of shoulder, 72f
of synovial joints, 65
Extensor carpi radialis brevis muscle, 85f–86f, 89f
Extensor carpi radialis longus muscle, 85f, 89f
of cat, 183f
Extensor carpi ulnaris muscle, 86f, 89f
of cat, 183f
tendon of, 89f
Extensor digiti minimi muscle, 89f
tendon of, 89f
Extensor digitorum brevis muscle, 93f
Extensor digitorum communis muscle, of cat, 183f
Extensor digitorum lateralis muscle, of cat, 183f
Extensor digitorum longus muscle, 85f–86f, 93f
of cat, 185f
tendons of, 11f
Extensor digitorum muscle, 85f–86f, 89f
tendon of, 89f
Extensor digitorum profundus muscle, 89f
Extensor hallucis longus muscle, 85f, 93f
Extensor indicis muscle, 89f
tendon of, 89f
Extensor pollicis brevis muscle, 86f, 89f
Extensor pollicis longus muscle, 85f, 89f
Extensor retinaculum, 89f
inferior, 11f
superior, 11f
Exteroceptive nervous impulses, 94
Extracellular fluid, 13f
Extrinsic muscle, of eye, 176f
Eye, 5f, 105f, 173, 175f
cavity of, 175f
chambers of, 175f–176f
choroid of, 173, 175f–176f, 178f
cross section of, 175f
layers of, 173
posterior chamber of, 175f–176f
supporting structures of, 175f–176f
Eyeball, 88f
muscles of, 88f
of sheep, 175f
Eyebrow, 176f
Eyelashes, 3f, 176f
Eyelid, 177f

Face, 3f, 5f
angle of, 3f
frontal region of, 3f
muscles of, 87f
nasal region of, 3f
temporal region of, 3f
Facial, definition, 5f
Facial (VII) nerve, 99f, 102f, 178f
Fallopian tubes. See Oviduct(s)
Fascia
deep, 166f
inferior, of urogenital diaphragm, 92f
superficial, 90f. See also Hypodermis
fat in, 166f
of testis, 157f
thoracolumbar, of cat, 185f
Fascia lata, 91f, 93f
of cat, 182f, 185f, 187f
iliotibial tract of, 85f
Fascial canal, genu of, 3f
Fasciculus cuneatus, 102f
Fasciculus gracilis, 102f
Fat cells. See Adipocytes
Fauces, 140f
Female reproductive system, 162–166
of fetal pig, 192f
in sagittal section, 163f
Femoral, definition, 4f–5f
Femoral artery, 117f
Femoral nerve, of cat, 187f
Femoral triangle
area of, 10f
of cat, 187f
Femoral vein, 118f
Femur, 40f, 56f, 70f, 77f, 93f
anterior view of, 61f
head of, 56f, 61f, 70f, 77f
intercondylar notch of, 61f
intertrochanteric crest of, 61f
at knee joint, 63f, 69f
muscles of, 91f
neck of, 56f, 61f, 70f
partially sectioned, 58f
patellar surface of, 61f
posterior view of, 61f
Fenestrated laminae, of elastic arteries, 112
Fetal development, 167–172
at 4 months, 171f
at 6 months, 171f
Fetal pig dissection, 191–192
Fetus
blood vessels, 172f
of cat, 190f
mesenchymal connective tissue in, 28f
Fibrils, 16f
Fibrinogen, 78
Fibroblast(s), 27
of areolar connective tissue, 28f
of stratified squamous epithelium, 23f
of tooth pulp, 127
Fibroelastic capsule, of prostate, 155
Fibromuscular coat, of gallbladder, 137f
Fibrous adventitia, 139
Fibrous capsule, of pineal gland, 151f
Fibrous cartilage (fibrocartilage), 31–32, 33f
Fibrous joints, 65–66, 67f–68f
Fibrous tissue, 26
Fibula, 40f, 77f, 93f
at ankle joint, 64f
articulation of, 62f
collateral ligament of, 77f
head of, 62f–63f, 68f, 93f
at knee joint, 63f, 69f
lateral malleolus of, 11f, 62f, 68f
notch of, 62f
syndesmosis joint of, 68f
Filaments, 12
Filiform papillae, of tongue, 126, 130f
Fimbriae, 163f–164f
Finger, 5f. See also Phalanx/phalanges
fetal, 38f
muscles of, 88f–89f
Fingernail. See Nail(s)
Fissure(s)
cerebral
longitudinal, 105f
transverse, 99f
of spinal cord, anterior median, 94, 103f–104f
superior orbital, 47f
superior sagittal, 105f
Flagellum, of sperm, 159f
Flexion
angular, 71f
of elbow, 72f
of head, 72f
of joints, 66
of knee, 72f
of synovial joints, 65
Flexor carpi radialis muscle, 85f, 88f
of cat, 186f
tendon of, 9f
Flexor carpi ulnaris muscle, 85f–86f, 88f
of cat, 183f, 186f

Flexor digitorum longus muscle, 85*f*–86*f*, 93*f*
Flexor digitorum profundus muscle, 88*f*
of cat, 183*f*, 186*f*
tendon of, 88*f*
Flexor digitorum superficialis muscle, 85*f*, 88*f*
tendon of, 88*f*
Flexor hallucis longus muscle, 86*f*, 93*f*
Flexor pollicis longus muscle, 85*f*, 88*f*
Flexor retinaculum muscle, 88*f*
Folia, 101*f*
Foliate (rare) papillae, of tongue, 126
Follicle(s)
hair. *See* Hair follicle(s)
ovarian, 168*f*
Follicular cells
ovarian, 14*f*
thyroid, 142*f*, 152*f*–153*f*
Follicular development, stages of, 168*f*
Folliculi, liquor, 164*f*
Fontanelle(s), of fetal skull
anterior, 41*f*–42*f*
mastoid, 42*f*
Foot, 5*f*, 11*f*
bones of, 64*f*
dorsiflexion of, 75*f*
in dorsoplantar position, 64*f*
heel of, 5*f*
joints of, 70*f*
lateral view of, 64*f*
longitudinal arch of, 11*f*
muscles of, 93*f*
plantar flexion of, 75*f*
sole of, 5*f*
eversion of, 74*f*
inversion of, 74*f*
superior view of, 64*f*
Foramen
caecum, 45*f*
infraorbital, 42*f*–43*f*
interventricular, 105*f*
jugular, 45*f*
lacerum, 43*f*, 45*f*
magnum, 43*f*, 45*f*, 48*f*
mandibular, 49*f*
mental, 42*f*–43*f*, 49*f*, 67*f*
nutrient, 58*f*
obturator, 55*f*
ovale, 43*f*, 45*f*, 47*f*
rotundum, 47*f*
spinosum, 43*f*, 45*f*, 47*f*
supraorbital, 42*f*–43*f*
transverse, 53*f*–54*f*, 69*f*
vertebral, 53*f*–54*f*
zygomaticofacial, 52*f*
Forehead, 3*f*
Foreskin, 156*f*
Fornix, 100*f*, 163*f*
anterior, 165*f*
Fossa
coronoid, of humerus, 58*f*
digastric, 49*f*
glenoid, 57*f*
iliac, 55*f*–56*f*, 68*f*, 70*f*
incisive, 43*f*
infraspinous, 56*f*–57*f*
mandibular, 48*f*
olecranon, 58*f*–59*f*
pituitary. *See* Sella turcica
popliteal, 11*f*, 86*f*, 93*f*
radial, 58*f*
subscapular, 57*f*
supraspinous, 56*f*–57*f*
Fovea
capitis, 61*f*
central, of macula lutea, 175*f*
Freeze-fractured cell, 14*f*
Frontal bone, 42*f*–45*f*, 67*f*, 87*f*–88*f*, 140*f*
disarticulated, 46*f*
of fetal skull, 41*f*–42*f*
orbital elevation of, 45*f*
Frontal eminence, of forehead, 3*f*
Frontalis muscle, 87*f*
Frontal lobe, 99*f*–100*f*, 105*f*
Frontal plane, 2, 8*f*
Frontal region, of face, 3*f*
Frontal sinus, 44*f*–45*f*, 138, 140*f*
Frontal tuberosity, of fetal skull, 41*f*
Fundic stomach, 133*f*
Fundus, 162
Fungiform papillae, of tongue, 126, 130*f*
Fusiform cells, 83, 95

Galea aponeurotica muscle, 87*f*
Gallbladder, 6*f*–7*f*, 126–127, 137*f*
of cat, 128*f*
spiral valves of, 127
Ganglion/ganglia
posterior (dorsal) root, 103*f*
spiral, 178*f*–179*f*
sympathetic, 98*f*
vestibular, 178*f*
Ganglionic cell layer, of eye, 178*f*
Gap junction, of cells, 13*f*
Gastric artery, 117*f*
Gastric gland(s)
of cardiac stomach, 132*f*
of fundic stomach, 133*f*
Gastric pit(s)
of cardiac stomach, 132*f*
of fundic stomach, 133*f*
of stomach-esophagus junction, 131*f*
Gastrocnemius muscle, 11*f*, 85*f*–86*f*, 93*f*
of cat, 185*f*
Gastrointestinal tract
ducts of, 132*f*
epithelium of, 131*f*–132*f*
lamina propria of, 120
layers of, 132*f*
plexus of, 132*f*
Germinal center(s)
of lymph nodes, 120, 123*f*
of lymphoid nodule, 135*f*
of palatine tonsil, 124*f*
Germinal disk, 169*f*
Gingiva, 126, 130*f*
Ginglymus joints, 65–66
Glabella, 3*f*, 42*f*–43*f*
Gland(s)
adrenal, 150, 151*f*, 154*f*
Brunner's, 134*f*
bulbourethral, 155, 156*f*, 161*f*
ceruminous, 173
Cowper's. *See* Gland(s), bulbourethral
cytogenic holocrine. *See* Gland(s), exocrine
duodenal, 134*f*
eccrine, 107
endocrine, 151*f*, 162
esophageal, 131*f*
exocrine, 127, 162
gastric, 132*f*, 133*f*
hypophysis, 150, 151*f*
intestinal, 133*f*, 135*f*
lacrimal, 87*f*
mammary, 162–163, 189*f*
medullary, 154*f*
meibomian. *See* Gland(s), tarsal
mixed, 141*f*, 150
mucosal, 126
parathyroid, 150, 151*f*–153*f*
parotid, 126, 129*f*, 188*f*, 191*f*
pineal, 102*f*, 106*f*, 150, 151*f*
pituitary, 100*f*, 150, 151*f*–152*f*
prostate, 155, 156*f*, 161*f*
salivary, 126, 188*f*, 191*f*
sebaceous, 107–108, 108*f*–110*f*
seromucous, 173
sublingual, 126
submandibular, 126, 128*f*
submaxillary, 24*f*, 126, 128*f*, 188*f*, 191*f*
suprarenal. *See* Gland(s), adrenal
sweat. *See* Sweat glands
tarsal, 176*f*–177*f*
thymus, 122*f*, 151*f*
thyroid. *See* Thyroid gland
tracheal, 32*f*, 142*f*
tuboalveolar, 126, 155
uterine, 165*f*
Glandular epithelium, of cecum, 135*f*
Glans penis, 156, 156*f*
in fetal pig, 192*f*
Glassy membrane, of hair follicle, 108, 110*f*
Glenoid cavity, 56*f*
margin of, 57*f*
Gliding joint(s), 65–66
in carpals, 69*f*
in wrists, 69*f*
Gliding movement
between sternum and clavicle, 71*f*
of synovial joint, 65–66
Globulins, 78
Glomerular capsule, 146*f*
Glomerulus/glomeruli
of Bowman's corpuscle, 22*f*
renal, 146*f*–147*f*
afferent arterioles of, 144
Glossopalatine arch, 120
Glossopharyngeal (IX) nerve, 99*f*, 102*f*
Glucagon, 127
Gluteal, definition, 4*f*–5*f*
Gluteal lines, 55*f*
Gluteus maximus muscle, 10*f*, 86*f*, 92*f*
of cat, 184*f*
Gluteus medius muscle, 10*f*, 86*f*
of cat, 184*f*
Glycogen, 12, 13*f*
Glycolipid, of plasma membrane, 13*f*
Glycoprotein(s), 150
of plasma membrane, 13*f*
Goblet cell(s), 23*f*
of colon, 136*f*
of simple columnar epithelium, 23*f*
of small intestine, 134*f*
of trachea, 25*f*, 141*f*
Golgi complex, 12, 13*f*
of freeze-fractured cell, 14*f*
TEM of, 15*f*
Golgi saccules. *See* Lamellar cisternae
Gomphosis, 65–66, 67*f*
Gonadal artery, 117*f*
Graafian follicle, in sagittal section, 164*f*
Gracilis muscle, 10*f*, 85*f*–86*f*, 91*f*, 93*f*
of cat, 187*f*
Granulocytes, 79
Granulosa cells, 14*f*
Gray matter
of cerebellum, 95
of cerebrum, 94
of spinal cord, 103*f*–104*f*
Groin, 5*f*
definition, 4*f*
Ground substance, of connective tissue, 27
Growth plate, 32*f*
epiphyseal, 37*f*
Gyri, 94, 99*f*, 105*f*

Hair, 107
cuticle of, 108*f*
papilla of, 108*f*, 110*f*
Hair cell(s)
of crista ampullaris, 180*f*
of ear, 179*f*
Hair follicle(s), 108, 109*f*
in axilla, 28*f*
bulb of, 108*f*, 110*f*
cortex of, 110*f*
cross section of, 110*f*
epithelial root sheaths of, 108
external root sheath of, 108*f*, 110*f*
internal root sheath of, 108*f*, 110*f*
layers of, 108
longitudinal section of, 110*f*
matrix of, 108*f*, 110*f*
medulla of, 110*f*
Hair line, 3*f*
Hair root
cortex of, 108*f*, 110*f*
cuticle of, 108*f*
medulla of, 108*f*, 110*f*
Hair root plexus, 108*f*
Hair shaft, 108*f*
Hamate bone, 60*f*
Hand, 5*f*
abduction of, 73*f*
adduction of, 73*f*
anterior surface of, 60*f*
bones of, 60*f*
muscles of, 88*f*–89*f*
posterior surface of, 60*f*
pronation of, 75*f*
supination of, 75*f*
Haploid egg nucleus, in meiosis, 19*f*
Haploid sperm nucleus, in meiosis, 19*f*
Hassall's corpuscle, 125*f*

Haversian canal(s), 35, 35*f*–36*f*
Haversian system(s), of bone, 34, 36*f*
H band(s), 16*f*
of skeletal muscle, 82
Head, 1, 5*f*
rotation of, 73*f*
Heart, 6*f*, 113, 114*f*–116*f*, 151*f*. *See also* Cardiac
apex of, 115*f*–116*f*
of cat, 128*f*, 188*f*, 189*f*
of fetal pig, 191*f*
Purkinje fibers in, 116*f*
of sheep, 115*f*–116*f*
Hematopoiesis, 78
Hematopoietic cords, of bone marrow, 81*f*
Hematopoietic tissue, 26, 35
Henle's layer, of hair follicle, 110*f*
Henle's loop
ascending limb of, 146*f*
descending limb of, 146*f*
Henson's node, 169*f*
Hepatic artery, 136*f*
common, 117*f*
Hepatic cells, plates of, 136*f*
Hepatic cords, 127
Hepatic portal vein, 118*f*
Hepatic prominence, embryonic, 169*f*
Hepatic sinusoids, of liver lobule, 136*f*
Hepatic vein, 118*f*, 136*f*
Hepatocytes, 127
of liver lobule, 136*f*
Heterochromatin
in lymphocyte, 14*f*
in nucleus, 15*f*
hGH. *See* Human growth hormone
Hilus
of kidney, 144, 146*f*
of lymph node, 123*f*
Hinge joints, 65
Hip, 5*f*
extension of, 72*f*
joints of, 65
Hipbone, 161*f*
Hip joint, 70*f*, 77*f*
Hofbauer cells, 171*f*
Homeostasis, 120
Hormones, 150. *See also specific hormone*
in morphogenesis, 167
production of, 127
Howship's lacunae, 34, 36*f*–37*f*
Human growth hormone, 167
Humeroulnar joint, 69*f*
Humerus, 6*f*, 40*f*, 76*f*, 89*f*
anatomical neck of, 58*f*
anterior surface of, 58*f*
articulation of, 59*f*
coronoid fossa of, 58*f*
deltoid tuberosity of, 58*f*
head of, 58*f*
intertubercular groove of, 58*f*
lateral epicondyle of, 58*f*, 89*f*
medial epicondyle of, 8*f*–9*f*, 58*f*, 88*f*–89*f*
posterior surface of, 58*f*
surgical neck of, 58*f*
tubercles of, 58*f*
Huxley's layer, of hair follicle, 110*f*
Hyaline cartilage, 31, 142*f*
perichondrium of, 142*f*
and surrounding tissue, 32*f*
Hyaline tissue, 26
Hyaluronic acid, in cartilage, 31
Hydrochloric acid, 127
Hyoid, definition, 3*f*
Hyoid bone, 139, 140*f*
body of, 49*f*
cornu (horns) of
greater, 49*f*
lesser, 49*f*
superior view of, 49*f*
Hypertrophic chondroblasts, in fetal bone, 38*f*
Hypochondriac region(s)
left, 7*f*
right, 7*f*
Hypodermis, 107. *See also* Fascia, superficial
Hypogastric artery, 117*f*
Hypogastric region, 7*f*
Hypogastric vein, left internal, 118*f*
Hypoglossal canal, 44*f*
Hypoglossal (XII) nerve, 99*f*, 103*f*
Hyponychium, 107, 111*f*
Hypopharynx. *See* Laryngopharynx
Hypophysis gland, 150, 151*f*
Hypothalamus, 100*f*, 151*f*
Hypothenar muscle, 85*f*

I bands, of skeletal muscle, 16*f*, 82
Ileocecal junction, 135*f*
Ileocecal valve, lumen of, 135*f*
Ileum, 125*f*, 126–127, 135*f*
termination of, 135*f*
Iliac artery
common, 117*f*
external, 117*f*
of cat, 187*f*
internal, 117*f*
Iliac crest, 55*f*, 68*f*, 70*f*, 91*f*
Iliac fossa, 55*f*–56*f*, 68*f*, 70*f*
Iliac node, 122*f*
Iliac region, 7*f*
Iliac spine
anterior superior, 90*f*–91*f*
anteroinferior, 55*f*
anterosuperior, 55*f*–56*f*, 68*f*, 77*f*
posteroinferior, 55*f*–56*f*
posterosuperior, 55*f*–56*f*
Iliacus, 91*f*
Iliacus muscle, 85*f*
Iliac vein
common, 118*f*
external, 118*f*
internal, 118*f*
Iliococcygeus muscle, 92*f*
Iliofemoral ligament, 77*f*
Iliopectineal line, 55*f*
Iliopsoas muscle, of cat, 187*f*
Iliotibial tract, 91*f*
of tensor fascia lata, 86*f*
Ilium, 55*f*, 68*f*, 70*f*
crest of, 56*f*
Immune system, 120
Incisive canal, 44*f*
Incisors, 42*f*–43*f*, 46*f*, 129*f*
central, 129*f*
lateral, 129*f*
Inclusions, 12, 13*f*
Incus, 173, 177*f*
Inferior, definition, 6*f*
Infraspinatus muscle, 86*f*
of cat, 183*f*
Infraspinous fossa, 56*f*–57*f*
Infundibulum, of pituitary gland, 152*f*
Inguinal, definition, 5*f*
Inguinal ligament, 90*f*–91*f*
site of, 10*f*
Inguinal node, 122*f*
Inguinal region, 7*f*
Inguinal ring, superficial, 90*f*
Innominate trunk, 115*f*
Insulin, 127
Integral protein, of plasma membrane, 13*f*
Integrating bipolar cell layer, of eye, 178*f*
Integumentary system, 107–111
Intercalacted disk, in cardiac muscle, 84*f*
Intercondylar eminence, 63*f*
of tibia, 68*f*
Interconnected cavities, 138
Intercostal muscles, 166*f*
Interlobar arteries, 144
Interlobar veins, 144
Interlobular artery(ies), renal, 144, 146*f*
Interlobular connective tissue, of pancreatic lobules, 137*f*
Interlobular duct, of excretory ducts, 137*f*
Interlobular septum, of parotid gland, 129*f*
Interlobular vein(s), renal, 144, 146*f*
Intermediate mass, cerebral, 100*f*, 106*f*
Interoceptive nervous impulses, 94
Interosseous margin, 59*f*
Interphase, 17*f*–18*f*
Interstitial lamellae
in compact bone tissue, 35*f*
osteocytes of, 36*f*
Interstitial systems, irregular, of dense bone, 36*f*
Intertrochanteric crest, 77*f*
Intertrochanteric line, 70*f*
Interventricular septum, 116*f*
Interventricular sulcus, 114*f*–115*f*
Intervertebral disk(s), 55*f*, 68*f*
cartilage of, 32, 33*f*
Intervillous space, 171*f*
of duodenum, 134*f*
Intestinal gland(s)
opening of, 133*f*
of vermiform appendix, 135*f*
Intestinal node, 122*f*
Intestinal tract, 127
Intestinal villi, of duodenum, 134*f*
Intestine(s). *See also* Colon; Large intestine; Small intestine
in fetal pig, 191*f*
Intracellular matrix, 15*f*
composition of, 31
Intralobular ducts, 166*f*
Inversion
of joints, 66
of sole of foot, 74*f*
Iris, 176*f*
anterior, 173
Ischial spine, 55*f*–56*f*, 68*f*
Ischial tuberosity, 55*f*–56*f*, 92*f*
Ischiocavernosus muscle, 92*f*
Ischiopubic ramus muscle, 92*f*
Ischium, 55*f*, 70*f*
Islets of Langerhans, 127, 137*f*, 150, 153*f*–154*f*
Isthmus, 165*f*

Jejunum, 126–127, 134*f*
JG cells. *See* Juxtaglomerular cells
Joints, 65–77, 76*f*–77*f*. *See also* Articulation; *specific joint*
abduction of, 66, 73*f*
adduction of, 66, 73*f*
classification of, 65
elevation of, 66
eversion of, 66
morphology of, 66
Jugular notch, 48*f*
Jugular vein
external, 118*f*
internal, 114*f*, 118*f*, 122*f*
of cat, 188*f*
Jugum, sphenoidal, 45*f*
Juxtaglomerular (JG) cells, 144

Keratinized epithelium, 21, 23*f*
Kidney(s), 147*f*, 151*f*. *See also* Renal
blood supply to, 144
of cat, 128*f*, 187*f*
cortex of, 147*f*, 149*f*
excretory functions of, 144
of fetal pig, 190*f*
medulla of, 147*f*
of sheep, 146*f*
simple columnar cells in, 22*f*
simple cuboidal epithelium in, 23*f*
Kinesiology, 66
Kinetochore
in meiosis, 19*f*
in prophase, 18*f*
Kinocilia, of crista ampullaris, 180*f*
Knee, 10*f*
anterior surface of, 5*f*
hollow behind of, 5*f*
joints of, 63*f*, 65, 69*f*, 77*f*
Kupffer cells, 127

Labia
majus, 162, 163*f*
minus, 162, 163*f*
Labrum, acetabular, 77*f*
Lacrimal bone, 43*f*
disarticulated, 46*f*
lateral surface of, 51*f*
medial surface of, 51*f*
Lacrimal canal, 46*f*
Lacrimal gland, 87*f*
Lacteal, of small intestine, 133*f*
Lactiferous duct, 166*f*
Lactiferous sinus, 166*f*
Lacunae
in bone, 34, 35*f*
developing, 37*f*

Lacunae *(continued)*
in cartilage, 31
elastic, 33*f*
fibrous, 33*f*
hyaline, 32*f*
chondrocytes in, 37*f*
Howship's, 34, 36*f*–37*f*
osteocytes in, 36*f*
Lambdoid suture, 67*f*
Lamellae, in compact bone tissue, 35*f*
Lamellar cisternae, 15*f*
Lamina, 69*f*
of axis, 54*f*
basal, 113
basement, 21
in trachea, 25*f*
of cervical vertebra, 54*f*
elastic
of arteries, 112
of veins, 113
fenestrated, 112
Lamina propria, 32*f*, 138, 142*f*–143*f*, 164*f*–165*f*
of anal mucosa, 136*f*
of cardiac stomach, 132*f*
of colon, 135*f*–136*f*
of epididymis, 25*f*
of esophagus, 131*f*
of gastrointestinal tract, 120, 132*f*
of ileocecal junction, 135*f*
of lingual tonsil, 124*f*
of palatine tonsil, 124*f*
of rectal mucosa, 136*f*
of simple ciliated columnar epithelium, 25*f*
of simple columnar epithelium, 23*f*
of small intestine, 133*f*–134*f*
of submaxillary gland, 24*f*
of tongue, 130*f*
in trachea, 25*f*
of ureters, 24*f*, 145
Large intestine, 122*f*, 127, 132*f*
ascending colon of, 7*f*
of cat, 128*f*, 187*f*
descending colon of, 7*f*
transverse colon of, 7*f*
Laryngeal sinus, 140*f*
Laryngopharynx (hypopharynx), 139, 140*f*
Larynx, 3*f*, 138–139, 141*f*
cartilage of, 31–32
of cat, 187*f*
of fetal pig, 189*f*
sagittal section of, 142*f*
Lateral, definition, 6*f*
Latissimus dorsi muscle, 9*f*, 85*f*–86*f*, 90*f*
of cat, 183*f*–184*f*, 186*f*
Leg(s), 1, 5*f*, 11*f*
muscles of
of cat, 185*f*
superficial, 93*f*
Lens, 175*f*
of eye, 175*f*–176*f*
suspensory ligament of, 175*f*
Leukocytes, 27, 78–79, 172*f*
Levator ani muscle, 92*f*
Levator labii superioris muscle, 87*f*
Levator palpebrae superioris muscle, 87*f*–88*f*, 176*f*
Levator scapulae muscle, 87*f*
Levator scapula ventralis muscle, of cat, 183*f*
Leydig cells, 150, 155
Ligament(s), 76*f*–77*f*. *See also specific ligament*
annular, 76*f*
broad, 165*f*
cartilage of, 32
collateral
of fibula, 77*f*
of radius, 76*f*
of tibial, 77*f*
of ulna, 76*f*
Cooper's, 166*f*
coracoacromial, 76*f*
coracohumeral, 76*f*
iliofemoral, 77*f*
inguinal, 90*f*–91*f*
site of, 10*f*
medial, 77*f*
oblique, 76*f*
ovarian, 165*f*
of patella, 77*f*
patellar, 10*f*, 85*f*, 91*f*, 93*f*
round, of uterus, 163*f*
sacrotuberous, 92*f*
suspensory
of breast, 166*f*
of lens, 175*f*
of penis, 156*f*
of uterine tube, 165*f*
trapezoid, 76*f*
uterosacral, 163*f*, 165*f*
vocal, 142*f*
Ligamentum arteriosum, 114*f*–115*f*
Ligamentum nuchae, dense regular connective tissue of, 29*f*
Limbic system, 101*f*
Limbs, 1. *See also* Lower extremity; Upper extremity
Limiting membrane, internal, of eye, 178*f*
Linea alba, 90*f*
Linea aspera, 61*f*
Lingual stratified squamous epithelium, 130*f*
Lingual tonsil(s), 120, 124*f*
Lingula, 49*f*
Lipid molecules, 12
Lips, 3*f*
Liquor folliculi, 164*f*
Liver, 6*f*, 126–127
of cat, 128*f*, 190*f*
of fetal pig, 191*f*
lobes of, 7*f*, 127
lobule of, 136*f*
Lobule(s), of liver, 166*f*
inactive, 166*f*
Lobuli testis, 155
Loin, 5*f*
Long bone(s)
cartilage of, 31
partially sectioned, 58*f*
shaft of, 35
terminal ends of, 35
Longitudinal arch, of foot, 11*f*
Longitudinal muscle, of gastrointestinal tract, 132*f*
Loop of Henle
ascending limb of, 146*f*
descending limb of, 146*f*
Loose areolar connective tissue, of hypodermis, 107
Loose tissue, 26
cells of, 27
Lower extremity, 5*f*
bones of, 40*f*
Lumbar, definition, 5*f*
Lumbar curvature, 52*f*
Lumbar region, 7*f*
Lumbar vertebra(e), 52*f*, 55*f*–56*f*, 68*f*
body of, 54*f*
fifth, lateral view of, 54*f*
inferior notch of, 54*f*
inferior view, 53*f*
posterior view, 56*f*
superior notch of, 54*f*
superior view, 53*f*
Lumen
of appendix, 135*f*
of artery, 119*f*
of cardiac stomach, 132*f*
of central vein, of liver lobule, 136*f*
of ductus (vas) deferens, 159*f*
of epididymis, 25*f*
of esophagus, 131*f*
of excretory duct, 24*f*
of fundic stomach, 133*f*
of gallbladder, 137*f*
of gastrointestinal tract, 132*f*
of ileocecal valve, 135*f*
of ileum, 135*f*
of jejunum, 134*f*
of prostate gland, 161*f*
of seminal vesicle, 159*f*
of simple ciliated columnar epithelium, 25*f*
of simple cuboidal epithelium, 23*f*
of ureter, 24*f*, 149*f*
of urinary bladder, 149*f*
of vein, 119*f*
Lunate bone, 60*f*
Lung(s), 6*f*, 138, 141*f*, 151*f*
blood supply of, 139
of cat, 128*f*
of fetal pig, 191*f*
Lung tissue, LM of, 143*f*
Lymphatic duct, 122*f*
Lymphatic follicle, aggregated. *See* Peyer's patch(es)
Lymphatic nodule(s), 120–121, 123*f*
of fundic stomach, 133*f*
of gastrointestinal tract, 132*f*
of small intestine, 133*f*
Lymphatic system, 120–125
Lymphatic trunks, 121
Lymphatic vessel(s), 120–121, 122*f*
in adipose connective tissue, 30*f*
afferent, 123*f*
in compact bone tissue, 35*f*
efferent, 123*f*
of small intestine, 133*f*
Lymph node(s), 120, 123*f*
of cat, 188*f*
cortex of, 123*f*
lymphocytes in, 29*f*, 123*f*
medulla of, 120, 123*f*
reticular connective tissue in, 29*f*
Lymph node cortex, 120
Lymphocyte(s), 79, 80*f*
B, 120
in bone marrow, 81*f*
in lymph node, 29*f*, 123*f*
in medullary cords, 120
T, 120, 124*f*–125*f*
TEM of, 14*f*
in thymus, 121
in tooth pulp, 127
Lymphoid nodule(s), 120, 123*f*, 135*f*, 142*f*, 152*f*
germinal center of, 135*f*
of lingual tonsil, 124*f*
of palatine tonsil, 124*f*
splenic, 125*f*
of vermiform appendix, 135*f*
Lymph organs, 122*f*
Lymph vessels, 122*f*
Lysosomes, 12, 13*f*

Macroglia, 94
Macrophage(s), 27, 79, 139
in areolar connective tissue, 28*f*
of medullary cords, 120
of tooth pulp, 127
Macula
within saccule, 179*f*
within utricle, 179*f*
Macula densa, of renal corpuscle, 148*f*
Macula lutea, central fovea of, 175*f*
Male reproductive system, 155–161
of fetal pig, 190*f*
sagittal plane of, 156*f*
supporting structures of, 161*f*
Malleolus
lateral, 11*f*, 62*f*, 68*f*
medial, 11*f*, 62*f*, 68*f*
Malleus, 174, 178*f*
Mammary, definition, 4*f*–5*f*
Mammary artery, of cat, 188*f*
Mammary duct, 166*f*
Mammary glands, 162–163
in pregnant cat, 189*f*
Mammillary body, 99*f*–100*f*, 106*f*
Mandible, 43*f*, 45*f*, 87*f*
angle of, 42*f*–44*f*, 49*f*
arch of, embryonic, 169*f*
body of, 42*f*, 44*f*, 49*f*
condyle of, 43*f*, 49*f*, 67*f*
depression of, 74*f*
disarticulated, 46*f*
elevation of, 74*f*
fetal, 42*f*
foramen of, 49*f*
fossa of, 48*f*
head of, 49*f*
left lateral view of, 49*f*
mental protuberance of, 44*f*
neck of, 49*f*
notch of, 43*f*, 49*f*, 67*f*
oblique line of, 49*f*
posterior medial view of, 49*f*
ramus of, 42*f*, 49*f*
posterior border of, 49*f*
symphysis of, 3*f*, 42*f*
Manual, definition, 5*f*
Manubrium, 56*f*
Marginal sinus, 172*f*
Masseter muscle, 85*f*–87*f*
of cat, 188*f*
of fetal pig, 191*f*

Mast cells, 27, 79
Maxilla, 42f–43f, 45f, 87f–88f, 140f
disarticulated, 46f
fetal, 41f–42f
medial view of, 46f
Maxillary sinus, 46f, 138
M bands, of skeletal muscle, 82
Medial, definition, 6f
Medial ligament, 77f
Medial rotation, of synovial joint, 66
Median eminence, cerebral, 102f
Median nerve, 88f
Median sagittal plane, 1–2
Medulla cells, of adrenal gland, 154f
Medulla oblongata, 99f–101f, 105f–106f
Medullary (marrow) cavity, 58f
in compact bone tissue, 35f
Medullary cord(s), 120, 123f
Medullary gland, 154f
Medullary sinus, of lymph node, 123f
Megakaryocyte(s), 79
of bone marrow, 81f
Meibomian glands. *See* Tarsal (meibomian) gland(s)
Meiosis, 19f
Meissner's corpuscles, 111f
Melanin stimulating hormone, 167
Melanocyte, of hair follicle, 108f
Membranous labyrinth, of ear, 174, 179f
Membranous urethra, 156f, 161f
Meninges, 94
spinal, 104f
Meniscus
cartilage of, 32
lateral, 77f
medial, 77f
Mental foramen, 42f–43f, 49f, 67f
Mentalis muscle, 87f
Mental protuberance, 49f
of face, 3f
Mesenchymal cells, of tooth pulp, 127
Mesenchymal tissue, 26
in fetus, 28f
Mesenchyme, 34
Mesenteric artery, 117f
Mesenteric vein, 118f
Mesentery, of gastrointestinal tract, 132f
Mesoderm, 150
Metacarpal, definition, 5f
Metacarpal bone(s), 6f, 40f, 60f, 69f–70f, 88f
base of, 60f
head of, 60f
shaft of, 60f
Metacarpophalangeal joints
abduction of, 73f
adduction of, 73f
Metamyelocyte, of bone marrow, 81f
Metaphase
in meiosis, 19f
in mitosis, 18f
micrograph of, 19f
Metaphase plate, 18f
in meiosis, 19f
Metaphysis, 58f
Metatarsal bone(s), 64f, 70f, 93f
Microfilament, 13f
Microglia, 94, 98f
Microtubule(s), 12, 13f
Microvilli, 13f
of simple columnar epithelium, 23f
of small intestine, 134f
Midbrain, 106f
cerebral peduncle of, 99f
Middle ear, 174, 178f–179f
Midline, definition, 6f
Midsagittal plane, 8f
Mitochondria, 12, 13f
in cellular cytosol, 15f
infiltrated with virus, 16f
in lymphocyte, 14f
Mitosis, 17f, 17–20, 18f, 168f
stages in, 19f
Mitotic spindle, 18f
Mitral valve, 115f
Mixed glands, 141f, 150
Moderator band, of heart, 116f
Molar(s), 43f, 46f
first, 129f
second, 129f
third, 129f
Monocyte(s), 79, 81f
nucleus of, 81f
Mons pubis, 163f
Morphogenesis, 167
Motilin, 127
Motor end plate, of skeletal muscle fiber, 97f
Motor neuron, 95f
axon of, 103f
cell body of, 103f
Mouth, 5f
MSH. *See* Melanin stimulating hormone
Mucosa
of bronchiole, 143f
of cardiac stomach, 132f
of colon, 135f
of duodenum, 134f
of epiglottis, 141f
of esophagus, 131f
of gallbladder, 137f
of gastrointestinal tract, 132f
of intestine, 127
of jejunum, 134f
of pyloric stomach, 133f
of rectum, 136f
of small intestine, 133f
of stomach, 127, 131f
of ureters, 145
Mucosal folds, of seminal vesicle, 159f
Mucosal glands, of alimentary tract, 126
Mucous acini
of esophageal glands, 131f
of submaxillary gland, 128f
of tongue, 130f
Mucous tissue, 26
Mucus, 165f
Muscle(s), 82–93, 83f–84f. *See also specific muscle*
actin of, 84f
circular, 133f
inner, 134f
intermediate, 159f
outer, 149f
cross section of, 84f
of ductus (vas) deferens, 159f
of eyeball, 88f
facial, 87f
of femur, 91f
of fingers, 88f–89f
of foot, 93f
of forelimb, cat, 183f, 186f
of hand, 88f–89f
of jejunum, 134f
junction with tendon, 84f
of leg, 93f
of cat, 185f
longitudinal, 133f
inner, 149f, 159f
outer, 134f, 159f
myosin of, 84f
of pelvic floor, 92f
of shoulder, cat, 183f–184f
skeletal, TEM of, 16f
of small intestine, 133f
superficial
anterior, 85f
of cat, 182f
posterior, 86f
of thigh, 91f
cat, 185f, 187f
of toes, 93f
of tongue, 130f
of ureter, 149f
of wrist, 88f–89f
Muscular arteries, 112–113
Muscularis, 164f
of cardiac stomach, 132f
of colon, 135f
of duodenum, 134f
of esophagus, 131f
of gastrointestinal tract, 132f
in ileum, 125f
of intestine, 127
of jejunum, 134f
of pyloric stomach, 133f
of rectum, 136f
of small intestine, 133f
of stomach, 127
of vermiform appendix, 135f
Muscularis mucosae
of cardiac stomach, 132f
of gastrointestinal tract, 132f
of jejunum, 134f
of small intestine, 133f
of stomach-esophagus junction, 131f
Muscularis oblique layer, of stomach-esophagus junction, 131f
Muscular tissue, 12, 21
Muscular wall, of seminal vesicle, 159f
Myelinated fiber
of nerve, 16f
sections through, 95f
Myelin sheath
of nerve, 16f
of neurolemmocyte, 95f
Myenteric plexus
of gastrointestinal tract, 132f
of small intestine, 133f
Mylohyoid groove, 49f
Mylohyoid line, 44f
Myocardium, 113
Myometrium, 165f, 170f
of uterus, 165f
Myosin, of skeletal muscle, 16f, 84f

Nail(s), 107, 111f
bed of, 38f, 107
fetal, 38f
fold of, 107
groove of, 38f
plate of, 38f, 107
root of, 107
Naris/nares
external, 140f
internal, 140f
posterior, 138
Nasal, definition, 4f–5f
Nasal aperture, posterior, 24f
Nasal bone(s), 42f–44f, 140f
disarticulated, 46f
of fetal skull, 41f
superior view of, 51f
Nasal bridge, 3f
Nasal cartilage, 31, 87f
Nasal cavity, 138, 140f
connective tissue, 140f
epithelial lining of, 138
section through, 140f
Nasal concha(e), 138, 140f
inferior, 42f, 140f
disarticulated, 46f
inferior view of, 50f
superior view of, 50f
middle, 42f, 50f, 140f
superior, 50f, 140f
Nasalis muscle, 87f
Nasal meatus(es), 140f
inferior, 140f
middle, 140f
superior, 140f
Nasal region, 3f
Nasal septum, of fetal skull, 41f
Nasal spine
anterior, 42f, 46f
posterior, 43f
Nasopalatine groove, 51f
Nasopharynx, 121, 138–139, 140f, 173
Navel, 5f
Navicular bone, 64f, 70f
Neck, 1, 5f
Nephron(s), 144–145, 146f
cortex of, 146f
medulla of, 146f
Nerve(s). *See also specific nerve*
cross section, 16f
longitudinal section, 97f
in neurovascular bundle, 119f
peripheral, 96f–97f
small, 96f, 119f
Nerve cells, 94
Nerve fibers, 16f. *See also* Axon(s)
of crista ampullaris, 181f
Nervous tissue, 12, 21
composition of, 94
Neural fibers, of pituitary gland, 152f
Neural groove, embryonic, 169f
Neural plate, of forebrain, 169f
Neurofibral node, 95f
Neurofibril, 95f
Neuroglia, 94, 95f
of pineal gland, 151f

Neurolemma, 95f
Neurolemmocyte
cytoplasm of, 95f
myelin sheath of, 95f
neurolemma of, 95f
nucleus of, 95f
Neuron(s), 94
cell body, 95f
efferent, 95f
motor, 95f, 103f
multipolar, 95f
giant, 96f
processes of, 95f
in spinal cord, 98f
in sympathetic ganglia, 98f
Neurovascular bundle, 119f
Neutrophil(s), 79, 80f–81f
Barr body of, 81f
granules of, 80f
multilobed nucleus of, 80f
stab cell of, 81f
TEM of, 16f
Neutrophil myelocyte, of bone marrow, 81f
Nipple, 166f
Nissi bodies, 95f
Nitrogenous compounds, elimination of, 144
Node of Ranvier, 95f
Nongranulocytes, 79
Nonkeratinized epithelium, 21, 24f
Nonkinetochore microtubule
in meiosis, 19f
in prophase, 18f
Nonmyelinated fibers, of nerve, 16f
Nose, 5f, 138
apex of, 3f
Nuchal line, inferior, 43f, 48f
Nuclear envelope, 13f
fragments of, in prophase, 18f
in interphase, 18f
Nuclear membrane, 15f
Nucleolus, 13f
in interphase, 18f
in meiosis, 19f
of neuron, 95f
parts of, 15f
Nucleopore, of freeze-fractured cell, 14f
Nucleus
in adipose connective tissue, 30f
bilobed, 80f
of cardiac muscle, 84f, 116f
of cell, 12, 13f
functions of, 12
TEM of, 15f
of egg
in meiosis, 19f
second mitotic division, 168f
of eosinophil, 80f
of freeze-fractured cell, 14f
geniculate
lateral, 102f
medial, 102f
of haploid egg, in meiosis, 19f
of haploid sperm, in meiosis, 19f
heterochromatin in, 15f
hypoglossal, 103f
inferior olivary, 103f
of monocyte, 81f
of neurolemmocyte, 95f
of neuron, 95f
of neutrophil, 16f, 80f
of oocyte, 164f, 168f
raphe, 103f
Schwann cell, 16f
of skeletal muscle cell, 16f, 84f
of sperm
formation of, 168f
in meiosis, 19f
vagus, 103f
Nutrient foramen, nutrient artery in, 58f

Oblique cord, 76f
Oblique ligament, 76f
Oblique muscle
external, 85f, 90f
anterior superior, 90f
aponeurosis of, 90f
of cat, 182f, 184f
inferior, 88f
internal, 90f
aponeurosis of, 90f
of cat, 184f
superior, 88f
Obturator foramen, 55f
Obturator internus muscle, 92f
Occipital bone, 43f–45f, 67f
basilar part of, 48f
disarticulated, 46f
external surface of, 48f
of fetal skull, 42f
Occipital condyle, 44f, 48f
Occipital crest, external, 43f
Occipitalis muscle, 87f
Occipital lobe, 99f–100f, 105f
Occipital protuberance
external, 43f–44f, 48f
internal, 45f
Oculomotor (III) nerve, 99f
Olecranon, of ulna, 59f, 89f
Olfactory bulb, 99f–100f, 106f
Olfactory epithelium, 140f
Olfactory foramina, 50f
Olfactory (I) nerve, 99f
Olfactory tract, 99f, 106f
Oligodendrocytes, 94
Oligodendroglia cell, 98f
Olive, 99f, 103f
Omentum
areolar connective tissue in, 28f
of cat, 128f
Omohyoid muscle, 85f, 87f
Oocyte(s), 168f
nucleus of, 164f, 168f
Optic canal, 47f
Optic chiasm, 100f, 106f
Optic disk, 174f
Optic (II) nerve, 88f, 99f, 105f–106f, 175f
Optic tract, 99f
Optic vesicle, embryonic, 169f
Oral, definition, 4f–5f
Oral cavity, 126, 140f
Oral opening, 3f
Oral pharynx, 126
Ora serrata, of eye, 175f
Orbicularis oculi muscle, 85f, 87f, 176f–177f
Orbicularis oris muscle, 85f, 87f
Orbital, definition, 4f–5f
Orbital fissure, superior, 42f, 47f
Organelles, 12, 13f
composition of, 12
Organ of Corti, 180f
Oropharynx, 139, 140f
Orthochromatic erythroblast, of bone marrow, 81f
Os
external, 165f
internal, 165f
Osseous labyrinth, of inner ear, 174
Ossification
endochondral, 32f
in developing bone, 37f
in fetal bone, 38f
zone of, in fetal bone, 38f
Osteoblasts, 34, 35f–37f, 79
Osteoclasts, 34, 35f–36f, 79
adhering to bony trabeculae, 37f
and bone reabsorption, 37f
Osteocytes, 34, 35f–36f
of interstitial lamellae, 36f
in lacunae, 36f
Osteogenesis, 79
Osteons
in compact bone tissue, 35f
in spongy bone tissue, 35
Osteoprogenitor cells, 34, 35f
Otic, definition, 5f
Oval window, of ear, 179f
Ovarian artery, 117f
Ovarian follicular cells, 14f
Ovarian ligament, 165f
Ovary(ies), 150, 151f, 162, 163f, 165f, 168f–171f
of cat, 128f, 189f
general structure of, 168f
sagittal section, 164f
Oviduct(s), 162, 163f–164f, 170f–171f
ampulla of, 165f
of cat, 128f
with ciliated epithelium, 164f
fimbriae of, 165f
infundibulum of, 165f
cross section, 164f
isthmus of, 165f
lumen of, 164f
simple ciliated columnar epithelium of, 25f
Ovum/ova, 162
formation of, 168f
prior to fertilization, 168f
sperm penetrating into, 168f
Oxygen, 138

Pachymeninx, 94
Pacinian corpuscle(s), 96f, 111f
cross section of, 96f
of penis, 160f
Palatine bone, 140f
conchal crest of, 51f
disarticulated, 46f
horizontal plate of, 51f
medial view of, 51f
perpendicular plate of, 51f
sphenopalatine notch of, 51f
Palatine tonsil(s), 120, 122f, 124f, 140f
Palatonasal canal, 43f
Palm, 5f
Palmar, definition, 2
Palmar arch artery
deep, 117f
superficial, 117f
Palmaris longus muscle, 85f, 88f
of cat, 185f
Palmar venous arch vein, 118f
Palpebral conjunctiva, 176f
Pancreas, 126–127, 150, 151f, 153f
Pancreatic acini, 137f, 153f–154f
Pancreatic gastrin, 127
Pancreatic lobules, 137f
Papilla(e)
dermal, 23f–24f, 108f–109f, 111f
hair, 108f, 110f
renal, 146f
of tongue, 126, 130f
Papillary duct, of nephron, 146f
Papillary layer, of dermis, 107
Papillary muscle, 115f
Parafollicular cells, of thyroid gland, 153f
Paranasal sinuses, 138, 140f
Parathyroid capsule, 153f
Parathyroid cells, 153f
chief, 153f
oxyphil, 153f
Parathyroid follicles, 152f
Parathyroid gland, 150, 151f–153f
Parenchyma, of spleen, 121
Parenchymal epithelium, 108
Parietal bone(s), 43f–45f, 67f
disarticulated, 46f
external surface of, 48f
of fetal skull, 41f–42f
frontal angle of, 48f
frontal border of, 48f
occipital border of, 48f
sagittal border of, 48f
Parietal cells, of Bowman's corpuscle, 22f
Parietal lobe, 99f–100f
Parietal pleura, 141f
Parietal tuberosity, 48f
of fetal skull, 41f
Parotid gland, 126, 129f
of cat, 188f
in fetal pig, 191f
Pars fibrosa, of nucleolus, 15f
Pars granulosa, of nucleolus, 15f
Pars intermedia, of pituitary gland, 152f
Patella, 10f–11f, 40f, 85f, 91f, 93f
apex of, 63f
base of, 63f, 77f
condyle facet of, 63f
at knee joint, 63f
medial facets of, 77f
posterior surface of, 63f
vertical ridge of, 63f
Patellar, definition, 4f–5f
Patellar ligament, 10f, 85f, 91f, 93f
Patellar surface, 77f
Pectineus muscle, 85f, 91f
Pectoantebrachialis muscle, of cat, 186f
Pectoralis major muscle, 8f–9f, 85f, 90f, 166f
of cat, 185f
Pectoralis minor muscle, of cat, 182f, 184f, 186f
Pectoral region, muscles of, cat, 186f
Pedal, definition, 5f

Pelvic cavity, 126
Pelvic curvature, 52*f*
Pelvic floor, muscles of, 92*f*
Pelvic (hip) girdle
 anterior view, 40*f*
 posterior view, 40*f*
Pelvic inlet (brim), 55*f*
Pelvis, 1, 5*f*
 female, superior view of, 55*f*
 male, superior view of, 55*f*
 posterior view of, 56*f*
 renal, 145, 146*f*
Penile urethra, 155
Penis, 92*f*, 155–156, 156*f*
 bulb of, 161*f*
 corona of, 156*f*
 crus of, 161*f*
 in fetal pig, 190*f*
 human, 160*f*
 of infant primate, 160*f*
 suspensory ligament of, 156*f*
Peptides, 150
Perichondrium
 of elastic cartilage, 33*f*
 of external ear, 173
 in fetal bone, 38*f*
 of hyaline cartilage, 32*f*
Perilymph, 179*f*
 of inner ear, 174
Perimetrium, 165*f*
Perimuscular connective tissue layer, of gallbladder, 137*f*
Perimysium, of cardiac muscle, 84*f*
Perineum, 1
 central tendon of, 92*f*
 deep transverse, 92*f*
 female, 92*f*
 male, 92*f*
 superficial transverse, 92*f*
Perineurium, of peripheral nerve, 97*f*
Periodontal membrane, 126
Periosteal blood vessels, in compact bone tissue, 35*f*
Periosteal bone, in fetal bone, 38*f*
Periosteum, 32*f*, 34, 58*f*
 of dense bone, 36*f*
 developing, 37*f*
 structure of, 35*f*
Peripheral protein, of plasma membrane, 13*f*
Peritubular capillaries, 144, 146*f*
Peroneal artery, 117*f*
Peroneal vein, 118*f*
Peroneus brevis muscle, 86*f*, 93*f*
Peroneus longus muscle, 11*f*, 85*f*–86*f*, 93*f*
 of cat, 184*f*
Peroneus tertius muscle, 86*f*, 93*f*
Peroxisome(s), 12, 13*f*
Petrous region, of inner ear, 174
Peyer's patch(es), 120–121, 122*f*, 125*f*, 135*f*
P granule, of neutrophil, 16*f*
Phalangeal, definition, 4*f*–5*f*
Phalanx/phalanges, 6*f*, 40*f*, 69*f*–70*f*
 of fingers
  abduction of, 73*f*
  adduction of, 73*f*
  developing, 111*f*
  distal, 38*f*, 60*f*
  fetal, 38*f*
  middle, 60*f*
  proximal, 38*f*, 60*f*
 of toes
  distal, 64*f*
  middle, 64*f*
  proximal, middle, 64*f*
Pharyngeal cavity, 138
Pharyngeal tonsil(s), 121, 139, 140*f*
Pharyngeal tubercle, 48*f*
Pharyngopalatine arch, 120
Pharynx, 138–139
Philtrum, 3*f*
Phospholipid bilayer, of plasma membrane, 13*f*
Phospholipid tails, of plasma membrane, 13*f*
Pia mater, 94
Pigmented cells
 of eye, 178*f*
 of stratified squamous epithelium, 23*f*
Pigments, cell, 12
Pineal body, 100*f*
Pineal gland, 102*f*, 106*f*, 150, 151*f*
Pinealocyte, of pineal gland, 151*f*
Pinna. *See* Auricle, of ear
Pisiform bone, 60*f*
Pituitary gland, 99*f*, 106*f*, 150, 151*f*–152*f*
 anterior, 152*f*
 infundibulum of, 100*f*
 posterior, 152*f*
Pivot joint, 65–66, 69*f*
Placenta, 170*f*-171*f*
 banded, of cat, 190*f*
 of cat, 190*f*
 gravid, 172*f*
 tissue, 172*f*
Plantar, definition, 2, 4*f*–5*f*
Plantar flexion
 of foot, 75*f*
 of joints, 66
Plantaris muscle, 86*f*, 93*f*
Plasma, 27, 78
 in neurovascular bundle, 119*f*
Plasma cells, 79
 of medullary cords, 120
Plasmalemma. *See* Cell membrane
Plasma membrane. *See* Cell membrane
Platelets, 78–79, 81*f*
Platysma muscle, 85*f*–87*f*
Pleura, blood supply of, 139
Pleural cavity, 141*f*
Plexiform layer, of eye
 inner, 178*f*
 outer, 178*f*
Plexus
 capillary, 139
 hair root, 108*f*
 myenteric
  of gastrointestinal tract, 132*f*
  of small intestine, 133*f*
 submucosal, of gastrointestinal tract, 132*f*
Plexus of Auerbach. *See* Plexus, myenteric
Plexus of Meissner. *See* Plexus, submucosal
Podocyte, of renal corpuscle, 148*f*
Polar body, 168*f*
 first, 169*f*
 second, 169*f*
Polar head, of plasma membrane, 13*f*
Pons, 99*f*, 105*f*–106*f*
Pons varolii, 106*f*
Popliteal, 10*f*
 definition, 4*f*–5*f*
Popliteal artery, 117*f*
Popliteal fossa, 11*f*, 86*f*, 93*f*
Popliteal vein, 118*f*
Popliteus muscle, 93*f*
Portal tract, between liver lobules, 136*f*
Portovaginalis, LM of, 165*f*
Posterior, definition, 2
Posterior arch, of axis vertebrae, 69*f*
Postoral arch, third, embryonic, 169*f*
Post-subscapularis nerve, of cat, 186*f*
Pouch of Douglas, 163*f*
Prechiasmatic groove, 45*f*
Premolar(s), 42*f*–43*f*, 46*f*
 first, 129*f*
 second, 129*f*
Prepuce, 156*f*
 epidermis of, 160*f*
Primitive streak, embryonic, 169*f*
Process
 acromion, 3*f*, 56*f*–57*f*
 alveolar, 67*f*
 bifid spinous, 54*f*
 ciliary, 174*f*
 clinoid, 47*f*
 coracoid, 56*f*–57*f*, 76*f*
 coronoid, 49*f*, 67*f*
  of ulna, 59*f*
 frontal, 42*f*, 46*f*
  of maxilla, 42*f*
  of zygomatic bone, 52*f*
 hamate, 60*f*
 inferior articular, 54*f*
 mamillary, 54*f*
 mastoid, 43*f*, 48*f*
  of fetal skull, 42*f*
 maxillary, 42*f*
 maxillary palatine, 43*f*
 of neurons, 95*f*
 odontoid, of axis vertebrae, 69*f*
 olecranon, 9*f*, 59*f*
 orbital, 51*f*
 palatine, 44*f*, 46*f*
  of maxilla, 44*f*
 pyramidal, 51*f*
 spinous, 54*f*
  of atlas, 69*f*
  of axis vertebra, 69*f*
 styloid
  of radius, 59*f*–60*f*
  of temporal bone, 43*f*, 48*f*
  of ulna, 59*f*
 superior articular, 54*f*
 transverse, 53*f*, 69*f*
 xiphoid, 56*f*
 zygomatic, 48*f*
Progesterone, 162
Pronation
 of hand, 75*f*
 of joints, 66
Pronator quadratus muscle, 85*f*, 88*f*
Pronator teres muscle, 88*f*
 of cat, 185*f*
 tendon of, 89*f*
Pronucleus
 female, formation of, 168*f*
 male, 168*f*
Prophase
 in meiosis, 19*f*
 in mitosis, 18*f*
  micrograph of, 19*f*
Prostate concretions, 161*f*
Prostate gland, 155, 156*f*, 161*f*
Prostatic urethra, 156*f*, 161*f*
Protein(s), of plasma membrane, 13*f*
Proteoglycans, in cartilage, 31
Prothrombin, 78
Protoplasm, 12
Protraction, of temporomandibular joint, 75*f*
Protrusion, of joints, 66
Proximal, definition, 2, 6*f*
Proximal tubules, renal, 147*f*–148*f*
 convoluted, 146*f*
Pseudostratified columnar epithelium
 ciliated, 25*f*, 142*f*
 of epididymis, 25*f*
 of trachea, 25*f*
Psoas major muscle, 85*f*, 91*f*
Psoas minor muscle, 91*f*
Pterion, 44*f*
Pterygoid canal, 47*f*
Pterygoid fovea, 49*f*
Pterygoid hamulus, 47*f*
Pterygoid plate
 hamulus of, 44*f*
 lateral, 47*f*
 medial, 47*f*
Pubic, definition, 5*f*
Pubic region, 7*f*, 10*f*
Pubic symphysis, 55*f*, 65, 68*f*, 156*f*, 163*f*
 bones, symphysis joints of, 68*f*
 cartilage of, 32, 33*f*
Pubis, 5*f*, 55*f*, 70*f*
 brim of, 68*f*
 crest of, 55*f*, 68*f*
 inferior ramus of, 92*f*
 tubercle of, 55*f*, 90*f*–91*f*
Pubococcygeus muscle, 92*f*
Pulmonary artery, 139
 left, 114*f*
 right, 114*f*
Pulmonary trunk, 114*f*–115*f*
Pulmonary veins
 left, 114*f*
 right, 114*f*
Pulp
 dental, 126, 130*f*
  lymphocytes in, 127
 splenic, 121, 125*f*
Pulp chamber, of teeth, 127
Pupil, 174*f*–175*f*
Purkinje cells, 96*f*, 102*f*
Purkinje fibers, cardiac, 116*f*
Pyloric stomach, 127, 133*f*
Pyramid(s)
 cerebral, 99*f*, 101*f*, 103*f*
  decussation of, 99*f*–100*f*, 103*f*
 renal, 146*f*
Pyramidal cortical cells, 95
Pyramidal (ganglionic) layers, of cerebral cortex, 101*f*

Quadrate membrane, 139
Quadratus lumborum muscle, 91*f*
Quadriceps femoris muscle, 93*f*
  tendon of, 85*f*, 91*f*, 93*f*

Radial artery, 117*f*
Radial collateral ligament, 76*f*
Radius, 6*f*, 40*f*, 59*f*–60*f*, 76*f*
  articulation of, 59*f*
  head of, 59*f*
  medial view of, 59*f*
  neck of, 59*f*
Rectal epithelium, 136*f*
Rectoanal junction, 136*f*
Rectum, 126–127, 136*f*, 156*f*, 163*f*
  of cat, 128*f*, 187*f*
Rectus abdominis muscle, 85*f*, 90*f*
  of cat, 181*f*, 183*f*
Rectus femoris muscle, 85*f*, 91*f*
  of cat, 186*f*
Rectus muscle
  inferior, 88*f*
  lateral, 88*f*, 175*f*
  medial, 88*f*, 175*f*
  superior, 88*f*, 176*f*
Rectus sheath, 90*f*
  anterior layer of, 90*f*
  posterior layer of, 90*f*
Red blood cells. *See* Erythrocytes
Red pulp, of spleen, 121, 125*f*
Renal artery, 144, 146*f*
  in fetal pig, 192*f*
  left, 117*f*
  right, 117*f*
Renal capsule, 146*f*
Renal corpuscle, 146*f*, 148*f*
  Bowman's capsule at, 22*f*
Renal cortex, 146*f*
Renal medulla, 146*f*
Renal papilla, 146*f*
Renal pelvis, 145, 146*f*
Renal pyramid, 146*f*
Renal vein, 144, 146*f*
  of cat, 189*f*
  of fetal pig, 192*f*
  left, 118*f*
Renin, 144
Reproductive system(s)
  female, 162–166
  male, 155–161
    supporting structures of, 161*f*
Respiration
  external, 138
  internal, 138
Respiratory system, 138–143
  functions of, 138
  and lymph nodules, 120
Respiratory tract
  lower, 138–139
  upper, 138, 140*f*
Rete testis, 155, 157*f*
Reticular connective tissue, in lymph node, 29*f*
Reticular fiber(s), 27
  in lymph node, 29*f*, 123*f*
Reticular layer, of dermis, 107
Reticular tissue, 26
Reticuloendothelial cells, of spleen, 121
Retina, 173, 175*f*–176*f*, 178*f*
Retinal artery(ies), 175*f*
  central, 175*f*
Retinal vein(s), 175*f*
  central, 175*f*
Retraction
  of joints, 66
  of temporomandibular joint, 75*f*
Retrouterine pouch, 163*f*
Rhomboideus major muscle, 9*f*
Rhomboideus muscle, of cat, 184*f*
Rib(s), 6*f*, 40*f*, 56*f*, 166*f*
  articular surface of, 57*f*
  cartilage of, 31
  costal cartilage articulation area, 57*f*
  head of, articular facets of, 57*f*
  second, 90*f*
  shaft of, 57*f*
  superior border of, 57*f*
  twelfth, 91*f*
  typical, 57*f*
Ribosomes, 12
  in cellular cytosol, 15*f*
  free, 13*f*
Rima glottidis, 140*f*, 142*f*
Risorius muscle, 85*f*–87*f*
Rods, 177*f*
Rotation
  at atlantoaxial cervical joint, 73*f*
  of central axis, 69*f*
  of head, 73*f*
  at synovial joint, 65–66
  at vertebral column, 73*f*
Round ligament, of uterus, 163*f*
Round window, of ear, 178*f*–179*f*
Rugae, 162
  of pyloric stomach, 133*f*

Saccular nerves, 178*f*
Saccule, 178*f*
Sacral crest, median, 56*f*
Sacral foramina, 56*f*
Sacroiliac joint, 55*f*–56*f*
Sacrotuberous ligament, 92*f*
Sacrum, 1, 52*f*, 55*f*, 91*f*, 156*f*, 163*f*
  position in pelvic region, 56*f*
Saddle joint, 65–66, 70*f*
Sagittal sinus
  inferior, 118*f*
  superior, 118*f*
Sagittal suture, 67*f*
Salivary glands, 126
  of cat, 188*f*
  of fetal pig, 191*f*
Saphenous vein(s)
  great, 118*f*
  small, 118*f*
Sarcolemma, 82
Sarcoplasmic reticulum, of skeletal muscle cell, 16*f*
Sartorius muscle, 10*f*, 85*f*–86*f*, 91*f*, 93*f*
  of cat, 182*f*, 184*f*–185*f*, 187*f*
Scala media, 179*f*–180*f*
Scala tympani, 179*f*–180*f*
Scala vestibuli, 179*f*–180*f*
Scalene muscle, 85*f*
  of cat, 184*f*
  middle, 87*f*
Scaphoid bone, 60*f*
Scapula, 6*f*, 40*f*, 56*f*, 76*f*, 90*f*
  anterior view of, 57*f*
  dorsal surface of, 57*f*
  inferior angle of, 57*f*
  infraspinous fossa of, 57*f*
  lateral border of, 56*f*–57*f*
  lateral margin of, 57*f*
  lateral view of, 57*f*
  medial border of, 57*f*
  neck of, 57*f*
  notch of, 57*f*
  superior angle of, 57*f*
  superior border of, 57*f*
  supraspinous fossa of, 57*f*
Scapular spine, 57*f*
Schwann cell, 95*f*
  nucleus, 16*f*
Sciatic notch
  greater, 55*f*–56*f*
  lesser, 55*f*–56*f*
Sclera, 174*f*–175*f*, 177*f*
Scleral venous sinus, 175*f*–176*f*
Scrotum, 151*f*, 155, 156*f*
Sebaceous gland, 107–108, 108*f*–110*f*
Secretory vesicle, 13*f*
Segmental arteries, 144
Segmental veins, 144
Sellaris joints, 65–66, 70*f*
Sella turcica, 45*f*, 47*f*
Semicircular canal(s), 178*f*–179*f*
  ampulla of, 179*f*
  anterior, 179*f*
  lateral, 179*f*
  posterior, 179*f*
Semicircular duct, ampulla of, 179*f*
  crista within, 179*f*
Semimembranosus muscle, 11*f*, 86*f*, 93*f*
  of cat, 186*f*
Seminal vesicle(s), 155, 156*f*, 159*f*, 161*f*
  duct of, 161*f*
Seminiferous tubule(s), 155, 157*f*–158*f*
Semitendinosus muscle, 86*f*, 93*f*
  of cat, 184*f*
Sense organs, 172–180
Sensory neuron, cell body of, 103*f*
Septal cells, 139
Septum, of testis, 157*f*
Septum pellucidum, 100*f*, 106*f*
Seromucous glands, 173
Serosa
  of colon, 135*f*
  of gastrointestinal tract, 132*f*
  of jejunum, 134*f*
  of rectum, 136*f*
  of small intestine, 133*f*
  of stomach, 127
  of uterus, 162
  of vermiform appendix, 135*f*
Serous acini
  of parotid gland, 129*f*
  of submaxillary gland, 128*f*
  of tongue, 130*f*
Serous glands of von Ebner, of tongue, 130*f*
Serratus anterior muscle, 85*f*, 90*f*
Serratus ventralis muscle, of cat, 184*f*
Sesamoid bone, 64*f*
S granules, of neutrophil, 16*f*
Shank region, muscles of, cat, 185*f*
Sharpley's fibers, 34
Shoulder, 5*f*, 9*f*
  extension of, 72*f*
  joints of, 65, 76*f*
  muscles of, cat, 183*f*–184*f*
Shoulder (pectoral) girdle, 40*f*
Sigmoid sinus, 118*f*
Simple columnar epithelium, 21, 23*f*
  ciliated, 25*f*
  of gallbladder, 137*f*
  in kidney, 22*f*
Simple cuboidal epithelium, 21, 23*f*
  in kidney, 23*f*
Simple squamous epithelium, 21, 22*f*
  of mesothelium, 22*f*
  of renal corpuscle, 148*f*
Sinus(es)
  coronary, 118*f*
  cortical, 123*f*
  ethmoidal, 50*f*, 138
  frontal, 44*f*–45*f*, 138, 140*f*
  lactiferous, 166*f*
  laryngeal, 140*f*
  marginal, 171*f*
  maxillary, 46*f*, 138
  paranasal, 138, 140*f*
  sagittal, 118*f*
  scleral venous, 175*f*–176*f*
  sigmoid, 118*f*
  sphenoidal, 44*f*, 47*f*, 138, 140*f*
    openings of, 47*f*
  straight, 118*f*
  transverse, 118*f*
  venous, of spleen, 121
Sinusoids, hepatic, of liver lobule, 136*f*
Skeletal muscle cell, nucleus of, 16*f*, 84*f*
Skeletal muscle fibers, 97*f*
Skeleton system, 40*f*
Skin, 90*f*, 107
  from abdominal region, 109*f*
  axillary, 109*f*
  subcutaneous layer of, 108*f*
Skull, 5*f*, 39*f*
  anterior view of, 42*f*
  base of, 43*f*
  bones of, disarticulated, 46*f*
  cranial portion of, 40*f*
  facial portion of, 40*f*
  fetal
    anterior view of, 41*f*
    lateral view of, 42*f*
    superior view of, 41*f*
  lateral projection of, 45*f*
  lateral view of, 43*f*
  respiratory structures in, 140*f*
  sagittal section of, 44*f*
  superior view of, 44*f*
  sutures of, 65
Small artery, of excretory ducts, 137*f*
Small intestine, 6*f*–7*f*, 121, 122*f*, 127, 132*f*, 134*f*, 151*f*
  arteriole of, 133*f*
  of cat, 128*f*
  epithelial lining of, 134*f*
  layers of, 133*f*
  lumen of, 125*f*
  mucosa of, 125*f*
  plexus of, 133*f*
Small veins, of excretory ducts, 137*f*

Smooth muscle, 83
in bronchiole, 143*f*
of epididymis, 158*f*
of prostate gland, 161*f*
visceral, 83*f*
nuclei of, 83*f*
Soft palate, 138, 140*f*
Sole, of foot, 2
Soleus muscle, 11*f*, 85*f*–86*f*, 93*f*
of cat, 185*f*
Somatostatin, 127
Sperm, 159*f*
flagellum of, 159*f*
head of, 159*f*
midpiece, 159*f*
penetrating into egg, 168*f*
tail of, 159*f*
Spermatic cord, 90*f*
in fetal pig, 192*f*
of testis, 157*f*
Spermatocytes, of seminiferous tubule, 158*f*
Spermatogenesis, 155
Spermatogonia, of seminiferous tubule, 158*f*
Spermatozoa, 155
in epididymis, 25*f*, 158*f*
of seminiferous tubule, 158*f*
Sphenoidal sinus, 47*f*, 138, 140*f*
openings of, 47*f*
Sphenoid bone, 42*f*–43*f*, 88*f*, 140*f*
anterior view of, 47*f*
body of, 47*f*
disarticulated, 46*f*
posterior view of, 47*f*
superior view of, 47*f*
wings of, 47*f*
Sphincter
anal, 92*f*
cardiac, 127
Spicules, 35, 37*f*
in fetal bone, 38*f*
Spinal, definition, 4*f*
Spinal cord, 94
anterior gray horn of, 103*f*
anterior median fissure of, 103*f*
anterior root of, 104*f*
anterior white column of, 103*f*
anterior white commissure of, 103*f*
central canal of, 103*f*–104*f*
fibers of, 104*f*
gray commissure of, 103*f*
gray matter in, 103*f*
lateral gray horn of, 103*f*
lateral white column of, 103*f*
LM cross section of, 104*f*
neurons in, 98*f*
origin of, 99*f*–100*f*, 103*f*, 105*f*–106*f*
posterior gray horn of, 103*f*
posterior root of, 104*f*
white matter in, 103*f*
Spinal nerve(s), 103*f*
anterior (ventral) root of, 103*f*
posterior (dorsal) root of, 103*f*
Spindle-shaped cortical cells, 95
Spinodeltoid muscle, of cat, 183*f*
Spinotrapezius muscle, of cat, 182*f*–183*f*
Spiral ganglion, of ear, 180*f*
Spiral organ, 180*f*
Spleen, 120–121, 122*f*, 125*f*
Splenic artery, 117*f*
Splenic cords, 121
Splenic nodule(s), lymphoid, 125*f*
Splenic vein, 118*f*
Splenius capitis, 87*f*
Spongy bone, 26, 34–35, 35*f*, 58*f*, 68*f*
trabeculae of, 35*f*
Spongy (cavernous) urethra, 156*f*, 161*f*
Squamous epithelium, 139
simple. *See* Simple squamous epithelium
stratified. *See* Stratified squamous epithelium
Squamous mesothelium, of spleen, 121
Squamous suture, border of, 48*f*
Stab cell, of neutrophil, 81*f*
Stapedius muscle, of middle ear, 174
Stapes, of middle ear, 174
Stellate cortical cells, 95
Stenson's duct
of cat, 188*f*
of fetal pig, 191*f*
Stereocilia
of crista ampullaris, 181*f*
in epididymis, 25*f*
Sternal region, of cat, 186*f*
Sternocleidomastoid muscle, 3*f*, 8*f*–9*f*, 85*f*–87*f*
Sternohyoid muscle, 85*f*, 87*f*
Sternum, 3*f*, 6*f*, 9*f*, 40*f*, 90*f*
body of, 56*f*
cartilage of, 31
and clavicle, gliding movement between, 71*f*
joints of, 65
Steroids, 150
Stomach, 6*f*–7*f*, 126–127, 132*f*, 151*f*
cardiac, 132*f*
in fetal pig, 191*f*
fundic, 133*f*
gastric epithelium of, 131*f*
junction with esophagus, 131*f*
pyloric, 127, 133*f*
Straight sinus, 118*f*
Straight tubule, of testis, 157*f*
Stratified columnar epithelium, 21
Stratified cuboidal epithelium, 21
of submaxillary gland, 24*f*
Stratified squamous epithelium, 21, 23*f*–24*f*, 141*f*
anal, 136*f*
esophageal, 131*f*
keratinized, 21, 23*f*
lingual, 124*f*, 130*f*
nonkeratinized, 21, 24*f*
palatine, 124*f*
Stratum basale, 107, 109*f*
Stratum corneum, 23*f*–24*f*, 107, 109*f*
Stratum germinativum, 23*f*–24*f*
Stratum granulosum, 23*f*, 107, 109*f*
Stratum lucidum, 23*f*, 107, 109*f*
Stratum spinosum, 23*f*, 107, 109*f*
Stroma
cortical, 164*f*
of lymph nodes, 120
Subacromial bursa, 76*f*
Subarachnoid space, 94
Subclavian artery(ies), 114*f*, 117*f*
of cat, 187*f*
Subclavian vein(s)
left, 114*f*, 118*f*, 122*f*
right, 114*f*, 122*f*
Subcutaneous muscles, 8*f*
Sublingual glands, 126
Submandibular gland, 126, 128*f*
Submandibular node, 122*f*
Submaxillary gland, 126, 128*f*
of cat, 188*f*
of fetal pig, 191*f*
stratified cuboidal epithelium in, 24*f*
Submucosa
of cardiac stomach, 132*f*
of colon, 135*f*
of duodenum, 134*f*
of esophagus, 131*f*
of fundic stomach, 133*f*
of gastrointestinal tract, 132*f*
in ileum, 125*f*
of intestine, 127
of jejunum, 134*f*
of pyloric stomach, 133*f*
of rectoanal junction, 136*f*
of rectum, 136*f*
of small intestine, 133*f*
of stomach, 127, 131*f*
of vermiform appendix, 135*f*
Submucosal plexus, of gastrointestinal tract, 132*f*
Subscapular fossa, 57*f*
Subscapular muscle, 76*f*
of cat, 184*f*
Substance P, 127
Substantia propria, of cornea, 177*f*
Sulcus/sulci, 94, 99*f*, 105*f*
cerebral, 94
coronary, 114*f*
interventricular, anterior, 114*f*–115*f*
lateral, 99*f*
palpebral, 3*f*
posterior median, 102*f*
posterolateral, 104*f*
temporal, 99*f*
Superficial, definition, 2
Superficial anterior muscles, 85*f*
Superficial vein, of penis, 160*f*
Superior, definition, 2, 6*f*
Supination
of hand, 75*f*
of joints, 66
Supinator muscle, 88*f*–89*f*
Supporting cell, of ear, 180*f*
Supporting tissue, bone, 26
Supraorbital margin, 42*f*
Suprarenal gland(s). *See* Adrenal gland(s)
Suprascapular notch, 57*f*
Supraspinatus muscle, of cat, 184*f*
Supraspinous fossa, 56*f*–57*f*
Sural, definition, 4*f*–5*f*
Suspensory ligament(s)
of breast, 166*f*
of lens, 175*f*
of penis, 156*f*
of uterine tube, 165*f*
Suture(s), 66
coronal, 43*f*, 67*f*
of fetal skull, 41*f*–42*f*
coronal (frontal), 44*f*
lambdoid, 44*f*, 67*f*
of fetal skull, 42*f*
median palatine, 43*f*
sagittal, 44*f*, 67*f*
of fetal skull, 41*f*
of skull, 65
fetal, 41*f*–42*f*
squamous, 43*f*
transverse palatine, 43*f*
Sweat glands, 107
apocrine, 107–108, 109*f*–110*f*
columnar epithelium of, 107
cuboidal epithelium of, 107
of eyelid, 177*f*
of Moll, 177*f*
Sympathetic ganglia, 98*f*
Symphysis, 65–66
mandibular, 3*f*, 42*f*
menti, of fetal skull, 41*f*
pubic, 55*f*, 65, 68*f*, 156*f*, 163*f*
cartilage of, 32, 33*f*
Symphysis joint, in pubic symphysis bones, 68*f*
Synapsis, in meiosis, 19*f*
Synaptic end bulb, 95*f*
Synaptic knobs (synapses), 98*f*
Synarthrosis joints, 65
Synchondrosis joints, 65–66
in long bones, 68*f*
Syncytiotrophoblast, 171*f*
Syndesmosis joints, 65–66
of tibia and fibula, 68*f*
Synovial condyloid joint, 70*f*
Synovial diarthrodial ginglymus joints, 69*f*
Synovial diarthrodial hinge joints, 69*f*
Synovial ellipsoidal joint, 70*f*
Synovial joint(s), 65–66, 69–70
abduction of, 65
adduction of, 65
circumduction of, 65–66
fetal, 38*f*
movements of, 65
types of, 65
Synovial movements, 71*f*–73*f*
Synovial pivot joint, 69*f*

Tail
of cat, 185*f*
embryonic, 169*f*
Talocalcaneal joint, 64*f*
Talus, 64*f*, 70*f*
Tarsal, definition, 5*f*
Tarsal bones
definition, 4*f*
joints of, 65, 70*f*
Tarsal (meibomian) gland(s), 176*f*–177*f*
Tarsal plate(s), 87*f*, 176*f*
Taste bud(s), 126, 130*f*–131*f*
epithelial ring of, 131*f*
type I cells, 131*f*
type II cells, 131*f*
Tectorial membrane, 180*f*
Tectum, 105*f*
Teeth, 42*f*–43*f*, 46*f*, 126, 129*f*–130*f*. *See also specific teeth*
cementum of, 126
deciduous, 41*f*
lower, 129*f*

Teeth *(continued)*
pulp chamber of, 127
upper, 129*f*
Telophase
in meiosis, 19*f*
in mitosis, 18*f*
micrograph of, 19*f*
Temporal bone, 42*f*, 173, 178*f*
disarticulated, 46*f*
of fetal skull, 41*f*–42*f*
lateral view of, 48*f*
mastoid region of, 48*f*
squamous region of, 48*f*
tympanic portion of, 48*f*
Temporal lobe, 100*f*
Temporal region, of face, 3*f*
Temporomandibular joint
protraction of, 75*f*
retraction of, 75*f*
Temporoparietalis muscle, 87*f*
Tendinous intersections, 90*f*
Tendo calcaneus, 11*f*
Tendons
cartilage of, 32
junction with skeletal muscle, 84*f*
Tensor fasciae latae muscle, 8*f*, 10*f*, 85*f*–86*f*, 91*f*
of cat, 182*f*, 184*f*
iliotibial tract of, 86*f*
Tensor tympani, of middle ear, 174
Tenuissimus muscle, of cat, 185*f*
Teres major muscle, 86*f*
Teres minor muscle, 86*f*
Testicular artery, 117*f*
Testicular capsule, 155
Testicular sac, 155
Testis/testes, 150, 151*f*, 155, 156*f*
of bull, 157*f*
of fetal pig, 192*f*
lobule of, 157*f*
rete, 155, 157*f*
sagittal plane, 157*f*
septum of, 157*f*
Testis interstitium, 150
Thalamus, 102*f*, 105*f*
Theca externa, 164*f*
Theca interna, 164*f*
Thenar muscle, 85*f*
Thigh, 5*f*, 10*f*
muscles of, 91*f*
Thoracic, definition, 4*f*–5*f*
Thoracic cavity, 126
Thoracic curvature, 52*f*
Thoracic duct, 122*f*
Thoracic vertebra(e), 52*f*, 56*f*
body of, 54*f*
costal facet of, 54*f*
inferior notch of, 54*f*
inferior view, 53*f*
lateral view of, 54*f*
pedicle of, 54*f*
superior view, 53*f*
Thoracolumbar fascia, of cat, 185*f*
Thorax, 40*f*
bony
anterior view of, 56*f*
posterior view of, 56*f*
Thrombocytes, 78–79
Thymus, 120–121, 122*f*, 151*f*
cortex of, 121
in fetal pig, 191*f*
lobules, 124*f*–125*f*
cortex of, 124*f*
medulla of, 124*f*
medulla of, 121, 124*f*
Thyrohyoid muscle, 85*f*
Thyroid cartilage, 87*f*, 139, 140*f*, 142*f*
Thyroid follicles, 152*f*
Thyroid gland, 140*f*, 150, 151*f*–153*f*
capsule of, 152*f*
in fetal pig, 191*f*
follicular cells of, 142*f*, 152*f*–153*f*
functions of, 167
Thyroid membrane, 139
Thyroxine, 167
Tibia, 11*f*, 40*f*, 77*f*, 85*f*, 93*f*
at ankle joint, 64*f*
anterior view of, 62*f*
condyles of
lateral, 62*f*, 68*f*
medial, 62*f*, 68*f*
fibular articular facet of, 68*f*
intercondylar tubercle of, 62*f*
at knee joint, 63*f*, 69*f*
medial malleolus of, 11*f*, 62*f*, 68*f*
posterior view of, 62*f*
syndesmosis joint of, 68*f*
Tibial artery
anterior, 117*f*
posterior, 117*f*
Tibial collateral ligament, 77*f*
Tibialis anterior muscle, 85*f*, 93*f*
of cat, 185*f*
Tibialis posterior muscle, 93*f*
tendon of, 93*f*
Tibial tuberosity, 68*f*
Tibial vein
anterior, 118*f*
posterior, 118*f*
Tibiofemoral joint, 69*f*
Tight junction, of cells, 13*f*
Tissue(s). *See also specific tissue*
functions of, 12
types, 21
T lymphocytes, 120, 124*f*–125*f*
Toes, 5*f*
muscles of, 93*f*
phalanges of, 64*f*
Tongue, 126, 130*f*. *See also* Lingual
and lymph nodules, 120
Tonsil(s), 120
lingual, 120, 124*f*
palatine, 120, 122*f*, 124*f*
pharyngeal, 121, 139, 140*f*
types, 120–121
Tonsillar crypt, 124*f*
of palatine tonsil, 124*f*
Toxic substances, elimination of, 144
Trabecula(e), 35, 94, 121, 125*f*
bony, osteoclasts adhering to, 37*f*
of cartilage, 32*f*
in developing bone, 37*f*
of developing skull bone, 36*f*
in fetal bone, 38*f*
of fetal finger, 38*f*
of lymph node, 123*f*
of spongy bone, 35*f*
Trabeculae carneae, 114*f*–115*f*
Trachea, 6*f*, 138–139, 140*f*–141*f*, 151*f*
basement membrane of, 141*f*
body wall of, 141*f*
cartilage of, 31
of cat, 188*f*
cross section of, 142*f*
goblet cells of, 141*f*
lumen of, 32*f*, 142*f*
pseudostratified columnar epithelium in, 25*f*
Tracheal glands, 32*f*
mixed, 142*f*
Transitional epithelium, 21
of ureters, 24*f*, 145, 149*f*
of urinary bladder, 145, 149*f*
Transport tissue, specialized, 26
Transverse plane, 2, 8*f*
through abdomen, 90*f*
Transversus abdominis muscle, 90*f*
aponeurosis of, 90*f*
of cat, 184*f*
Transversus costarum muscle, of cat, 184*f*
Trapezium, 70*f*
joints of, 65
Trapezium bone, 60*f*
Trapezius muscle, 9*f*, 85*f*–87*f*
Trapezoid bone, 60*f*
Trapezoid ligament, 76*f*
Triad, of skeletal muscle cell, 16*f*
Triceps brachii muscle, 9*f*, 85*f*–86*f*, 89*f*
lateral head, of cat, 182*f*–183*f*
long head, of cat, 183*f*
long heads of, 8*f*
medial heads of, 8*f*
Triceps muscle, 9*f*
lateral head, 183*f*
long head, 183*f*
Tricuspid valve, 114*f*, 116*f*
Trigeminal (V) nerve, 99*f*
Triquetrum (triangular) bone, 60*f*
Trochanter
greater, 56*f*, 61*f*, 70*f*, 77*f*
lesser, 56*f*, 61*f*, 70*f*, 77*f*
Trochlea, 88*f*
of humerus, 58*f*–59*f*
Trochlear (IV) nerve, 99*f*, 102*f*
Trochlear notch, 59*f*
Trochoid joint, 65–66, 69*f*
Trunk, 1, 5*f*
Tuber cinereum, 99*f*
Tubercle(s)
of atlas
anterior, 53*f*–54*f*
posterior, 53*f*
of calcaneus, anterior, 64*f*
of humerus
greater, 58*f*
lesser, 58*f*
pharyngeal, 48*f*
pubic, 55*f*, 90*f*–91*f*
of tibia, intercondylar, 68*f*
Tuberosity
calcaneal, 70*f*
conoid, 57*f*
costal, 57*f*
deltoid, 58*f*
frontal, of fetal skull, 41*f*
gluteal, 61*f*
ischial, 55*f*–56*f*, 92*f*
parietal, 48*f*
of fetal skull, 41*f*
radial, 59*f*
tibial, 62*f*, 68*f*
Tuboalveolar glands, 126, 155
Tubule(s), 147*f*
inactive, 166*f*
secondary, 166*f*
Tunica adventitia, 112–113
Tunica albuginea, 155
of corpus cavernosum penis, 160*f*
of corpus spongiosum, 160*f*
of testis, 157*f*
Tunica dartos, of penis, 160*f*
Tunica intima, 112–113
Tunica media, 112–113
Tunica vaginalis, 155
of testis, 157*f*
Tunica vasculosa, 155
Turbinate bones, 138
Turbinates. *See* Nasal concha(e)
Tympanic antrum, of ear, 178*f*
Tympanic cavity, of middle ear, 174
Tympanic membrane, 178*f*
of middle ear, 174
secondary, 178*f*–179*f*

Ulna, 6*f*, 40*f*, 59*f*–60*f*, 76*f*, 88*f*
articulation of, 59*f*
medial view of, 59*f*
notch of, 59*f*
olecranon of, 59*f*, 89*f*
Ulnar artery, 117*f*
Ulnar collateral ligament, 76*f*
Umbilical, definition, 5*f*
Umbilical arteries, in fetal pig, 191*f*–192*f*
Umbilical cord, 170*f*–171*f*
of cat, 190*f*
Umbilical region, 7*f*
Umbilical vein, 172*f*
Undifferentiated granules, 12
Upper extremity, 5*f*, 9*f*
bones of, 40*f*
Ureter, 146*f*, 149*f*
Ureter(s), 145, 161*f*, 165*f*
of cat, 189*f*
of fetal pig, 192*f*
lumen of, 24*f*
transitional epithelium of, 24*f*, 145, 149*f*
Urethra, 145, 155, 160*f*, 163*f*
of cat, 187*f*
membranous, 156*f*, 161*f*
penile, 155
prostatic, 156*f*, 161*f*
spongy (cavernous), 156*f*
Urethral orifice, external, 92*f*, 156*f*
Urethral wall, of penis, 160*f*
Urinary bladder, 6*f*–7*f*, 145, 149*f*, 155, 156*f*, 161*f*, 163*f*
of cat, 128*f*, 189*f*
epithelial folds of, 149*f*
of fetal pig, 191*f*–192*f*
Urinary system, 144–149
Urine, 145, 146*f*
Urogenital diaphragm, 156*f*, 161*f*
inferior fascia of, 92*f*
Urogenital system, of cat, 189*f*
Urothelium, of urinary bladder wall, 149*f*
Uterine cavity, 165*f*, 169*f*
Uterine glands, 165*f*
Uterine horns, in pregnant cat, 189*f*
Uterine tubes. *See* Oviduct(s)
Uterine wall, of cat, 190*f*
Uterosacral ligament, 163*f*, 165*f*

Uterus, 151*f*, 162, 163*f*
  body of (corpus uteri), 162, 163*f*, 165*f*
  of cat, 128*f*, 189*f*
  cervix of, 165*f*
  cross section, 165*f*
  endometrium, 162
  exposed, 169*f*
  fundus of, 162, 163*f*, 165*f*
  horns of
    in cat, 189*f*
    in fetal pig, 192*f*
  lumen of, 162, 165*f*
  myometrium (muscularis), 162
  perimetrium of, 165*f*
  portio vaginalis, 162
  round ligament of, 163*f*
  serosa (perimetrium), 162
Utricle, 179*f*
Utricular nerve, 179*f*
Uveal layer, of eye, 173

Vacuole(s), 12
  in lymphocyte, 14*f*
Vagina, 92*f*, 162, 163*f*, 165*f*, 169*f*
  in fetal pig, 192*f*
  lumen of, 165*f*
  posterior fornix of, 163*f*
Vagus (X) nerve, 99*f*, 102*f*–103*f*
Valve(s)
  cardiac
    bicuspid (mitral), 115*f*
    tricuspid, 114*f*, 116*f*
  of gallbladder, 127
  ileocecal, 135*f*
  lymphatic, 123*f*
Vasa recta, 144
Vasa vasorum, of elastic arteries, 112
Vascular system tissue, 26
Vasculature
  arterial, 112–113
  venous, 113
Vas deferens. *See* Ductus (vas) deferens
Vasoactive intestinal peptide (VIP), 127
Vastus lateralis muscle, 10*f*, 85*f*–86*f*, 91*f*
  of cat, 186*f*
Vastus medialis muscle, 10*f*, 85*f*, 91*f*
  of cat, 187*f*
Vein(s), 113. *See also specific vein*
  body wall of, 119*f*
  in bronchiole, 143*f*
  cross section, 119*f*
  of jejunum, 134*f*
  large, 113
  lumen of, 119*f*
  medium, 113
  in neurovascular bundle, 119*f*
  in skin, 108*f*
  small, 113
Vena cava
  inferior, 114*f*, 118*f*
    of cat, 188*f*
  superior, 114*f*, 118*f*
    of cat, 188*f*
Venous arch vein, dorsal, 118*f*
Venous sinuses, of spleen, 121
Venous system, 118*f*
Ventral, definition, 2
Ventricle(s)
  cardiac
    left, 114*f*–115*f*
      body wall of, 115*f*
    right, 114*f*–115*f*, 139
      body wall of, 114*f*, 116*f*
  cerebral
    ciliated ependymal cells lining, 96*f*
    fourth, 100*f*, 103*f*, 105*f*
      floor of, 102*f*
    lateral, 105*f*–106*f*
    third, 102*f*
  laryngeal, 142*f*
Ventricular fold, 140*f*
Venule(s), 113
  of small intestine, 133*f*
Vermiform appendix, 135*f*
Vermis, 95, 105*f*
Vertebra(e). *See also specific vertebrae*
  axis of
    odontoid process of, 69*f*
    spinous process of, 69*f*
  joints of, 65
Vertebral artery, 117*f*
Vertebral column
  anterior view, 40*f*
  curvature of, 52*f*
  lateral view, 52*f*
  posterior (dorsal) view, 40*f*, 52*f*
  rotation at, 73*f*
  synovial movements of, 71*f*
Vesicorectal pouch, 156*f*
Vesicouterine pouch, 163*f*
Vestibular labyrinth, of inner ear, 174
Vestibular membrane, of ear, 179*f*
Vestibule(s)
  of ear, 179*f*
  of nasal cavity, 140*f*
Vestibulocochlear (VIII) nerve, 99*f*, 102*f*, 179*f*
  cochlear branch of, 178*f*–179*f*
  vestibular branch of, 178*f*–179*f*
Villus/villi
  of gastrointestinal tract, 132*f*
  of ileum, 135*f*
  of intestine, 127
  of jejunum, 134*f*
  of small intestine, 133*f*–134*f*
VIP. *See* Vasoactive intestinal peptide
Virus, in mitochondria, 16*f*
Visceral cells, of Bowman's corpuscle, 22*f*
Visceral pleura, 141*f*
Visual axis, 175*f*
Vitreous body, of eye, 175*f*–176*f*
Vitreous chamber, of eye, 175*f*
Vocal cord, 140*f*
Vocal fold
  false. *See* Ventricular fold
  truce. *See* Vocal cord
Vocal ligaments, 142*f*
Volkmann's canal, 36*f*
Vomer bone, 51*f*
  ala of, 51*f*
  disarticulated, 46*f*
  lateral view of, 51*f*
  posterior border of, 43*f*

Water, elimination of, 144
White blood cells. *See* Leukocytes
White matter
  cerebellar, 95, 102*f*
  cerebral, 101*f*
  of spinal cord, 103*f*–104*f*
White pulp, of spleen, 121, 125*f*
Wisdom tooth, 129*f*
Wrist(s), 5*f*
  bones of, 60*f*
  joints of, 65
  muscles of, 88*f*–89*f*

Yolk sac, 169*f*

Z bands, 16*f*
  of skeletal muscle, 82
Zona fasciculata, 154*f*
Zona glomerulosa, 154*f*
Zona pellucida, 164*f*, 168*f*
Zona reticularis, 154*f*
Zygomatic arch, 43*f*, 87*f*
Zygomatic bone, 42*f*, 87*f*
  disarticulated, 46*f*
  frontal process of, 52*f*
  lateral view of, 52*f*
  maxillary border of, 52*f*
  orbital border of, 52*f*
  temporal border of, 52*f*
Zygomaticofacial foramen, 52*f*
Zygomatic protuberance, of face, 3*f*
Zygomaticus major muscle, 87*f*
Zygomaticus minor muscle, 87*f*
Zygote, 167
  mitotic division in, 169*f*